Environmental Impact in the United States

Environmental impact assessment is now firmly established as an important and often mandatory part of proposing any development project. *Environmental Impact Assessment in the United States* provides foundational knowledge of environmental review in the United States as carried out at federal, state, and local levels, with detailed information about the National Environmental Policy Act (NEPA) and its applications, and other relevant federal and state legislation. This book will aid planners, architects, engineers, project managers, or consultants who work with environmental impact statements to assess the effects of a proposed activity on the environment and who develop and assess measures to avoid or minimize those impacts. It will serve as a desk reference for professional environmental planners as well as a core textbook for students who intend to work in the fields of environmental policy, civil engineering, environmental law, resources management, or other areas of environmental management.

Robert M. Sanford (Rob) is Professor Emeritus of Environmental Science and Policy at the University of Southern Maine in Gorham, Maine. He received a BA in anthropology (archaeology) at SUNY Potsdam, and his MS and PhD in environmental science from SUNY College of Environmental Science and Forestry. He served as an environmental hearing officer in Vermont for 9 years. Before and after that, he was an environmental impact assessment and planning consultant. His research interests include environmental impact assessment, environmental planning, and environmental education. He was a co-director of the SENCER (Science Education for New Civic Engagements and Responsibilities) New England Center for Innovation (SCI) from 2012 to 2022. In 2019, he was appointed to the Maine Board of Environmental Protection. His recent books include *Reading Rural Landscapes: A Field Guide to New England's Past*; *Environmental Science: Active Learning Laboratories and Applied Problem Sets* (3rd Ed.); *River Voices, Practicing Archaeology* (3rd Ed.); and *Environmental Site Plans and Development Review*.

Donald G. Holtgrieve earned his PhD degree from the University of Oregon in 1973. He taught geography, environmental studies, and environmental planning at California State University campuses at Hayward and Chico before moving back to Eugene at the University of Oregon. He created two environmental planning consulting firms and supervised the preparation of environmental impact reports, resource management plans, and various other community planning documents for federal, state, and local agencies. He planned environmentally sensitive land development projects and created three wildlife preserves for non-profit watershed groups. He is involved with volunteer service to the McKenzie River Trust and the Hatfield Marine Science Center in Newport, Oregon.

Environmental Impact Assessment in the United States

Robert M. Sanford and
Donald G. Holtgrieve

Illustrations by Patricia Boyle

Cover image: Robert M. Sanford

First published 2023
by Routledge
605 Third Avenue, New York, NY 10158

and by Routledge
4 Park Square, Milton Park, Abingdon, Oxon, OX14 4RN

Routledge is an imprint of the Taylor & Francis Group, an informa business

© 2023 Robert M. Sanford and Donald G. Holtgrieve

The right of Robert M. Sanford and Donald G. Holtgrieve to be identified as authors of this work has been asserted in accordance with sections 77 and 78 of the Copyright, Designs and Patents Act 1988.

All rights reserved. No part of this book may be reprinted or reproduced or utilised in any form or by any electronic, mechanical, or other means, now known or hereafter invented, including photocopying and recording, or in any information storage or retrieval system, without permission in writing from the publishers.

Trademark notice: Product or corporate names may be trademarks or registered trademarks, and are used only for identification and explanation without intent to infringe.

Library of Congress Cataloging-in-Publication Data
Names: Sanford, Robert M., author. | Holtgrieve, Donald G., author. | Boyle, Patricia, illustrator
Title: Environmental impact assessment in the United States / Robert M. Sanford and Donald G. Holtgrieve ; illustrations by Patricia Boyle.
Description: New York, NY : Routledge, 2023. | Includes bibliographical references and index.
Identifiers: LCCN 2022030743 (print) | LCCN 2022030744 (ebook) |
  ISBN 9780367467319 (hardback) | ISBN 9780367467326 (paperback) |
  ISBN 9781003030713 (ebook)
Subjects: LCSH: Environmental impact statements—United States.
Classification: LCC TD194.5 .S26 2023 (print) | LCC TD194.5 (ebook) |
  DDC 333.71/40973—dc23/eng/20221011
LC record available at https://lccn.loc.gov/2022030743
LC ebook record available at https://lccn.loc.gov/2022030744

ISBN: 978-0-367-46731-9 (hbk)
ISBN: 978-0-367-46732-6 (pbk)
ISBN: 978-1-003-03071-3 (ebk)

DOI: 10.4324/9781003030713

Typeset in Bembo MT Pro
by Apex CoVantage, LLC

Contents

List of figures   vii

List of tables   ix

Foreword   xiii

Acknowledgments   xv

1   Introduction   **1**

2   The EIA process   **21**

3   Screening, scoping, and related aspects   **46**

4   Assessing environmental impacts   **65**

5   Geology, topography, and earth resources   **80**

6   Hydrology, water quality, and water supply   **92**

7   Biological: Species and habitats   **103**

8   Air quality and climate change   **115**

9   Archaeology and historic preservation   **127**

10	Energy	138
11	Noise impact analysis	150
12	Aesthetics and visual impact analysis	161
13	Social impacts and environmental justice	171
14	Infrastructure, fiscal impacts, and community services	182
15	Traffic and transport systems	193
16	Writing the report	205
17	Making and implementing the decision	216

Appendices

National Environmental Policy Act Implementing Regulations	*237*
Preparation of students as environmental practitioners	*302*

Index   *309*

Figures

1.1	A public notice of Finding of No Significant Impact (FONSI) commonly associated with the dispensation of federal funding	4
1.2	How EIA relates to environmental planning	7
1.3	Stages in the EIA process	9
2.1	Spheres of US global involvement	31
3.1	Land development process for a private sector project	50
3.2	Public participation in EIA and EIS	60
4.1	Accuracy and precision illustrated	71
5.1	Basic soil properties	82
5.2	Callahan Mine site	88
6.1	Well plume	97
8.1	Air quality assessment cycle	116
8.2	Air quality model	120
9.1	Area of potential effect for a transmission line	132
9.2	NEPA and Section 106 flow chart (based on Neumann and Sanford, 2010: 43)	134
10.1	Three energy aspects for EIA review: exterior, interior, and production	139
11.1	Airport noise contours	155
12.1	Viewshed area	163
14.1	Projects can involve a variety of services	183
17.1	Decision tree for impact significance in Brazil (Tagliani and Walter, 2018)	218
17.2	Value tree analysis for decisions	224

Tables

2.1	EIA hierarchy and the branches of US government	25
2.2	State equivalent NEPA laws ("Little NEPA") as of 2022	29
2.3	Hypothetical display of how potential environmental impacts may vary depending on the phase (life stage) and nature of the project	34
2.4	Sample job plan outline created by students in an EIA class (budget column and other details omitted)	39
2.5	Another sample job plan outline	40
2.6	One-year timeline for an EIA	41
3.1	Screening matrix for a project to construct an erosion control and stormwater barrier dike	48
3.2	Sample matrix for offshore wind turbine project impact assessment	57
3.3	Project and environment matrix	58
3.4	Stakeholder exercise	61
3.5	Master plan stakeholder matrix	63
4.1	Two-month timeline for EIA project management	68
4.2	Categories of impact matrix	73
4.3	EIA report timeline	78
5.1	Earth resource components of an environmental setting	83
5.2	Projects associated with significant earth resource issues	84
5.3	Sample checklist for a soil engineering report	85
5.4	Sample checklist for erosion control plans and details	86
5.5	Sample California checklist for sites in unstable areas	87
5.6	Factors to consider in control of erosion through vegetation	89
6.1	Sample wetland functions and values checklist for an EIA on a site that contains wetlands	94
6.2	Major pollutants affecting water management	96
6.3	Water impact categories	98
6.4	Checklist of water resource categories of impact for project phases in an EIA	99
7.1	Checklist of items for assessing an environmental setting	105
7.2	Brief comparison of Habitat Evaluation System and Habitat Evaluation Procedure	107

7.3	A checklist for avian impact assessment	108
7.4	Checklist of biological impacts from the California Environmental Quality Act for your state	109
7.5	Common types of biological impact mitigation measures	111
8.1	AQI for ozone and particle pollution effects on humans	117
8.2	Types of air pollutants for a gravel pit	122
8.3	Climate change and waterfront development	124
9.1	NEPA and three intersecting federal laws involving cultural resources	129
9.2	Contents of an archaeological or historical site assessment or survey report	130
10.1	EIA energy component checklist based on LEED green building certification	142
10.2	Energy policy and program impact assessment checklist	143
10.3	Sample landscape checklist for energy efficient sites	144
10.4	Checklist of energy-related emission reduction categories based on LEED green building certification	145
11.1	EPA/OSHA standards for permissible noise exposure limits on construction sites	154
11.2	HUD noise preference levels	154
11.3	Sample options for noise mitigation at a gravel pit with a crusher and screener	157
11.4	Potential noise for a national park	158
12.1	Items to address in assessing a landscape plan	164
12.2	Lighting impact and mitigation	165
12.3	Visual Impact Assessment worksheet	167
12.4	Henning odor classification	167
13.1	Variables for social impacts	172
13.2	Checklist of steps for evaluating the environmental justice component of an EA report	175
13.3	Potential social impacts of a near-shore aquaculture station	177
13.4	United Nations social impact assessment good practices in impact mitigation	179
14.1	Fire and emergency department questions for project review	184
14.2	Police department checklist for project review	185
14.3	Highway department review questions	186
14.4	Sewer and water checklist	187
15.1	Sample checklist for applications involving transportation review or permits	194
15.2	Sample checklist for transportation construction plans	195
15.3	Questions to ask about a traffic study	198
15.4	What Smart Growth transportation site planning principles does the project meet?	198
15.5	General transit-related questions to ask about a project	199
15.6	Sample threshold trip levels based on type and size of project	200
16.1	Checklist for EIA report completion	206
16.2	EIA writing sample critique	210

16.3	Response letter critique	213
17.1	Some ways in which a project can be modified	217
17.2	Sample risk matrix based on probability of event and severity of event	220
17.3	Program risk management assessment scale for federal defense contractors	221
17.4	Environmental Situation Rapid Environmental Assessment Response Form	229
17.5	Environmental Impacts of Relief Activities Rapid Environmental Assessment Response Form	230

Foreword

This text takes an applied perspective on the process of environmental impact assessment (EIA). It presents emphasis on the most significant piece of environmental legislation of the 20th century: the four-page National Environmental Policy Act (NEPA). Signed into law by President Richard Nixon, NEPA became the cornerstone for environmental legislation, planning, and public participation in the United States and for countries around the world. As environmental civil rights legislation, it also marked a major change in the way the US federal government does business. The public was finally given a mechanism to help force environmental accountability in a systematic way. Yet many members of the public are unaware of what a powerful mechanism the EIA process can be. This text provides an overview of how to conduct environmental assessments, systems analysis, review of environmental impact statements, and use of various regulatory review processes. We will also look at state and some international aspects of environmental assessment.

Users of this text will gain experience in applying principles and theories of environmental impact assessments (EIAs) and environmental impact statements (EISs) under state and federal laws. We cover most kinds of impacts, ranging from air pollution and archaeology to traffic and educational services. A primary task is to "scope" and consolidate these impacts, make sense of them, and use that knowledge to promote effective decision-making. EIA is a planning law that makes good decisions happen. While it is not perfect, it has evolved to be a major policy instrument emulated by dozens of other countries.

The EIA/EIS is required by its authors and users to be the result of an interdisciplinary team effort. It represents a combination of law, science, organization, and art. We attempt to summarize attributes and retain solid, useful information for enabling a successful environmental review of proposed physical construction and policy-focused projects.

This text is intended to be used by students and practitioners in the various fields of environmental science, environmental studies, civil and environmental engineering, land use planning, political or governmental sciences, environmental policy, geography, and other disciplines. It is also for citizens who want a guide to help understand the EIA process and how to participate in it. The text, like an environmental impact statement, is interdisciplinary.

Environmental impact analysis (EIA) is an ever-changing process that is dependent upon internet content to answer questions and to guide the practitioner toward successful project decision-making. To that end, this text offers internet and print resources to support and expand knowledge for preparing or critically reviewing environmental impact documents. The intended user readership in this field also includes non-professional but very concerned citizens who wish to know more about proposed projects that may affect them.

Each topic could be expanded to be an entire book, but we have chosen to limit them to what it takes to get the reader started in the EIA process. The entire text could be spent in simply defining terms and concepts. That would be overwhelming, and there are environmental dictionaries and agency websites that accomplish this quite well. Accordingly, we are limiting our terms and concepts to a brief section within each topic whenever we have not simply defined the term or concept in context. There is some overlap in the concepts provided within the topics, as we have designed them to function somewhat independently.

In this book we have stayed away from the subject of NEPA in the courts. The judicial system is vital in the checks and balances of the government. However, in some ways, the EIA process has let us down by the time an issue comes to the courts for litigation. The EIA process becomes something for lawyers to argue about rather than for environmental scientists and specialists to use as a guide in environmental planning and sustainability searching. Court decisions do become recorded in agency guidelines and regulations so that they remain useful for working professionals. The EIA process is a tool for incremental progress in balancing the needs of society with environmental resource management. Each EIA has the potential to improve the overall process, especially as we continue in a dynamic world of climate and environmental policy changes.

The Council on Environmental Quality (CEQ, 2022) provides a desk reference that contains the National Environmental Policy Act of 1969, the CEQ regulations for implementing NEPA 40 CFR Parts 1500–1508), the Environmental Quality Improvement Act of 1970, Section 309 of the Clean Air Act, and Executive Order 11514. We have included this in our appendix along with advice on entering the environmental impact assessment field as an environmental professional.

Acknowledgments

Thank you to our publisher, who suggested this collaboration. We greatly appreciate our students who have motivated us and helped us hone our work. Nathan Hamilton, Daniel Martínez, Lee Ann McLaughlin, Michelle Radley, Joe Staples, Karen Wilson, and Afton Trotter read portions and provided helpful comments. Patricia (Tricia) Boyle, an artist and 2021 graduate of the University of Southern Maine, Department of Environmental Science and Policy, helped create the figures and deserves our deep appreciation. The late giant in EIA, Dr. Larry Canter, was a gracious correspondent for many years. Thank you to the library staff of the University of Southern Maine. Thank you Dan Sanford for helping with the cover photo. Thank you Robin Sanford for your support. Kristine Holtgrieve is very appreciated as a project manager and amazing idea person when things seem stalled. Of course, any mistakes, missteps, and omissions are our own.

TOPIC 1

Introduction

1.1	Brief history of environmental impact analysis	1
1.2	EIA and public awareness	2
1.3	EIA as research	5
1.4	EIA as a decision-making process	6
1.5	Comparisons with Strategic Environmental Assessment	13
1.6	Current issues with EIA and NEPA effectiveness in the United States	14
1.7	EIA in national and global affairs	16
1.8	Concepts and terms	16
1.9	Implementing EIA	18
1.10	Selected resources	19
1.11	Topic references	20

1.1 BRIEF HISTORY OF ENVIRONMENTAL IMPACT ANALYSIS

Modern techniques in paleontology, archaeology and mapping technology have confirmed that humankind not only occupied multiple biomes in the distant past, but their members left considerable evidence of adaption to environmental differences. The use of fire and hunting implements are long known, as are their consequences. While the "Pleistocene over-kill hypothesis" that humans caused the loss of megafauna lacks conclusive evidence (Nagaoka et al., 2018), it is clear that humans influenced and changed their environments. We did more than eliminate or enhance populations of animal and plant species. We changed entire ecosystems. Direct modification of river systems can be tracked back to the Fertile Crescent's Tigris and Euphrates Rivers 6,000 years ago. Control of water there and in other centers of plant and animal domestication triggered a widespread agricultural revolution. Farming allowed higher birth rates and greater survival for humans, with corresponding environmental changes. That and other technological and social changes led to a gradual awareness of the need for environmental accountability. The United States has a long history of human-altered ecosystems, beginning with Native Americans over 20,000 years ago (Pyne, 1997).

This book is based on one question of public environmental accountability. That question is, "What would happen to our environmental resources, and to us, if we do this. . . ?" You can fill in the blank for the action. It could be burning the Amazon rainforest or landscaping your front yard. The idea is that all decision makers are responsible for their actions that impact our common environment. The developments of earth moving, water management, materials burning, and waste creation can be traced through centuries into the 19th-century industrial revolution and 20th-century age of overconsumption at the expense of almost all things we call "resources." (Note: *Resource* is a term used to identify something that is useful to people, and *waste* refers to something without value to humans.)

In the United States, concerns about the effects of human activity on the environment go back at least to the 19th and early 20th centuries in the works of Henry David Thoreau, George Perkins Marsh, John Muir, Rachel Carson, Aldo Leopold, Ian McHarg, and others. George Perkins Marsh authored *Man and Nature or Physical Geography as Modified by Human Action* in 1864. We can call that book the first Environmental Impact Statement (EIS)!

Most of us now understand and accept that human activity was and is a direct cause of changes to our planet's entire ecosystem made up of air, water, earth material, and living things. What is less widely recognized is that these actions can be measured, evaluated, and presented to political, economic, social, and personal decision makers. The first step in the education of decision-making about environmental issues is to identify harmful impacts and learn how to correct them to benefit the ecosystem and its human inhabitants. This process is currently being carried out and is proving to be very effective wherever it is made an official part of land or resource development. Environmental Impact Assessment (EIA) is a process of thinking things through, discussing, evaluating, and recommending approval or denial of a project or action that may have effects on environmental resources. The first formal experiment with the environmental impact assessment process was the passage of the National Environmental Policy Act (NEPA) in the United States in 1969. NEPA assigned environmental accountability to every federal agency in addition to their other responsibilities. An Environmental Assessment (EA) and its more in-depth version, an Environmental Impact Statement (EIS), are two documents that can result from the assessment process. (Note: An EIS outside of the United States is generally referred to as an EIA, whereas in the United States, an EIA is a process rather than a document. However, the terms can overlap, muddying the distinction.) Topic 2 presents a summary of information you should know to begin working with EIA projects and documents.

Assignment

1. Why should all responsible people be aware of the process of environmental review in public decision-making?
2. What were some of the first ecosystem alterations by humans in what would become the United States?

1.2 EIA AND PUBLIC AWARENESS

Any media search with the words "environmental impact" will likely generate many thousands of hits. For use in this text, we identify two major kinds of environmental impact assessments. They are impact assessments of *ongoing* actions or processes such as mining, fishing, transporting, or consuming resources, or they are assessments of *proposed* development of new projects. The process or system of assessing ongoing actions is a form of monitoring. The actions may or may not have effects on local ecosystems or other environments. The second category of impact assessment—of proposed actions—is what usually draws attention from the public at large and those proposing or opposing a "new" project. The most visible examples of such projects are land development, infrastructure improvements, energy growth, and resource extraction.

This distinction is identified by our legal systems in the way government agencies deal with the participants. The ongoing environmental assessments usually include a research component and an environmental compliance requirement that involves meeting "standards of practice." This might be compliance with emissions levels into the atmosphere or quotas of allowable fish catches in specific watersheds.

Development impacts are often, but not always, predicted and evaluated before agencies will issue an approval to go forward with the project. This kind of approval process is guided by government requirements. The focus here is on the process of gathering existing and potential impacts, then reporting them. That is what this book is about.

Projects that have drawn considerable attention in North America are things like building natural gas pipelines, highways, wind turbine farms, placement of international border barriers, mining in critical wildlife areas including salmon habitats, "fracking" oil extraction, dam relicensing, and projects that imperil threatened or endangered species of plants and animals. In most of these circumstances, and others, the "players" (stakeholders) need information to discuss the merits of projects and to make decisions about them. The best resource for information in these cases should be and almost always is politically neutral, prepared by a very knowledgeable team, and is available to any and all who wish to read it. Seldom do we hear a major media presenter say the words "Environmental Impact Statement," which belies the influence and power of EIA in environmental accountability as part of government decision-making. Fortunately, detailed information is available to those who want it. This common knowledge collected through environmental assessments and independent research is available to everyone. The EIA process mandates public notices and public involvement. Figure 1.1 is a typical public notice that an agency has determined that a federal action will not have a significant impact. Therese are relatively common. We can expect orders of magnitude more of these public notices of Finding of No Significant Impact (FONSI) in comparison with project that might actually have a significant impact and therefore receive an EIS. For every EIS, we might expect around 200 EAs (Gerrard, 2009).

Assignment

1. Identify an informational news link that refers to or contains the use of an Environmental Impact Statement.
2. What is being done in your geographic area (or agency) to promote awareness of EISs and the EIA process?

PUBLIC NOTICES

NOTICE OF FINDING OF NO SIGNFICANT IMPACT AND NOTICE OF INTENT TO REQUEST RELEASE OF FUNDS

March 14, 2021, City of Portland, Maine
389 Congress Street Room 313, Portland, Maine 04101
(207)874-8711

These notices shall satisfy two separate but related procedural requirements for activities to be undertaken by the City of Portland.

REQUEST FOR RELEASE OF FUNDS

On or about March 30, 2021, the City of Portland will submit a request to the HUD Region 1, Boston, MA, for the release of:

Project: Middle Street Housing Partners LP.
1. $193,266 HOME funds under Title II of the Cranston Gonzalez Affordable Housing Act of 1996, as amended;
2. CDBG entitlement funds under Title I of the Housing and Community Development Act of 1974, as amended;
3. $368,676 Section 8 Project Based Voucher Program authorized by Section 8(o)(13) of the U.S. Housing Act of 1937 (42 U.S.C. § 1437f(o)(13)) as amended.

To undertake a project described as 83 Middle Street Apartments for the purpose of providing construction of new rental housing development at 83 Middle Street, Portland Maine. Forty-five new rental units will be made available to seniors age 55 and older and qualified income at 80% AMI or less. The Property is located in the City of Portland and owned by Community Housing of Maine. Estimated $193,266 HOME funds, $200,000 CDBG funds, and $368,676 Section 8 Project Based Voucher Program from Portland Housing Authority, total project cost of $10,483,388.

FINDING OF NO SIGNIFICANT IMPACT

The City of Portland has determined that the projects will have no significant impact on the human environment. Therefore, an Environmental Impact Statement under the National Environmental Policy Act of 1969 (NEPA) is not required. Additional project information is contained in the Environmental Review Record (ERR) on file at City of Portland, 389 Congress St. Portland, ME 04101, and may be examined or copied weekdays 8 A.M to 4 P.M.

FIGURE 1.1 A public notice of Finding of No Significant Impact (FONSI) commonly associated with the dispensation of federal funding.

PUBLIC COMMENTS

Any individual, group, or agency may submit written comments on the ERR to the City of Portland, Division of Housing and Community Development Department 389 Congress St. Room 313, Portland, ME 04101. All comments received by March 29, 2021, will be considered by the City of Portland and Portland Housing Authority prior to authorizing submission of a request for release of funds. Comments should specify which Notice they are addressing.

ENVIRONMENTAL CERTIFICATION

The City of Portland certifies to HUD Region 1 Boston that Jon Jennings in his official capacity as City Manager consents to accept the jurisdiction of the Federal Courts if an action is brought to enforce responsibilities in relation to the environmental review process and that these responsibilities have been satisfied. HUD's approval of the certification satisfies its responsibilities under NEPA and related laws and authorities and allows the City of Portland and the Portland Housing Authority to use HOME & CDBG Program funds.

OBJECTIONS TO RELEASE OF FUNDS

HUD Region 1, Boston will accept objections to its release of funding in the City of Portland certification for a period of 15 days following the anticipated submission date or Its actual receipt of the request . . . [TEXT OMITTED]

Jon Jennings, City Manager, 389 Congress Street, Portland, ME 04101

Date of Notice: March 14, 2021

FIGURE 1.1 (Continued)

1.3 EIA AS RESEARCH

EIA is a form of applied research because it involves finding things out though an active investigation. In addition to evaluating impacts, project descriptions and environmental settings are informed by gathering information. EIA practitioners in disciplines such as archaeology and wildlife biology have made interesting discoveries in the field, but most data gathering is project specific and limited in scope. Nevertheless, data acquisition is critical to describing existing conditions for a project site and is used as a starting point for attempting to predict future circumstances. Accordingly, the EIA produces environmental assessment reports in the form of Environmental Assessments, Environmental Impact Statements, Findings of No Significant Impact, Environmental Impact Reports, Phase I, II, and II Reports, and other documents that result from collecting data and producing research for a public decision-making process. All of this refers back to the concept of environmental accountability.

Readers of impact statements often are confronted with appendices placed at the end of the volume at hand. These commonly occupy entire additional volumes for large, complex projects. These report appendages may be bulky and distracting but are probably necessary for understanding the primary text. Agency requirements sometimes limit allowable primary content to a specific number of pages. At the federal level, page limits are specified in the NEPA Council on Environmental Quality (CEQ) regulations, 40 CFR § 1502.7. The CEQ sets up guidelines in these regulations for each federal agency to use in making its own implementation regulations and rules.

Information gathering and analysis in support of a formal EIA is required to be carried out by an interdisciplinary team (40 CFR § 1502.6) which is intended to help ensure comprehensive and balanced treatment of all environmental impact topics. Bachelor's degree level training is usually sufficient to establish expertise, although of course many participants have advanced degrees.

Environmental assessments often act as a collective gathering of information and requirements from multiple sources. Thus, you may find the term "incorporate by reference." This is where previous environmental assessments, historic documents, legal opinions, or other such information are referred to in your document and are made available (usually online) to the current reader. It is a useful way to expedite review of a recent or ongoing impact study. It improves brevity of documents but carries legal implications or regulatory force of the referenced documents.

Assignment

1. Where can you find EIA reports locally?
2. Can an ongoing EIA process be incorporated into an academic research program? If so, give a specific example of how it could be done. If not, why not?

1.4 EIA AS A DECISION-MAKING PROCESS

EIA is a process of thinking things through and is a critical component of environmental planning for new projects. Environmental planning is a broad field of study and practice that has three major subcomponents: traditional land use planning, pollution control, and environmental impact assessment (Ortolono, 1984), as illustrated in Figure 1.2. Many cities, counties and other agencies prepare an Environmental Action Plan (EAP). Environmental plans arise from policy decision-making arenas that may contain related goals and overlapping implementations—that is, they may be part of a larger framework. Ideally, the Environmental Action Plan for policy implementation would include provisions for use of the EIA process (e.g., Daniels, 2014).

Conceptually, the need to conserve environmental resources yet utilize these resources to support our economy results in a balancing act. To responsibly use resources (*conservation*) is quite different than just saving these resources (*preservation*). Conservation practices involve some consumption, but it is informed and is expected to be sustainable. If we need to consume

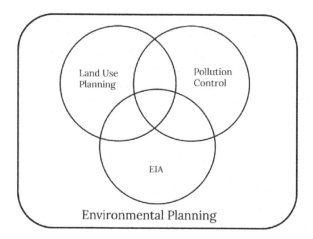

FIGURE 1.2 How EIA relates to environmental planning.

a resource, we should at least know the consequences of that action so that we can make appropriate long-term and short-term trade-off decisions. Sound decision-making requires thorough, objective analysis of all relevant data and complete awareness of trade-offs. This is the rational basis for NEPA. After the NEPA process concludes for a project, the planning process is not done. It can continue to a yes or no decision, which presumably will be made based on the NEPA documents. Thus, after the EIA process has been concluded, the environmental impact forecasting has been done and other, more political and financial processes take over. A rational and sound scientific basis provides the foundation for good public policy decision-making.

Assignment

1. Why is EIA generally considered as one of the three subsets of environmental planning?
2. What contribution can the EIA process make for achieving sustainable land development?

In the United States, the National Environmental Policy Act (NEPA) was enacted "to promote the general welfare," cited in the Preamble of the US Constitution. After years of creating environmental agencies and practicing resource conservation, a new mandate for all federal agencies arose—to undertake a second mission for each of them: environmental accountability. We argue that NEPA is the most important federal environmental law for the 21st century because it applies to all other laws and all agencies whether environmental or not. NEPA requires demonstration of environmental accountability as a part of all environmentally related decision-making by every federal entity. Over five decades, the rest of the developed world followed with similar pronouncements, and most countries at least attempt to do environmental assessments before imposing final decisions on the landscape. NEPA's EIA process is not just for governments. Today, international agencies like the World Bank and International Monetary Fund, most commercial banks, and many investors require an environmental review of some sort as part of their approval process for new developments.

The EIA process follows guidelines promulgated by the lead agency. A major strength of NEPA is improving buy-in by agencies through having them craft their own regulations, with some assistance. The federal Council for Environmental Quality issues guidelines for use by all other federal agencies in creating their own, more specific, guidelines. The (online) guidelines contain mandated and recommended procedures, some of which are summarized here (CEQ regulations and related laws are provided in the appendix). In addition, the states, counties, and municipalities will also have their own regulations and procedures for environmental review: some are NEPA equivalents (so-called little NEPAs) while others such as Vermont's Act 250 are comprehensive like NEPA but result in a yes or no decision, unlike NEPA.

A basic sequence of steps is used in almost all EIAs in the United States and internationally (Figure 1.3). Within the first step, the primary starting factor for environmental planners is the location, description, and condition of the proposed site for development. This is sometimes called *baseline information*. Such information is often in the form of maps with an existing conditions report that inventories what is on the site now from an environmental perspective.

The project description is really a summary of what decisions will be made to carry out the proponent's goals. These decisions may include approvals, location, design, construction, and so on—in short, what will be done, how will they do it, where, and so on. In most cases, it is an ongoing process where teams of participants develop a vision for project managers. The project description must also contain the purpose of and need for the project. The description is the basis from which the EIA flows. A failure to adequately define the project represents a fundamental flaw and a significant source of legal challenges.

Project screening answers the basic question: should an environmental analysis be undertaken for this newly introduced project? A reason for *not* doing the EIA might be circumstances such as the project has already demonstrated that it will not be environmentally harmful (within a Strategic Environmental Impact Assessment). Another reason might be that it is exempted from environmental review by statute or executive order. A third reason might be the fact that the proposed project is similar to others already completed.

Scoping might be visualized as a room full of experts trying to decide if their project *may* have environmental impacts. Of all the possible impacts, those considered to be of concern or perhaps significantly harmful are identified. This is also a time where staffing the project team with qualified people will begin. Scoping is figuring out where and how to put your energies in assessing environmental impacts. Scoping will continue to occur as part of individual categories of impact assessment and various other steps in the process.

With continued reference to the project description, alternatives to the project and alternative ways of carrying out the project are considered. A scoping process includes deciding what is and is not going to be considered among the possible alternatives. An Environmental Impact Statement (EIS) is required to assess and report on alternatives, including the alternative of *not* doing the project. The agency may identify a preferred alternative in this array, based on impacts and the agency's project objective.

The core of almost any environmental impact statement is, obviously, the impacts that the project will create. The prediction of impacts and an evaluation of their significance to the natural environment is what will become an information aid to all the parties that may be involved in the project from day one to final completion. It is sometimes used as an informational source for community groups, political participants, and even the court system.

The most important part of impact analysis is deciding what to do about them. Significant impacts may be mitigated in a variety of ways such as relocation, downsizing, redesign, and so

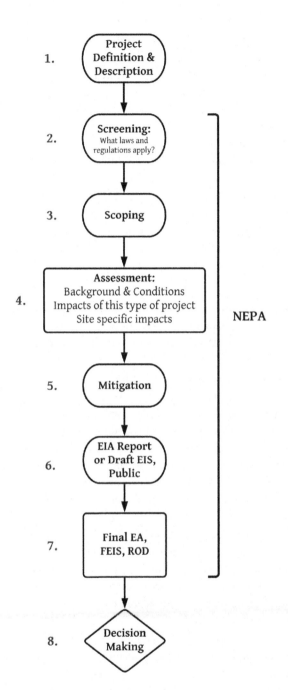

FIGURE 1.3 Stages in the EIA process.

on. Mitigation of impacts to an acceptable level is required in almost all cases. Other ways of dealing with significant impacts are by use of the alternatives option mentioned above. Also, somewhere within the mitigation rubric is the necessity for a monitoring plan to insure that the mitigation measures are actually carried out and that the impacts are tolerable.

The final step in this process is filing a Record of Decision (ROD), which explains why required alternatives, mitigation measures, or monitoring were included in the decision to approve completeness of the EIS. In the case of an Environmental Assessment, the final document upon issuance of the EA is a Finding of No Significant Impact (FONSI). A FONSI may also occur when a project falls below the threshold of requiring an Environmental Assessment, as reflected in Figure 1.1.

Once an environmental document is prepared, it must go to its readers. This stage of the decision-making process is a mandatory part of public participation. It is where the comprehensive project is reviewed and modified by all of the groups and individuals involved. These will include the document preparers, agency representatives, project proponents from the public, project opponents, legal practitioners, media representatives, and students. Public participation is expected for multiple stages of the EIA process and usually carried out by electronic communication and personal or public conferences.

Assignment

1. Define the EIA context for the following terms: judicial review, lead agency, cooperating agency, Environmental Assessment (EA), adequacy, scoping, major action, standards, criteria.
2. Why should public participation occur before the EIS is prepared?

Without physical, political, social, legal, and administrative hindrances, an EIA approval process could take less than a year. In complex cases such as nuclear waste disposal, mining, or petroleum extraction, the process could take multiple years and even a decade.

EIA is a process that inventories information and then prepares for analysis and assessment for making decisions in the overall planning process. One hundred or more countries now have environmental impact assessment laws. These usually contains three perspectives or components. These are a legal perspective based on statutory authority, a science perspective where information is gathered and impartially analyzed, and a social perspective focused on positive outcomes for all parties.

Good decision-making should not hinder the economy, yet there is a persistent myth that EIA wastes time, increases costs, and causes job losses. The opposite is true: EIA creates jobs directly through assessment and related consulting. Further, maintaining a healthy environment is good for society and the economy. The cost of environmental cleanup is reduced through good planning and through making decisions that mitigate environmental harm. Thus, EIA and jobs, time, and costs are not in opposition.

Environmental stability is a social asset and an economic asset. So too is the promise of economic stability *because* of environmental sustainability. Environmental, economic, and social objectives all fold together and integrate in the EIA process. Effective coordination among government agencies is needed to develop and carry out environmental policy. Openness in decision-making gives communities and the public a direct voice in government decisions.

Assignment

1. What did the passage of NEPA do for each federal agency?
2. Identify a federal agency that has publicly available information on its decision-making processes pertaining to NEPA. Summarize the information and reference its source.

When discussion about almost any development project begins, there are process-related questions from a large number of interested people (potential stakeholders). They represent neighborhood groups, companies, public agencies, community planners, and, of course, the developers. Eventually, potential participants will be introduced to the environmental impact review process though the public notification and engagement components. Here are some common questions (with answers) that we have been asked, that have been responded to in "Forty Most Asked Questions Concerning CEQ's National Environmental Policy Act Regulations" compiled by the CEQ in 1986, and that are still valid to this day. This list is posted on the CEQ's website, the US Department of Energy website, and other places. Much of the context for these questions will be provided in subsequent topics, and thus, the questions may make more sense there, but it is useful to start with them here as a quick introduction to NEPA.

What are the steps or stages of the EIA process? The stages follow the same progression, although they may be broken down into more components depending on how detailed we want to get. First we start with a description of the project. Next is screening, the act of determining what laws and regulations apply and if an EIA may be required. The third step is called scoping, where the parameters of the EIA are decided upon and which agencies and other organizations will be invited to participate. Step four involves team information assembly—collecting data—and production of an internal draft of the eventual Environmental Analysis or draft Environmental Impact statement. Step five for an Environmental Assessment is approving, editing, and distributing the DEIS or EA, as well as collecting all of the observations and comments from those who have read it. The next step involves incorporating all of the input from all of the participants into a Final Impact Statement or, in the case of an EA, a final project approval. The approval, in whatever form it takes, will contain all of the decisions about project alternatives, identified impacts, mitigation measures, and monitoring procedures that were incorporated into the project development program. As far as NEPA goes, the acceptance of the final EIS is the end of the process, and approvals are subsequent agency decisions upon the conclusion of NEPA compliance. The distinction is significant: NEPA is a process law, not a permit decision law.

What is a lead agency? Most major development projects involve many decision makers, each representing their own or their agency's interests. Among these interested parties, an agreed-upon lead agency serves as project manager. That agency usually makes the final decisions on project approval.

Who prepares the project description for the EA (Environmental Assessment) or EIS (Environmental Impact Statement)? Usually, the primary or sponsoring entity (lead agency) provides plans for whatever is proposed to be done as a "project" (or "undertaking" or "action"). This information may be modified as necessary during the scoping process. A proper project description sets the stage for the EA/EIS.

Who pays for the costs of an environmental review? Government agencies usually include environmental review cost as part of their project budget. Privately sponsored projects are usually charged a fee for preparation and processing the environmental documents.

How may citizens or special interest groups get involved in the EIA process? It is strongly recommended that individuals or organizations contact the lead agency as soon as possible after the project is introduced. Doing so creates an ongoing flow of information as it becomes available. There will be several required opportunities for public participation to take place once you are "on the list" or through general public notification.

Does approval of the Record of Decision (ROD) mean that the project is approved? No. The ROD only assures that the EA or EIS is adequate for decision-making and is ready for approval or denial of the project application. Once the final Environmental Impact Statement (FEIS) and ROD are issued, the NEPA process is essentially concluded.

Do all of the impacts identified within the project have to be mitigated? No. The keyword is "significance." If it is determined that an action may create a significant impact on environmental conditions, it has to be addressed and changed to a less than significant situation. The impacts after mitigation are residual impacts, upon which the acceptability of the action is based.

What can be done if a reviewing party does not agree with the conclusions in the ROD? If any of the participating parties who followed and participated in the EIA process are in disagreement with a final decision, they may initiate a legal challenge. This is often a manifestation of the concept of "environmental justice."

What happens if changes in the project description occur during the review process or after the FEIS or EA is approved? It is possible that a new EA or EIS would be required, but, most likely, a supplemental document would be required, reviewed, and approved.

How are project alternatives managed? There may be alternative design features, alternative locations, or other alternative functions such as mitigation features within a project description. Three alternatives are a "no alternative" (the no-action alternative, which is required to be addressed), a "preferred alternative," and the most environmentally "sustainable alternative." Of course, an alternative may fall into two of these three categories.

Who recommends or determines the "preferred alternative"? The lead agency's project manager usually makes that decision in consultation with team members. The agency may start out with a preferred alternative, or it may use the assessment to point to one.

Who recommends or determines what is environmentally preferable? Again, this is usually a team decision of the lead agency based on internal expertise and comments from other agencies and the public.

What if lead and cooperating agencies disagree on the EIA scope or level of analysis? The lead agency has ultimate responsibility for the content of the document, but it must demonstrate cooperation in making those decisions.

What is meant by the term "third party contracts," and how may they be used? The term is used by applicants who contract with a consulting firm to prepare an EIS. The lead agency has direct oversight over research and production of the document even if the consultant fee is paid by the applicant.

What is the scope of mitigation measures that must be discussed in an EA or EIS? The mitigation measures must cover the range of all impacts presented in the document. The mitigation measures must be considered even if, by themselves, they would not be considered significant.

May state and federal agencies serve as joint lead agencies? Yes. Collaboration encourages efficiency and shared responsibility.

When are EISs required on policies, plans, or programs? An EIS must be prepared if an agency proposes to implement a specific policy, to adopt a plan for a group of related actions, or to implement a specific statutory program or executive directive. Often these are Strategic Environmental Assessments.

When is an area-wide or overview EIS appropriate? The preparation of an area-wide document may be useful when similar actions, viewed with other foreseeable or proposed agency actions, share common timing or geography.

What is tiering, and how is it used? Tiering is a procedure that allows an agency to avoid duplication of paperwork through the incorporation by reference of an environmental statement of broader scope or vice versa. Site-specific EISs might follow more general documents already prepared and incorporate them by reference.

When is it appropriate to use appendices instead of including information in the body of an EIS? Lengthy technical discussions of methodology, baseline studies, or other such work are best reserved for the appendix. All findings and conclusions should not be reserved to appendices.

How are responses to comments from reviewers of Draft EAs and EISs treated? Normally the responses should result in changes to the text of the EIS, not simply a separate answer at the back of the document. If the agency decides that no substantive response to a comment is necessary, it must briefly explain why.

Under what circumstances do old EISs have to be supplemented before taking an action on a proposed project? As a rule of thumb, if the proposal has not yet been implemented after 5 years, it should be carefully reexamined to determine if new circumstances compel preparation of an EIS supplement.

How long or detailed must an Environmental Assessment be? Since the EA is a concise document, it should not contain long descriptions or detailed data that the agency may have gathered. Rather, it should contain a brief discussion of the need for the proposal, alternatives to the proposal, the environmental impacts of the proposed action, mitigation measures, and a list of agencies and persons consulted.

What is the level of detail of information that must be included in a Finding of No Significant Impact (FONSI)? The finding must succinctly state the reasons for deciding that the action will have no significant environmental effects. An EA might accompany the FONSI in support of the findings.

Can an EA with a finding of no significant impact still require mitigation measures? Yes, mitigated EAs are fairly common. Mitigation is about reducing impacts and good environmental planning is about properly managing impacts.

Assignment

1. Are FONSIs issued for EISs?
2. Can a project receive both an EA and an EIA?
3. Do private corporations prepare EISs?
4. Explain why an EA is more common than an EIS.
5. Why might it good public policy to have an agency prepare its own EIS?

1.5 COMPARISONS WITH STRATEGIC ENVIRONMENTAL ASSESSMENT

Strategic Environmental Assessment (SEA) is a process and a tool for evaluating the effects of proposed policies, plans, and programs on natural resources, social, cultural, and economic conditions. Thus, SEA takes a very wide foreseeable scope of possible development projects. The focus is deliberately wide because many of the pressures on both the physical environment and social conditions are a result of custom, tradition, and institutional factors such as the ownership and management of land. It is not possible to analyze the effects of policies on physical resources without understanding the social, cultural, economic, and institutional context in which resource exploitation takes place. It is a decision support process in which impacts and environmental sustainability aspects of a development idea are considered in policy, planning, and program creation. Thus, it is anticipatory and a planning tool rather than a reaction, hence the "strategic" designation.

Planning at the policy level for environmentally based programs will sometimes incorporate a Programmatic EIS, which usually resembles an SEA for a series of similar projects or actions. An SEA may also be conducted before a corresponding EIA is undertaken. This means that information on the environmental impact of a plan can cascade down through the tiers of decision-making and can be used in an EIA at a later stage.

Riki Therivel, a globally known environmental impact analyst, defines SEA as a process that aims to integrate environmental and sustainability considerations in making decisions (Therivel, 2010). In a perfect world the SEA would be undertaken parallel with the physical plan-making. Obvious problems with this procedure are many including data availability, differing scales of design and decision-making, different time frames, and the volatility of each component of the plan. Often there are significant plan changes up to the final go or no-go decision. The frequency of having to do supplemental impact statements demonstrates this reality. SEAs are less frequently used in the United States as in many other countries for the above and other reasons.

Another way to address the potential effects of an ongoing project is *tiering*, the use of different layers of planning, with shared information (Therivel and González, 2021). This procedure may utilize three or more EISs as a regional plan, then a site plan, down to an actual development plan. A benefit to an agency that uses this process is that cumulative or otherwise connected projects are addressed.

Assignment

1. What are the main reasons for use of Strategic Environmental Assessments (SEAs) in environmental planning?
2. How can "tiering" connect SEA to cumulative impact analysis?

1.6 CURRENT ISSUES WITH EIA AND NEPA EFFECTIVENESS IN THE UNITED STATES

This subject could have easily been placed in the back of this book, where you would have already learned how to create environmental documents to be used by decision makers. It is placed here so that you can be on the alert for ways to make the EIA process more effective throughout your experiences in environmental decision-making.

The issues we have listed will appear in several places throughout this book. By far, the most important impact in most environmental reviews is global warming and climate change. It is the ultimate cumulative impact.

It is possible to calculate changes in carbon concentrations, but thus far, no adequate placement of responsibility for those emissions has been agreed upon. The techniques of measuring such impacts and their significance are presented in *Drawdown. The Most Comprehensive Plan Ever Proposed to Reverse Global Warming*, edited by Paul Hawken (2017). Potential alternatives and mitigation measures are known, but placing and accepting responsibility for remediation are not agreed upon. Related catastrophic events are likewise under-acknowledged even though occurrences can reasonably be predicted.

Environmental changes such as global sea level rise, ocean acidification, and plastic pollution have affected the world's oceans, but maritime laws are wholly inadequate for addressing responsibility or mitigation actions.

At the national level, NEPA has survived decades of challenges and changes, mostly for the better. But there are continued suggestions and proposals for what some call improvements and others describe as weakening of jurisdiction and effectiveness. Jurisdiction of local, state, and federal roles in large projects needs to be clarified, especially when natural resources are at issue. Infrastructure projects, which are usually federally funded, also may have very profound effects on local circumstances. Length of time required for project approval (or denial) seems to be complained about by all involved parties in an environmental review, yet adequate public involvement demands long time frames.

Some commonly expressed concerns about the effectiveness of environmental review processes are listed here; watch for them in your reviews of EAs or EISs.

1. Determining the geographic limits of a project anywhere between micro to global levels is important but often overlooked. Transportation projects usually draw attention to this issue.
2. Implementation of mitigation measures sometimes get neglected. Monitoring and enforcement of commitments is sometimes neglected. This would be particularly important for long term mitigation such as reforestation.
3. Measuring thresholds, limits, capacities, and compliance are sometimes used as indicators of impacts without explanation. Terms like restoration, sustainability, ecosystems services, and so on are descriptive but often not really measured.
4. Some important factors in project development such as profitability, commercial competitiveness, compliance with some planning regulations, or the identities of some persons involved are not impacts and are not required to be included in an impact review. Other marginal topics such as disaster planning, health hazards, strategic plans, and social issues are not often required by statute but have frequently been included through the scoping process of a project proposal.

Some other major issues for EIA effectiveness as pointed out by Larry Canter (1996) are:

1. Adverse environmental impacts may be underestimated because they have not occurred yet, or the science has not progressed adequately, or because of preparer bias or inadequate review.
2. NEPA is usually a case-by-case review, and cumulative impacts can be missed or improperly addressed.

3. NEPA has a limited scope of applicability and does not usually apply to state or private actions that do not trigger federal jurisdiction.
4. Periodic attempts to streamline the environmental documentation process can affect its range of coverage and its thoroughness.
5. Misunderstanding that NEPA is only a process in a larger arena of planning, and that EISs are a part of this process, not the end of planning.

Assignment

1. What do you think forms a basis for the majority of challenges to a DEIS?
2. What are three reasons why EISs tend to underestimate impacts?
3. Overall, is NEPA working? Is it effective? Support your answers.

1.7 EIA IN NATIONAL AND GLOBAL AFFAIRS

After NEPA was adopted in 1969 and shown to be a valuable aid to environmental decision-making in the United States, similar requirements were implemented in most developed countries, starting with Canada, Australia, Germany, and France. By 1996, more than 100 countries had adopted EIA systems similar to NEPA (Sadler, 1996), and this number has continued to increase (Glasson and Therivel, 2019). In addition, EIA procedures are provided by non-profit corporations or various other agreements or protocols. Trade agreements, maritime laws, and political treaties also contain EIA components. In many cases, funding of new projects is dependent on environmental review. This is the case for the World Bank and the International Monetary Fund. Most of these systems are in the form of mandatory regulations, while others are in the form of guidelines but not requirements. Enforcement also varies by country.

The UN Environment Programme (UNEP) provides publications and training for developing countries. An important private sector global initiative was the publication of *Equator Principles*, which provides guidelines for financial institutions in relation to funding decisions on major international projects. As our book title indicates, we are primarily addressing EIA practices in the United States, but note that there are many American sources of expertise in this field that assist groups in other countries working in this realm. An example is *Environmental Law Alliance Worldwide*, located in Eugene, Oregon, which has advised organizations in over 80 countries.

Assignment

1. How does the EIA process in a country of your choice differ from NEPA in the United States?
2. How are Canada's EIA processes similar to NEPA?

1.8 CONCEPTS AND TERMS

We could easily devote hundreds of pages to a dictionary of EIA terms, but instead we focus on a few of the most important ones to get started.

Baseline information: Project criteria and environmental background including existing levels of impact already occurring.

Categorical exclusion (CX): An action an agency has listed as exempt from an EIA because the action has only an insignificant level of potential impact.

Cooperating agency: One of the other agencies or organizations that participate in the EIA process along with the lead agency.

Council on Environmental Quality (CEQ): This federal body exists within the executive branch of government. It is tasked with implementing NEPA through development of regulations, advising the president, and dealing with policy matters.

Environmental Assessment (EA): A concise public document that analyzes the environmental impacts of a proposed federal action and provides sufficient evidence to evaluate the significance of impacts. An EA may evolve into an EIS as information increases. Environmental assessment (in lowercase letters) refers to the general process of assessing environmental impacts rather than a specific document.

Environmental Impact Assessment (or **Analysis**; EIA): The process of evaluating the effects of an action or plan that may cause a negative impact on environmental resources or circumstances. Outside of the United States, EIA tends to mean a document equivalent to an EIS in the United States.

Environmental Impact Statement (EIS, DEIS, FEIS): A detailed statement prepared when an agency proposes an action that may significantly affect the environment. It may be in draft or final form.

Environmental setting: The site or parcel that contains an existing or potential project. May include off-site property if part of the same ecosystem or if associated with potentially significant impacts.

Finding of No Significant Impact (FONSI): A public document that briefly presents reasons why an action will not have a significant effect on the environment.

Lead agency: The primary governmental party that serves as a project manager or sponsor/author for the preparation of environmental documents.

Mitigation measures: A discussion of the means to mitigate (reduce or eliminate) all adverse environmental impacts identified in an environmental review document.

Monitoring: A program or stage in the EIA process or project approval conditions in which environmental impacts are tracked, reported, and addressed.

National Environmental Policy Act (NEPA): The first and most often used law that requires study and analysis of potential impacts from proposed projects. This requirement placed a burden of environmental accountability upon each federal agency by charging them with environmental impact assessment procedures.

Phase 1 Report: Document that seeks to investigate whether or not a particular environmental resource (such as an archaeological site) or environmental impact (such as a contamination from an oil spill) exists. It asks the question, "Is anything there?"

Phase II Report: Document that determines the extent of an environmental resource (such as an archaeological site) or environmental impact (such as an oil spill) at a site or setting. It asks the questions, "Something is there, where is it, how much of it?"

Phase III Report: Document addressing recovery or salvage of an environmental resource or environmental impact of known type and distribution, often as determined by Phase I and II Reports.

Project alternatives: A required component of any environmental review document. Alternatives could include locations, time frames, design differences, size or scale, or other factors within the project description.

Project description: A detailed and required component of any environmental review which includes plans, policies, and programs.

Project phases: Projects typically have four phases or stages, each with potential impacts: planning, construction, operation, and closure/post-operation.

Record of Decision (ROD): A public document that reflects the agency's final decision about a proposed action. It should include the rationale behind that decision and commitments to monitoring and mitigation.

Scoping: The process of determining the types of actions to be included within the study and determining the responsibilities of each participating party.

Screening: A process that determines whether an environmental analysis should be undertaken for a newly introduced project. Includes determining what rules and regulations apply.

Strategic Environmental Assessment (SEA): Sometimes called a programmatic EIS, it is effectively used in policy-oriented decision-making for plans or programs.

Tiering: The process of preparing multiple levels of environmental review, usually from general to more specific, as new information is available. Attempts accurate transfer of information among the levels.

Assignment

1. Select and explain three of the five terms provided by your instructor from the above list.
2. What are the top two or three terms you might suggest be added to the list in this section?

1.9 IMPLEMENTING EIA

EIA is a live planning and development process. It is not about financing, permits, or construction. It is about the inclusion of environmental thinking in planning and development of new features in local to global landscapes. If you go through the EIA process properly, you will have complied with most environmental legal requirements. However, this is different from permit laws, which generally deal with yes or no decisions. This difference arises from a philosophical approach to environmental compliance on several levels. EIA participants should not hide the impacts. They should find out what they are, acknowledge them, and see what mitigative measures are available. Be fair to all decision makers. That will keep everyone out of the courtroom. A reasonable assumption for all is that a good project will stand up to the light of day through a rigorous formal review process. That said, the EIS process is complex, long, and often involves numerous embedded and related permits as part of going through the environmental assessment.

Assignment

1. What components or characteristics do you think should be in a definition of the ideal EIA?
2. Why does being a process law rather than a permit decision law (e.g., yes or no) help make NEPA successful?

Under perfect conditions with unlimited resources, the EIA final product is a tool for making inspired decisions. It should always aspire to such. As described by Canter (1996), the EIS document should:

- Capture all significant impacts
- Compare reasonable alternatives
- Read clearly and exhibit organization
- Present understandable graphics
- Contain strong documentation
- Demonstrate decision-making processes
- Encourage and record public participation
- Provide for monitoring of outcomes
- Remain timely and provide feedback
- Be legally and practically enforceable.

The review of the draft EIS (DEIS) provides a self-correcting mechanism so that the final EIS is an improved document of environmental accountability. Still, NEPA is met once the final EIS is issued. But to ensure the policy continues to improve in effectiveness for conserving environmental resources and reducing environmental impacts, the overall process needs to feed back into the policy-making aspects of EIA so that future assessment reports are improvements over previous ones.

Assignment

1. An EA does not necessarily need to consider alternatives to the project, but an EIS does. What are some characteristics an EA shares with an EIS?
2. If you were on an impact assessment team working for a government agency, what tasks would be most interesting to you: fieldwork, writing, geographic information systems, graphics, project management, lab science, data collection, public relations, legal research and regulatory compliance, historical research, other? Circle or write in the ones you would choose.

1.10 SELECTED RESOURCES

Bass, Ronald, Albert Herson and Kenneth Bogdon. (2001). *The NEPA Book: A Step-By-Step Guide on How to Comply with the National Environmental Policy Act*. Point Arena, CA: Solano Press.
Canter, Larry. (1996). *Environmental Impact Assessment*. New York: McGraw-Hill.
Canter, Larry. (2015). *Cumulative Effects Assessment and Management*. Horseshoe Bay, TX: EIA Press.
Counsel on Environmental Quality. (CEQ) www.whitehouse.gov/ceq/. A clearing house for all sorts of NEPA information.
Counsel on Environmental Quality (CEQ). (1986). Forty Most Asked Questions Concerning NEPA. www.energy.gov/nepa/downloads/forty-most-asked-questions-concerning-ceqs-national-environ-mental-policy-act.
Daniels, Tom. (2014). *The Environmental Planning Handbook for Sustainable Communities and Regions*, 2nd ed. Chicago: Planners Press.
Eccleston, Charles. (2000). *Environmental Impact Statements: A Comprehensive Guide to Project and Strategic Planning*. New York: Wiley.

Eccleston, Charles. (2017). *Environmental Impact Assessment: A Guide to Best Professional Practices*. Boca Raton: CRC.

Environmental Impact Assessment Review (EIA Review) Is a Refereed, Interdisciplinary Journal Serving a Global Audience of Practitioners, Policy-Makers, Regulators, Academics and Others with an Interest in the Field of Impact Assessment. Elsevier.

Environmental Law Alliance World Wide. www.elaw.org/.

Environmental Practice, Is the Official Journal of the National Association of Environmental Professionals. It Provides an Open Forum to Members and Other Concerned Individuals for the Discussion and Analysis of Significant Environmental Issues. The Journal Is Available at Libraries or by Becoming a NAEP Member.

Environmental Protection Agency. Environmental Impact Statement Database. https://cdxnodengn.epa.gov/cdx-enepa-public/action/eis/search.

Environmental Protection Agency. www.epa.gov/.

Glasson, John and Riki Therivel (Eds.). (2019). *Introduction to Environmental Impact Assessment*, 5th ed. New York: Routledge.

Hawken, Paul (Ed.). (2017). *Drawdown. The Most Comprehensive Plan Ever Proposed to Reverse Global Warming*. The Drawdown Project. New York: Penguin Books.

Morgan, Richard. (1998). *Environmental Impact Assessment: A methodological Perspective*. New York: Kluwer Academic Publishers.

Morris, Peter and Riki Therivel (Eds.). (2009). *Methods of Environmental Impact Assessment*, 3rd ed. New York: Routledge.

Morrison-Saunders, Angus. (2018). *Advanced Introduction to Environmental Impact Assessment*. Cheltenham: Edward Elgar.

Noble, Bram. (2015). *Introduction to Environmental Impact Assessment: A Guide to Principles and Practice*, 3rd ed. Don Mills, Ontario: Oxford University Press.

Petts, Judith (Ed.). (1999). *Handbook of Environmental Impact Assessment*, Volumes 1 and 2, New York: Wiley (Blackwell Science).

1.11 TOPIC REFERENCES

Canter, Larry. (1996). *Environmental Impact Assessment*, 2nd ed. New York: McGraw-Hill.

Council on Environmental Quality. (CEQ). (1986). Forty Most Asked Questions Concerning NEPA. www.energy.gov/nepa/downloads/forty-most-asked-questions-concerning-ceqs-national-environmental-policy-act.

Daniels, Tom. (2014). *The Environmental Planning Handbook for Sustainable Communities and Regions*, 2nd ed. Chicago: Planners Press.

Gerrard, Michael B. (2009). *The Effect of NEPA Outside the Courtroom*. 39 ELR 10615–10617. Washington, DC: Environmental Law Institute.

Glasson, John and Riki Therivel. (2019). *Introduction to Environmental Impact Assessment*, 5th ed. New York: Routledge.

Hawkin, Paul (Ed.). (2017). *Drawdown: The Most Comprehensive Plan Ever Proposed to Reverse Global Warming*. The Drawdown Project. New York: Penguin Books.

Nagaoka, Lisa, Torben Rick and Steve Wolverton. (2018). The Overkill Model and Its Impact on Environmental Research. *Ecology and Evolution*, 8(19), 9683–9696. https://onlinelibrary.wiley.com/doi/full/10.1002/ece3.4393.

Ortolono, Leonard. (1984). *Environmental Planning and Decision Making*. New York: John Wiley & Sons.

Pyne, Stephen J. (1997). *Fire in America: A Cultural History of Wildland and Rural Fire*. Seattle: University of Washington Press.

Sadler, Barry. (1996). *Environmental Assessment in a Changing World: Evaluating Practice to Improve Performance*. UNECE Final Report. https://unece.org/DAM/env/eia/documents/StudyEffectivenessEA.pdf.

Therivel, Riki. (2010). *Strategic Environmental Impact Assessment in Action*. New York: Routledge.

Therivel, Riki and Ainhoa González. (2021). "Ripe for Decision": Tiering in Environmental Assessment. *Environmental Impact Assessment Review*, 87. March 2021, 106520.

TOPIC 2

The EIA process

2.1	History of NEPA and the EIA process	21
2.2	Environmental law and legal basis for NEPA	25
2.3	Judicial review and environmental justice	26
2.4	Roles in the environmental review process	27
2.5	Integrating with other environmental laws and procedures	27
2.6	EIA process in the various states and other jurisdictions	28
2.7	International EIA and globalization	30
2.8	Managing the NEPA process	33
2.9	Concepts and terms	42
2.10	Selected resources	44
2.11	Topic references	45

2.1 HISTORY OF NEPA AND THE EIA PROCESS

This topic covers the history of NEPA and the variety of projects and circumstances that are within its purview. It also discusses the legal foundations for use of environmental review documents. Legal concepts such as environmental justice, judicial review, public domain, and public trust are introduced. We address the roles of major stakeholders—agencies, the public at large, elected officials, document preparers, planners, and of course, environmental protection advocates. Finally, we step through the general Environmental Impact Assessment (EIA) process from inception to final approval.

The National Environmental Policy Act (NEPA; www.epa.gov/nepa) was the first comprehensive US environmental law. Enacted in 1969 under the Nixon administration and signed into law in 1970, NEPA requires all federal agencies to go through a formal process before taking any action anticipated to have substantial impact on the environment. Jim Kershner (2011) provides a history of its political origins; the EPA contains articles on the history of NEPA on its website (www.epa.gov/history/epa-history-national-environmental-policy-act); and of course there are numerous other sources.

DOI: 10.4324/9781003030713-2

The EPA reviews almost all federal environmental documents and many state environmental impact studies. It also sets standards by which impacts are evaluated, curates EISs as a repository, and it prepares them for the EPA's own projects. These multiple roles make EPA a key player in almost all NEPA-related decisions. The EPA is thus a useful resource for NEPA-related actions that may affect you. Anyone can obtaining EISs at the NEPA website (www.epa.gov/nepa/how-obtain-copy-environmental-impact-statement).

Historically, approval of development projects came from an agency's need or desire for it, its cost, the feasibility of getting it done, and political will or desire. Environmental impact was not a primary concern. After 1970, NEPA made environmental consequences part of the equation, defining environment—and environmental impacts—in a large, inclusive manner. Most projects now requiring environmental review are changes in land use, although policies and other actions may still qualify. The way that EIA fits into the land development process and land-altering actions by a federal agency is addressed by the respective agency and, writ large, in a Council on Environmental Quality (CEQ) booklet (https://ceq.doe.gov/docs/get-involved/Citizens_Guide_Dec07.pdf; CEQ, 2020a). NEPA as an environmental planning policy exerts a jurisdictional pull through licensing, permits, federal funding, federal ownership of involved land, specific legislation, and direct actions or proposals by a federal agency.

In addition to federal agency undertakings, many state and local agencies and even some private organizations are required to go through the NEPA or similar process due to a federal connection. Each city or other jurisdiction has its own land development requirements and most, but not all, include some form of an environmental review process. Some cities such as New York have their own specific NEPA equivalent, as do more than 20 states. Thus, there are NEPA "tiers" of potential NEPA jurisdiction and state or local NEPA equivalents.

The National Environmental Policy Act (NEPA) process begins when a federal agency receives or develops a proposal to take a "major" federal action (also known as an "undertaking"). These actions are defined at 40 CFR § 1508.18. The environmental review under NEPA can involve three different levels of analysis:

Level 1. Categorical Exclusion determination (CATEX or CE or CX)
Level 2. Environmental Assessment/Finding of No Significant Impact (EA/FONSI)
Level 3. Environmental Impact Statement (EIS)

The depth of analysis progresses through each level of analysis. The determination of what level to undertake is itself a contestable decision. Public disbursement of the documents and decisions are a key component in allowing participation by project proponents and opponents and epitomize the meaning of civic involvement in environmental actions.

At the first level, an action may be "categorically excluded" from a detailed environmental analysis if the federal action will not have a significant effect on the human environment. The reason for the exclusion and a list of excluded actions are generally detailed by each federal agency and posted on its website. An obvious concern for all parties involved is how the exclusion decision is made, but it is the principle of agency autonomy that lets each agency make its own list of exemptions. Professionalism, peer review, and public scrutiny all help make the decision process accountable.

A federal agency can determine if a proposed action is on its list of Categorical Exclusions (CATEX) that do not require environmental assessment. In these cases, the process is complete. If the project is not exempt, the agency may then prepare an Environmental Assessment (EA) or launch into a full Environmental Impact Statement (EIS) if the project clearly has significant impact potential. The EA determines whether an action has the potential to cause significant environmental effect. In the case of fairly minor impacts, the EA may be the concluding step, accompanied by a Finding of No Significant Impact (FONSI), otherwise the EA is a preliminary scoping process for a full draft Environmental Impact Statement (DEIS), and subsequently a final Environmental Impact Statement (FEIS). In conjunction with the CEQ, each federal agency has adopted its own NEPA procedures for the preparation of EAs (https://ceq.doe.gov/laws-regulations/agency_implementing_procedures.html, 2020b).

An EA may be expected to include a brief discussion of need, alternatives, impacts, and information sources. First is the need for the proposal or action (i.e., the project). Alternatives should be discussed when there are unresolved issues concerning ways to utilize resources. An EA should present the predicted environmental impacts of the proposed action and their alternative solutions. Finally, the EA should contain a listing of agencies and persons consulted.

The EA is roughly equivalent to a "Phase 1," which is used in environmental engineering for the analysis of a site for pollutants and asks the question, "Is there anything there?" Based on the EA, one of two actions will occur. If the agency determines that the action will not have significant environmental impacts, it will issue a Finding of No Significant Impact (FONSI). A FONSI is a document that presents the reasons why the agency has concluded that there are no significant environmental impacts projected to occur upon implementation of the action. Federal agencies will prepare an Environmental Impact Statement (EIS) if a proposed major federal action is determined to significantly affect the quality of the human environment. The regulatory requirements for an EIS are more detailed and rigorous than the requirements for an EA.

If the EA determines that the environmental impacts of a proposed federal action will be significant, the EIS process begins, resulting in a DEIS is prepared, distributed, and commented on. Next, a FEIS is issued.

The EIS decision-making process may be quite lengthy and complex. Here we can only provide a summary. An overview of the process timelines is given by the CEQ (2020a) at https://ceq.doe.gov/nepa-practice/eis-timelines.html.

1. An agency publishes a Notice of Intent (NOI) in the Federal Register. The Notice of Intent informs the public of the upcoming environmental analysis and describes how the public can become involved in the EIS preparation. This Notice of Intent starts the scoping process, which is a period in which the federal agency and the public collaborate to define the range of issues and possible alternatives to be addressed.
2. A draft EIS is prepared and published for public review and comment for a minimum of 45 days. Upon close of the comment period, agencies consider all substantive comments and, if necessary, conduct further analyses.

3. A final EIS is then published. It provides responses to substantive comments. Publication of the final EIS begins the minimum 30-day "wait period," in which agencies are generally required to wait 30 days before making a final decision on a proposed action. EPA publishes a Notice of Availability in the Federal Register announcing the availability of both draft and final EISs to the public. The EPA maintains a list of EISs with comment letters and open comment periods.
4. The EIS process ends with the issuance of a Record of Decision (ROD) and a Notice of Availability. The ROD explains the agency's decision, describes the alternatives the agency considered, and discusses the agency's plans for mitigation and monitoring, if necessary.

The EIA is many things: it is a physical document that discloses information, it is a planning tool that is part of the planning process, it is a reflection of participatory democracy that helps to reveal and solve problems, and it is a statement of environmental accountability that forces action. Canter (1996) provided a thorough discussion of what an EIS should contain, and of course, it addressed by federal agencies along with the CEQ regulations (40 CFR Parts 1500–1508).

Most EISs contain these components:

Cover sheet: Includes, among other things, the name of the lead agency and any cooperating agency contact information, the title of the proposed action and its location, a paragraph abstract of the EIS, and the date by which comments must be received.

Summary: A summary (or "executive summary") of the EIS, including the major conclusions, areas of controversy, and major issues to be resolved.

Table of contents: Assists the reader in navigating through the EIS.

Purpose and need statement: Explains the reason the agency is proposing the action and what the agency expects to achieve.

Alternatives: Consideration of a reasonable range of alternatives that can accomplish the purpose and need of the proposed action.

Affected environment: Describes the physical and human environments of the area to be affected by the alternatives under consideration. Federal agencies often refer to this setting as the "Area of Potential Effect (APE)."

Environmental consequences: A discussion of the direct and indirect environmental effects and their significance.

List of preparers: A list of the names and qualifications of the interdisciplinary team members primarily responsible for preparing the EIS.

List of agencies, organizations, and persons to whom the EIS were sent.

Index: The index focuses on areas of reasonable interest to the reader.

Appendices: Background materials prepared in connection with the EIS.

A supplement to a draft or final EIS is required if substantive changes to the action were made or when new information arises that could affect the action or its impacts. If an agency decides to supplement its EIS, it prepares, circulates, and files the supplemental EIS in the same fashion as a draft or final EIS.

Assignment

1. What are the purposes of an EIA?
2. What are the major steps to take for federal agencies to evaluate the environmental impacts of a proposed action?
3. What should an EIS contain?

2.2 ENVIRONMENTAL LAW AND LEGAL BASIS FOR NEPA

Governmental hierarchy of authority is the basis for compliance where everything springs from the US Constitution (Table 2.1).

The roots of environmental impact assessment law can be seen in the history of federal legislation. For much of our early history the federal government spent a great deal of effort getting rid of federal land that it had acquired from foreign interveners (the Louisiana Purchase comes to mind), inappropriate takings from native peoples, and other sources so it could be put to work by the private sector, generating capital, promoting commerce, and perhaps even improving the general welfare. Yes, we protected special areas, such as anything that involved George Washington, who was venerated as a near saint. But there was not much concept of protecting non-historic land. Europeans and others coming to America were fascinated by our seemingly endless natural treasures: the Grand Canyon, Niagara Falls, Yellowstone, and many other places. Our growing recognition of them and of the need to protect Civil War sites led to an awakening of national conscience. An awareness of the hazards of pollution to health and to navigation also contributed to a nascent sense of value and accountability.

The Refuse Act, a portion of the Rivers and Harbors Act prohibited dumping into navigable waters without a permit. The Reclamation Act of 1902 used public funds from land sales for irrigation and other reclamation of arid and semiarid lands. New Deal legislation brought into being federal programs that cleaned up lands and built roads, parks, and other infrastructure. The 1939 Omnibus Flood Control Act allowed construction of flood control projects and federal acquisition of various state and local dams. These any many other laws marked the growth of federal power and authority. But the federal government continued to grow unwieldy in the management of its own agencies, not to mention the land and waters it was also managing.

TABLE 2.1 EIA hierarchy and the branches of US government.

Authority	Description	Branch
Highest authority	US Constitution	Legislative branch
Must align with above	Statutes	Legislative branch
Implements the statute	Regulations	Executive branch
Carries out the regulations	Rules	Executive branch
Must meet the rules	EIA and other products that arise from the rules and regulations	Executive branch
Checks and balances	Court systems	Judicial branch

The 1968 Intergovernmental Coordination Act was one attempt to correct this, and NEPA, coming upon the heels of an environmental awakening, was another attempt. NEPA attempts to balance the environment, commerce, and needs of the nation in "productive harmony," to paraphrase the NEPA statute, Section 2, 42 USC § 4331 (Purpose).

Assignment

1. What are the major historical legal influences on EIA law?
2. Why is a historical understanding important for the use of environmental laws today?

2.3 JUDICIAL REVIEW AND ENVIRONMENTAL JUSTICE

The judicial branch of American government provides checks and balances to EIA and to the other two branches of government, executive and legislative. NEPA itself is a quasi-judicial process that operates within the executive branch. The judicial branch is a final check; it uses "arbitrary and capricious" or "reasonableness" standards to examine federal decisions on EIS preparation.

Challenges under NEPA are usually made on procedural grounds rather than on the merits of particular environmental impacts. It is easier to make a claim that a step was not properly followed than to get into the weeds about the content of the project or its potential impacts. The procedures reflect aspects of "fairness" in government decisions and actions, which must be equitable for all.

As part of the EIA process, every US federal agency must achieve environmental justice as part of its mission by identifying and addressing disproportionately high and adverse human health or environmental effects of its programs, policies, and activities on minority populations and low-income populations, including tribal populations. Six principles for environmental justice analyses are followed to determine any of these circumstances:

1. Consider the composition of the affected area to determine whether low-income, minority, or tribal populations are present and whether there may be disproportionately high and adverse human health or environmental effects on these populations.
2. Consider relevant public health and industry data concerning the potential for multiple exposures or cumulative exposure to human health or environmental hazards in the affected population, as well as historical patterns of exposure to environmental hazards.
3. Recognize the interrelated cultural, social, occupational, historical, or economic factors that may amplify the natural and physical environmental effects of the proposed action.
4. Develop effective public participation strategies.
5. Assure meaningful community representation in the process, beginning at the earliest possible time.
6. Seek tribal representation in the process.

Assignment

1. Briefly explain how use of an environmental law such as the Endangered Species Act can be affected by a federal court?

2. Provide an example of how a federal project location might discriminate against a particular cultural or social group.

2.4 ROLES IN THE ENVIRONMENTAL REVIEW PROCESS

Agencies and organizations have particular roles in the EIA process. These individuals and groups are stakeholders. Of course, there are other stakeholders, but first we focus here on agencies.

Every federal agency already has the NEPA-assigned mission of environmental accountability. There are five potential roles that an agency may play in the EIA process. An agency may have all, some, or none of the following roles.

1. Setting standards for indices and other environmental data
2. Collecting and providing data and other forms of information
3. Reviewing and commenting on EIA/EIS documents
4. Preparing assessments for its own projects
5. Managing EIA processes and EIA reports (curation).

The EPA will conduct all five of these roles and will be an active player in all aspects of an EIA. Many people think the EPA enforces NEPA, but each agency does its own monitoring and enforcement. The CEQ may perhaps be said to have overall authority, and it reports directly to the president.

The Army Corps of Engineers and the Department of Energy prepare a proportionally large number of impact statements. The Department of Transportation also processes a large number of them. Some agency components, such as the Fish and Wildlife Service, are particularly known for commenting on impact statements with regard to environmental impacts to resources under their purview, such as threatened and endangered species. The Department of Energy probably has the most diverse menu of NEPA-related projects.

Assignment

1. Choose a federal agency other than the EPA to study online. What are the roles that agency has played in NEPA in the past 5 years?
2. What are the roles of the CEQ?

2.5 INTEGRATING WITH OTHER ENVIRONMENTAL LAWS AND PROCEDURES

Integration is a movement to combine the EIA process with other laws. The reasons for this interest are that many laws have been in effect for some time with established procedures and programs related to environmental resources. They also may be able to reduce time and costs or do a better job of estimating, evaluating, and commenting on impacts. They may also reduce duplication.

Environmentally related laws that often interact with NEPA include:

Clean Water Act (CWA)
Clean Air Act (CAA)
Army Corps of Engineers permits (Section 404 of CWA)
Rivers and Harbors Act
Endangered Species Act
National Historic Preservation Act (especially Section 106)
Federal Highway Administration projects
FERC (Federal Energy Regulatory Commission)
Federal Hazardous Materials Laws
Executive Orders (have the effect of law for affected agencies)

Projects that are subject to NEPA typically require multiple permits and approvals. Some of the ways the integration process of NEPA and other federal laws and agencies may work are:

1. Preliminary constraints analysis
   Purpose of the action, conditions of the environment, likely regulatory requirements, preliminary site evaluation
2. Consultation with regulatory agencies
3. Comprehensive environmental compliance strategy
4. Proposal and signing of Memoranda of Understanding
5. Technical studies and impact reviews
6. Results into draft NEPA document
7. Coordinated public and interagency review
8. Final NEPA document
9. Adopt NEPA document
10. Use the NEPA document in regulatory decisions
11. Policy reviews. Is this making government more efficient and effective?
12. Cooperative agreements
13. Making formal agreements.

These examples illustrate the importance of environmental factors in almost all planning and management of resources.

Assignment

Select one of the laws or executive orders referenced above and summarize the way it interacts with NEPA in terms of the EIA process.

2.6 EIA PROCESS IN THE VARIOUS STATES AND OTHER JURISDICTIONS

State equivalents of NEPA are prepared jointly among the state and federal lead agencies. This allows for efficient environmental review. Further, there can be collaborations between different

TABLE 2.2 State equivalent NEPA laws ("Little NEPA") as of 2022.

1. California Environmental Policy Act (CEQA)
2. Connecticut Environmental Policy Act (CEPA)
3. District of Columbia Environmental Policy Act of 1989 (DCEPA)
4. Georgia Environmental Policy Act (GEPA)
5. Hawai'i Environmental Policy Act (HEPA)
6. Indiana Environmental Policy Act (IEPA)
7. Maryland Environmental Policy Act (MEPA)
8. Massachusetts Environmental Policy Act (MEPA)
9. Minnesota Environmental Policy Act (MEPA)
10. Montana Environmental Policy Act (MEPA)
11. New Jersey Executive Order 215 (EO 215)
12. New York State Environmental Quality Review Act (SEQRA or SEQR)
13. New York City Environmental Quality Review (CEQR)
14. North Carolina State Environmental Policy Act (SEPA)
15. Puerto Rico Environmental Public Policy Act (EPPA)
16. South Dakota Environmental Policy Act (SDEPA)
17. Tahoe Regional Planning Compact (TRPC)
18. Virginia Environmental Impact Report Procedure (VAEIR)
19. Washington State Environmental Policy Act (SEPA)
20. Wisconsin Environmental Policy Act (WEPA)

state, local agencies, and federal agencies that share a common responsibility. A successful example is the Lake Tahoe management area lying in four counties of the states of California and Nevada. EIA work can also be carried out on an international basis. This applies to projects on or near US borders or that extend into or from other countries. Pipelines, transmission lines, and oceanic projects are common examples.

When a proposed action requires a NEPA review and compliance with state or local environmental reviews, it is important for practitioners to understand their requirements to achieve common goals and avoid duplication or conflict.

Some states, notably California and Washington, have even more detailed requirements. Hawai'i, Massachusetts, New York, and Washington include some private developments in addition to public ones. Table 2.2 shows NEPA equivalents—mostly states—below the federal level (CEQ, 2021).

Assignment

1. Search the internet to see if any other states, US territories, or state equivalents have adopted a version of NEPA.
2. Select one of the state-level NEPA equivalents above and compare it to NEPA. Summarize any differences.

The most common environmental effects that local and state agencies seek to mitigate or avoid in their actions include impacts on air and water quality, disruption to waste management, interruptions to animal or plant species and habitats (especially endangered species), impacts on wetlands, changes to land due to cultivation or construction, and impacts on public infrastructure. Such issues are also environmentally addressed in city or county comprehensive plans.

2.7 INTERNATIONAL EIA AND GLOBALIZATION

Many books and articles describe the successes of EIA in developed and developing countries. Canter (1996), Morris and Therivel (2009), Noble (2015), Eccleston (2017), Glasson and Therivel (2019), and others have compared NEPA with equivalents in various countries. A number of differences exist, including the elements of public participation, enforcement, formats of the EIS, cost allocations, project viability, stability of projects, dealing with cumulative impacts, responses to climate change, project and jurisdictional boundaries, and where EIA falls on the planning continuum.

Although this text is primarily concerned with EIA in the United States, we can learn from the experiences of how other countries handle assessment. Further, the United States is both the recipient and the source of some environmental impacts from other countries. Acid rain, for example, does not respect territorial borders. Not only are there multinational projects, but there are multinational impacts. Treaties and trade agreements can be another source of connection. The sharing of impact assessment data is enhanced by international cooperation. In this era of globalization and climate change the world has grown smaller, and we can expect an increase in the international connection aspects of EIA. Examples of this trend are in global financing systems, joint venture development projects among nations, and political commitments to global warming impacts.

Assignment

1. How could another nation become a participating party to a US federal action under NEPA?
2. Why might an international agency doing work in another country require an EIA?

There are four "spheres" of US involvement planet-wide (globalization; Figure 2.1). These spheres are not mutually exclusive. They are ways to organize potential impacts for analysis and comparison.

The first sphere comprises direct physical impacts. For air, this can include acid rain, carbon dioxide, ozone, toxins, and particulate matter. For water, impacts can include loss of fisheries, nautical shipping, water quality/quantity, and ocean acidification. We suggest that water-related issues are the number one environmental problem for the world's population. Other direct global physical impacts include desertification, deforestation, wildfires, transport of hazardous materials, food production and distribution, waste disposal, and other activities that affect the biosphere including extinctions, extirpation, and various aspects of ecosystems services. Climate change—global warming—is a major factor linking many of these physical impacts and their environmental, social, and political consequences.

A second sphere is identified with trade and economic growth issues. The treadmill of production, the three horsemen (more, faster, and cheaper), fuels demands on natural resources.

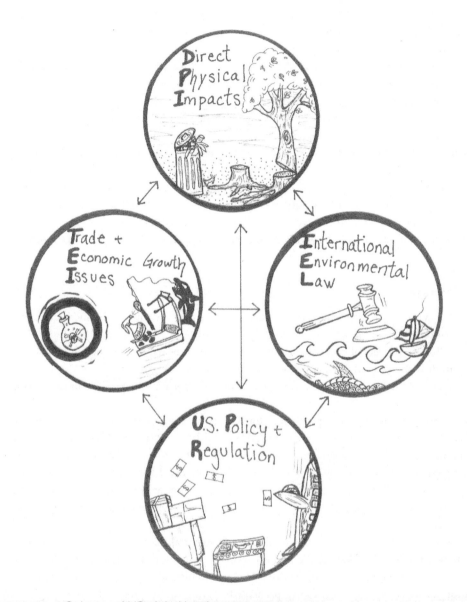

FIGURE 2.1 Spheres of US global involvement.

Most trading nations have export/import policies that can cause environmental harm; the use of pesticides, illegal fishing, and wildlife poaching come to mind. One consequence is the "circle of poison" (Weir and Shapiro, 1981). If we as a nation are not careful, we get our toxins back in food—coffee, pineapples, and other exotic imports, and on things we could grow ourselves but want out-of-season, like melons, grapes, and tomatoes. The issue is we want diverse choices and continual availability but we do not necessarily acknowledge or want the consequences.

The third sphere is international environmental law, which dates back as long as there have been nations. Maritime law is a classic example of this. Treatment of migratory or transboundary endangered species is another aspect, as is management of fisheries commons. In manufacturing, ISO 14000 is another example: its standardizations include industrial ecology and other environmental aspects of the life cycle of products. The system of accountability reflected in Environmental Impact Assessment is the most relevant manifestation of international environmental law other than trade aspects. The United Nations has played a major role in the use of EIA, particularly in developing countries.

Finally, we have our fourth sphere, that of US policy and regulation. The consequences of our own values and behaviors are felt by other countries. We can explicitly recognize this in our environmental impact assessments and implicitly through economic import/export, including the manifestation of US values. Areas of policy influence include market structures and tools that affect natural resources, issues of population and of birth control, packaging, manufacturing, and food. The typical food product on our table has traveled 1,300 km. Environmental impacts result from growth, harvest, manufacturing, and travel of our food. Unfortunately, often the cheapest sources of goods are from those countries and sites where accountability is the least.

Assignment

1. For what reasons might an EIS be required for a project within the boundaries of another nation?
2. What are the global consequences if we export or externalize our environmental impacts in the form of pesticides, emissions, and waste products?
3. How are these spheres important in responding to global problems like climate warming?

Over 180 countries have adopted some form of EIA system and rules or regulations about environmental protection. Most are modeled after NEPA. In 2018, the UN Environment Programme released a report that provides an overview of national legislation and institutional arrangements relevant to environmental impact assessments (EIAs) and strategic environmental assessments (SEAs). The report, global in its focus, underscores the importance of such assessments in achieving a 2030 Agenda for Sustainable Development and a Strategic Plan for Biodiversity.

Some countries have strengthened their regulatory frameworks, but other countries have exhibited a trend of weakening the EIA process. There seems to be significant global variation in the role of public participation in EIAs. In some countries, requirements for public participation for EIAs have been expanded, particularly during the scoping and review stage. A few countries include provisions related to indigenous peoples.

Another need is to improve the collection and measurement of data, with an increased focus on climate change and human health. Also needed is a means of anticipating and avoiding impacts on biodiversity and ecosystem services. Impact assessment laws are in a position to support these actions by promoting government accountability.

Other obvious global environmental impacts that require serious attention are ocean pollution and acidification, measurement and evaluation of carbon emissions, and environmentally sensitive urban development. This list could be expanded to book length. A takeaway here is that cumulative impacts that are required for analysis at local levels probably contribute to global environmental issues.

John Glasson and Riki Therivel in *Introduction to Environmental Impact Assessment* (2019) explain how many of the major international funding institutions such as the World Bank have established EIA procedures.

Assignment

If the three horsemen of the treadmill are *more, faster,* and *cheaper,* perhaps the lone knight is *sustainability*. What would be the role of EIA in promoting sustainability?

Environmental impact assessment and its components, especially environmental justice, can form a response to the pressures (the horsemen of *more, faster,* and *cheaper*) of the treadmill of production. One of the biggest things that can be done is to promote gender equity, especially in improving women's status through increased education and the resultant economic opportunity. Child labor in agriculture, mining, and manufacturing is another huge factor, and is affected by female status in society. Other EIA-related responses include using an ecosystem concept of trade and resource management, and an ecosystem concept of development.

Cradle-to-grave accountability of products, food, construction, and government processes are also tools. Full-cost pricing (products reflect cost of disposal, environmental degradation) can improve the accuracy of environmental impact assessment. Happiness and quality of life indices are a measurement tool that can reflect the harmonious balance of resource use, environmental conditions, and economic status.

2.8 MANAGING THE NEPA PROCESS

Management of NEPA is left to individual agencies. The CEQ can be consulted as an advisor, but the autonomy of individual agency accountability reflects a fundamental principle of self-regulation that helped NEPA gain acceptance. The process of peer review and public participation helps keep the system accountable and manageable. EIA and NEPA management are discussed further in the next two topics of this text, but here we present the essentials of getting started: the phases of a project build-out and the aspects of soliciting EIA project services.

2.8.1 The four phases of a project

Most of us naturally assume the major impacts of a development project are going to be associated with its construction and operation. However, every project has four potential phases: planning, construction, operation, and closure. Consequently, every project's potential impacts should be examined in the context of these four phases to see if they engender specific impacts, particularly environmental ones (Table 2.3). The planning phase would include exploration and testing, site assessments, and work-associated pre-construction to determine the viability of a project. Construction includes site work and the building of infrastructure and related work. Impacts will be forecast for the expected life of the project—its operation or use. Closure may involve very little beyond reclamation, or in the case of a landfill, extensive long-term monitoring.

TABLE 2.3 Hypothetical display of how potential environmental impacts may vary depending on the phase (life stage) and nature of the project.

Project	Planning	Construction	Operation	Closure
Coal mine	Low	Low	High	High
Oil well	Moderate	Low	High	Low
Fisheries	Low	Low	High	Low
Landfill	Low	Moderate	High	High
Highway	Low	High	High	Low
Timber harvest	Low	Moderate	High	Low

Assignment

What would you estimate to be the potential major impacts for each of the four phases for a 200,000 square foot waste-to-energy plant built in your community?

2.8.2 The RFQ and the RFP

The planning phase of a project engenders the need for a Request for Qualifications (RFQ) or a Request for Proposals (RFP) for EIA services. An RFQ streamlines the efficiency of the RFP process by narrowing down the recipients to qualified entities. An RFP is a solicitation of service opportunity. An agency may outsource an EIA for three potential reasons:

1. Lack of sufficient in-house expertise
2. A lack of time
3. A contract/consultancy mandate.

We can find contract information such as RFQs and RFPs at government agency websites, in the System for Award Management (https://sam.gov) and at Professional Associations. Trade journals, newspapers, and agency pre-qualified notification lists may be helpful. GSA eBuy! (www.ebuy.gsa.gov/ebuy/) provides information about the federal, state, and local government marketplace and can communicate business opportunities, requirements, and quotations via web and email. The Federal Register (www.federalregister.gov) is a public document announcing all code and regulation changes and proposals for changes, as well as funding allocations and needs.

If an RFQ for a project is posted, entities that successfully respond to the RFQ and meet the qualification criteria will be included in the subsequent RFP solicitation. Other agencies, such as the Department of Housing and Urban Development (HUD), develop pre-qualified lists of vendors during an open solicitation period (24 CFR § 85.36(c)(4)). Consultant qualifications can be submitted at any time to an agency that maintains a pre-qualified vendor list. When a solicitation announcement comes up, the consultant gets a copy of the RFP.

Abstracting service providers and other commercial firms continually search for opportunities and alert clients and others that maintain websites for users to search.

Sample abstracting services are:

1. www.findrfp.com. Membership allows searching for RFPs from government entities, as well as bids and contracts from all over the United States. This is not a free service.
2. www.rfpdb.com lists government, corporate, and non-profit RFPs. This site has many search query and filtering options to narrow down your search; however, you need to create an account for full access to this page.
3. www.stateandfederalbids.com. The State and Federal Bids site is a free site that grants the user access to RFPs and other government bids and contracts. You are required to create an account in order for the site to send you specialized reports with RFPs that you are specifically interested in.
4. FedConnect allows anyone to view public opportunities that agencies post. Users registered with FedConnect can see public and directed opportunities for a specific vendor or applicant, or a limited group of vendors or applicants.

2.8.2.1 Responding to RFPs

There are two main types of RFPs: prescriptive (detailed) and subscriptive (general).

A prescriptive RFP tends to be fully specified, where everything is detailed. Subscriptive RFPs leave it to the responders to provide details. Many are a combination. Usually, the high-risk health-based or safety-based arenas yield prescriptive rather than subscriptive specifications and requirements. However, the sponsoring agency may simply have a fully articulated vison or set of requirements. A response to this type of RFP is usually very engineering oriented.

Some RFPs are conceptual, largely based on performance standards or a general mission statement. For example, "Find out if I can build here" reflects a place in search of a project.

Such a subscriptive—"here is what I need," or "here is where I need something"—can be outcome-based, or exploratory. Responding to this requires care in the bidding process so you do not give away your intellectual property but fail to get the contract.

Some proposals submitted in response to subscriptive RFPs are very detailed and some are purely conceptual. Some are policy related and some are specific construction projects. Again, some RFPs proposals are quite prescribed, and some are subscriptive, as are what they call for in their responses.

2.8.2.2 Considerations in writing and responding to development RFPs

1. What is the *project*? What needs to be done? Context of agency mission and authority. Is it a general, subscriptive project or a detailed, prescriptive project? A full and proper description of a project makes the screening and scoping much more efficient.
2. *Screening*. What laws and regulations apply? This is easy to miss or leave out for a preparer that assumes everyone is familiar with the regulations, but it is an important aspect for peer and public review and for determining jurisdictional authority.
3. *Statement of Qualifications* (SAQ). Ability to do the job. May be established in a general manner ahead of time through Request for Qualifications (RFQ). What is the ability of the agency, and what are the abilities of the preparers?

4. *Scope of Work* (SOW). Scoping process: identifying focal points and areas of emphasis. This is about efficiency since we cannot assess everything but do not want to miss the essentials.
5. *Time constraints.* Requirements: field conditions, timing, notices.
6. *Equipment.* In-house and rented.
7. *Personnel.* Specific to the project rather than general SAQ. In-house expertise, consultants, subcontracting for overall job plan (carrying out the SOW).
8. *Ethics.* What do you need to do the job right? (This is not a formal section, but it is reflected in the meaning of "professional.")
9. *Report*: requirements, formats, notices, publication, and distribution.
10. *Budget*: function of time, personnel, equipment, sensitivity.

In project planning, keep in mind the rule of thirds for your time and effort: one-third for outreach and recruitment for new projects, one-third for current project work in the field, and one-third for wrap-ups involving laboratory, writing, and legal/public for "old" projects.

Another planning concept or mantra in selecting and responding to RFPs is "BRAD": budget, risk, accountability, and deliverables. Essentially, this manifests in what goes into the scope of work.

2.8.2.3 Scope of Work (SOW)

The sponsoring agency may identify the scope of work in the RFP (a prescriptive approach). It may also require the bidder to respond with its own intentions to carry out the scope of work (a subscriptive approach). Many are a combination of these approaches. The guide to what is expected will be in the RFP. A subscriptive Scope of Work for general consulting services in urban redevelopment might contain something similar to the following.

Description/scope for professional services in real estate transactions and redevelopments: Provide property research, investigation, and analysis services to support long-term growth and economic development. Services will include, but are not limited to:

1a. Phase I environmental site assessments;
1b. Phase II environmental site assessments (including geophysical surveys and sample collections);
1c. Identification and sampling of hazardous materials of construction conditions and environmental conditions (including, but not limited to, lead-based paint, asbestos-containing materials, universal waste, radiation, mercury, mold, and radon);
1d. Preparation of remedial cost estimates;
1e. Litigation support;
1f. Waste management (hazardous, special, solid, universal, or miscellaneous), excavation or removal incidental to testing;
1g. Removal of hazardous materials, incidental to testing (including, but not limited to, lead-based paint, asbestos-containing materials, universal waste, radioactive materials, mercury, and mold);

1h. Evaluation, recommendation, and implementation of innovative technologies (including, but not limited to, soil vapor extraction systems and stabilizing sealants and coatings);
1i. RCRA permitting;
1j. Hazardous waste record keeping and reporting;
1k. Grant and research services;
1l. Training, program development, and public outreach.

The Scope of Work may also list key personnel. In the above case, these might include principal, project manager, project engineer, project scientist, licensed professional engineer, licensed structural engineer, licensed professional geologist, licensed/certified planner (AICP preferred), licensed surveyor, field manager, field engineer, field scientist, licensed asbestos building inspector, licensed asbestos project designer, licensed asbestos project manager, licensed asbestos air sampling professional, and licensed asbestos supervisor.

2.8.2.4 Factors to consider in writing a request for proposals

The RFP may be quite detailed about what to submit in response. In general, something along the lines of the following will be expected.

1. What is the *project*? What needs to be done? Context of agency mission and authority.
2. *Screening*: What laws and regulations apply? (Federal EIS under NEPA means federal undertaking, funding, land, license/permit, and combinations of these.)
3. *Statement of Qualifications* (SOQ): ability to do the job. May be established in a general manner ahead of time through Request for Qualifications (RFQ).
4. *Scope of Work* (SOW). Scoping process: identifying focal points. Prescriptive or subscriptive approach/requirements.
5. *Time constraints*: These may be dictates of the contract, field conditions, timing, notices, and personnel.
6. *Equipment*: In-house and rented.
7. *Personnel*: Specific to the project rather than general SOQ. In-house expertise, consultants, subcontracting for overall job plan (carrying out the SOW).
8. *Ethics*: What do you need to do the job right? (This is not a formal section but it overlies whatever you do and how you do it). The proposal may also require environmental justice to be addressed as a separate component.
9. *Report*: Requirements. This will include electronic and hardcopy formats, notices, publication, and distribution.
10. *Budget*: Think of this as a function of time, personnel, equipment, and professional evaluation of what it takes to get the job done right.

Assignment

Choose one of the following potential projects and outline an RFP that addresses the ten factors from above. Keep in mind the RFP is to *do* the environmental impact assessment of a project and generate a Phase 1 Report (it is not an RFP to *build* the project).

Projects to choose from:

1. Wind farm turbines
2. Marine fisheries management plan for commercial harvest
3. Urban pedestrian bridge
4. Land-based aquaculture project 200,000 tons/year
5. Interstate highway 50-mile realignment
6. State park expansion: 50 acres with 100 new campsites
7. Railway expansion/spur of 100 miles
8. Airport expansion to accommodate 20% capacity increase
9. Dam relicensing
10. Dam removal
11. Sea level rise stabilization of 1 mile of beachfront
12. Affordable housing, 200 units
13. Invasive species (zebra mussel) treatment
14. Solar farm 1 MW on 100,000 square feet
15. Tidal generator 24 turbines, average 57 MW (24% capacity)
16. Historic fort restoration Colonial/Revolutionary War era
17. Medium-sized brownfield smelter expansion
18. Timber harvesting 1 million board feet annually
19. Environmental policy proposal for hazardous waste disposal
20. Wildlife restoration.

2.8.2.5 Bids

The federal RFP may arrive as a large document subfile. Another document subfile will contain a "Solicitation, Offer and Award" form (SF-33 [Standard Form]). The one-page SF-33 is for the bidder to fill in, sign, and return as part of the submitted bid.

Some bids are open; others are sealed. Sealed bids will have costs/pricing and related schedule submitted under one cover, and a technical proposal addressing the SOW under another. Reviewers at the federal agency will examine and rank each response.

Evaluation and award follow a two-step process. The first is a technical review in which the firm's technical expertise and qualifications, the technical merit of what it proposes to do, and the proposed personnel's already committed time are all considered. The idea is to evaluate what the firm would do, how it would do it, and realistically be able to do, without being distracted by cost considerations. The second step in the process treats contractual and award issues, specifically if the firm satisfies any contracting restrictions (e.g., small business or minority-owned small business set-asides) and what cost is being proposed. The budget should specify how much of each person's time is going to be spent on *this* federal project.

If the proposals are equivalent in terms of technical merit and if the firms otherwise qualify under any contracting restrictions, then the firm with the lowest bid will be offered the contract to do the requested work. Most government RFPs reserve the right to make awards to the next lowest bidder if, in their judgment, the lowest bid is too low to actually perform the work.

Generally, the agency will negotiate with the lowest first-round bidders—usually a maximum of three—to see if an even lower cost can be arranged. This is the "best-and-final" offer made by the firm.

2.8.3 The job plan

The job plan is what will be used in responding to the RFP. It is more than the scope of work; it is how the agency or consultant will structure and carry out the scope of work and meet the requested project deliverable, presumably the EA or EIS.

Consultants and agencies may have a range of structured procedures, checklists and approaches to develop a job plan. Among the factors to consider in developing the job plan and proposal response are:

1. Agency mission
2. RFP outcomes/deliverables
3. Screening: what regulations apply
4. Personnel: in-house expertise
5. Subcontractors and the need for contracting
6. Ethics: what you need to do the job right
7. "Scoping" process: sorting out what needs to be done and how to focus on it.

Job plans are a management tool and may be required for specific bid processes. There are many ways to organize them (Tables 2.4 and 2.5 provide two examples).

Assignment

Assume you have to write an EIA report and a final EIS for a project to convert an existing industrial or commercial building in your city or town into 30 units of affordable housing. Write a job plan to accomplish this assessment. Present it in a table of your own design. Table 2.6 is a sample timeline format.

TABLE 2.4 Sample job plan outline created by students in an EIA class (budget column and other details omitted).

Expected Date of Delivery	Component	Description	Owner (Lead Responsibility)
	Management plan		
	Site visit		
	Screening		Group
	Scoping	Cultural	
	Scoping	Legal and regulatory	
	Scoping	Environmental	
	Scoping	Fiscal impact	
	Assessment description, deliverables	Background information and site proposal	
	Draft base map		
	Public notice	Cultural	
	Public notice	Legal and regulatory	
	Public notice	Environmental	

(continued)

TABLE 2.4 (Continued)

Expected Date of Delivery	Component	Description	Owner (Lead Responsibility)
	Public notice	Economical	
	Project update		
	Draft IA assessment matrix	Risk ranking of impact assessment	
	Research	Environmental	
	Draft SEQR form	Standardized EA	
	Research	Legal and regulatory	
	Research	Environmental justice	
	Research	Cultural/social	
	Draft outline	Draft submission	
	Legal and regulatory summary and application	Draft submission	
	Research	Economical	
	Impact categories and tables update	Graphics and data	
	Individual updates	Cultural	
	Individual updates	Legal and regulatory	
	Individual updates	Environmental	
	Individual updates	Economical	
	Project update		
	Individual drafts	Cultural, environmental, justice	
	Individual drafts	Legal and regulatory	
	Individual drafts	Environmental	
	Individual drafts	Economic/fiscal	
	Final draft of report section		
	Final draft EA report		
	Peer review		
	Multimedia presentation		
	Final EIA report		

TABLE 2.5 Another sample job plan outline.

	Step	Personnel	Timing	Potential Issues	Budget
1	*Describe project*				
2	*Describe institution*/proponent/ applicable standards *and regulations*				
3	*Potential impacts* ("portable" that go with this type of project)				

TABLE 2.5 (Continued)

	Step	Personnel	Timing	Potential Issues	Budget
4	*Describe setting*: what is the affected environment?				
5	*Predict impacts* (combination of 3 and 4)				
6	*Assess* the predicted impacts (compare 5 with 2				
7	Mitigate impacts and determine *residual impacts* (RI = 6–7)				
8	*Choose* preferred action (Least RI)				
9	*Write* EA/EIS				
10	Build/project with *monitoring*				
11	Operate project with *monitoring*				
12	*Feedback* (metadata)				

TABLE 2.6 One-year timeline for an EIA.

Task/Responsibility	Who (what job title, such as engineer, planner, manager, architect)	When (time line and duration of task)
Respond to RFQ		
Respond to RFP		
List of supplies and services		
Screening (determine what permits and laws apply)		
Site testing and evaluation		
Initial site visit		
Scoping of impacts		
Matrix/checklist of impact categories		
Literature review		
Write EA		
Description of environmental background		
Description of the project		
First site reports due		
Review project with client		
Public meeting on environmental impacts		
Complete and file a draft EIS		
Hearing on draft EIS		
Write and file final EIS		

2.9 CONCEPTS AND TERMS

Here we provide just a few of the many terms available. It is easy to get overwhelmed in the lexicon of government regulation. These terms are selected for their importance to understanding how environmental impact analysis is performed in the United States. Some of the definitions are from the Department of Energy's *Glossary of Terms Used in NEPA Documents* (1998). Other federal agencies have their own guides and glossaries.

Affected environment: A description of the existing environment to be affected by the proposed action.

Alternatives: Reasonable ways to fix an identified problem or to satisfy the stated need.

Best management practices: Decisions, actions, prohibitions, maintenance, and other practices to prevent or reduce maintenance procedures and other management means to prevent or reduce environmental impacts.

Categorical exclusion (CX): Exempt from NEPA. A category of actions that do not individually or cumulatively have a significant effect on the human environment.

Cumulative effects: The incremental environmental impact or effect of the proposed action, together with impacts of past, present, and reasonably foreseeable future actions.

Draft Environmental Impact Statement (DEIS): A completed environmental impact statement intended for circulation and review by interested parties.

Ecosystems services: A functional and rather utilitarian perspective that examines the benefits to humans that ecosystems provide. Many ecosystem services are non-market but still considerable.

Environmental Assessment (EA): A concise public document that briefly discusses the purpose and need for an action and provides sufficient evidence and analysis of impacts to determine whether to prepare an environmental impact statement or finding of no significant impact. In lowercase letters, "environmental assessment" is a general term referring to the generic process of determining environmental impacts.

Environmental Audit: A type of evaluation intended to identify environmental compliance and management system implementation gaps, along with related corrective actions. An Environmental Audit may also be a formal document similar to an EA.

Environmental Impact Statement (EIS): A detailed written statement analyzing the environmental impacts of a proposed action, adverse effects of the project that cannot be avoided, alternative courses of action, short-term uses of the environment versus the maintenance and enhancement of long-term productivity, and any irreversible and irretrievable commitment of resources.

Environmental justice: The fair treatment and meaningful involvement of all people regardless of race, color, national origin, or income with respect to the development, implementation, and enforcement of environmental laws, regulations, and policies.

Environmental Management System (EMS): A set of processes and practices that enable an organization to reduce its environmental impacts and increase its operating efficiency.

Environmental Protection Agency (EPA): A US federal government agency whose mission is to protect human and environmental health. It oversees programs to promote energy efficiency, environmental stewardship, sustainable growth, air and water quality, and pollution prevention.

Estoppel: A legal concept binding on government. An agency must follow its own established/published procedures. If it does not, it may be legally "estopped" from an enforcement action.
Existing conditions report: An analysis of a site that is proposed for development within a NEPA procedure. The report includes background conditions of the site.
Ex parte: Improper contact. This is binding upon agencies in administrative and legal procedures. Essentially, deciders cannot have sidebar conversations about a project. Two environmental hearing board members cannot be chatting together outside of the hearing about a project under their review. Similarly, they cannot be discussing the project with the applicant or another party.
Final Environmental Impact Statement (FEIS): A complete environmental analysis including all of the questions and reviews of the DEIS.
Finding of No Significant Impact (FONSI): A document supported by an environmental assessment that analyzes whether a federal action will have no significant effect on the human environment and for which an environmental impact statement, therefore, will not be prepared.
International Organization for Standardization (ISO): An international standard-setting body composed of representatives from various organizations. The organization promotes worldwide proprietary, industrial, and commercial standards.
Life Cycle Assessment (LCA): A methodology for assessing environmental impacts associated with all the stages of the life cycle of a commercial structure, product, process, or service.
Laches: Diligence in making a legal claim. This legal concept is binding on government pertaining to unreasonable delay. For example, an agency cannot ignore an environmental violation, then decide much later to enforce against it.
Mitigation measures: Planning actions taken to avoid an impact or to minimize the degree or magnitude of the impact, or to compensate for the impact
National Environmental Policy Act (NEPA): Requires all federal agencies to examine the environmental impacts of their actions, incorporate environmental information, and utilize public participation in the planning and implementation of all actions.
Party: A participant in the EIA process. May be a consulting or commenting entity or an entity with particular legal standing in regard to a project.
Proposed action: A plan or "undertaking" that contains sufficient details about the intended actions to be taken, or that will result, to allow alternatives to be developed and its environmental impacts analyzed.
Program EIS: Evaluates the environmental impacts of broad agency actions such as the setting of national policies or the development of programs. A program EIA may be followed by a more specific operational EIs.
Rapid Environmental Assessment (REA): A methodology for rapidly assessing and analyzing the environmental context of a particular crisis or disaster.
Record of Decision (ROD): Statement of a regulatory process history. Once protests are resolved, the agency issues a ROD, which includes its final action prior to implementation. If members of the public are still dissatisfied with the outcome, they may sue the agency in federal court.
Request for Proposals (RFP): A government document specifying agency work to be done and inviting proposals to do that work.

Request for Qualifications (RFQ): A government document seeking to build its list of entities who are qualified to perform agency work in response to RFPs.

Scope of Work (SOW): Describes what work will be done as part of an agency undertaking or contract. May be specific and detailed or general, as provided for in an RFP.

Scoping: An early and open process for determining the extent and variety of issues and impacts to be addressed and for identifying the significant issues related to a proposed action.

Screening: Determining what laws and regulatory processes apply to a project review.

Significance: Requires consideration of both context and intensity of an action and must be analyzed in its current and proposed short-and long-term effects on the whole of a given resource. "Significance" has specific legal connotations in some processes, in addition to its statistical implications of meaningfulness.

Strategic Impact Statement: A systematic decision support process aiming to ensure that environmental and possibly other sustainability aspects are considered effectively in policy, plan, and program making.

Sustainability Assessment (SA): A complex appraisal method for supporting decision-making and policy in a broad environmental, economic and social context, and transcends a purely technical/scientific evaluation indicators.

Tiering: The coverage of general matters in broader environmental impact statements with subsequent narrower statements of environmental analysis, incorporating by reference.

Unavoidable adverse effects: Refers to effects that cannot be avoided due to constraints in alternatives. These effects do not have to be avoided by the planning agency, but they must be disclosed, discussed, and mitigated, if possible.

Assignment

Assume you have a team of five people to do an EIA for a 100-unit residential project in the Bayside neighborhood of Portland, Maine. Your first step is to see what is meant by "Bayside" in Portland. An online search via Google or other engine will tell you that there is an East Bayside and a West Bayside. What else can you find out? You can also get an idea of the geographic boundaries, the socioeconomic demographics, the buildings, and the infrastructure. This will form the area and context for the EIA project planning exercise.

1. Describe Bayside.
2. What skill set and expertise does the team need?
3. Create a 1-year timeline that conducts an EIA and produces an EIS. Assign roles and deadlines for tasks/responsibilities (Table 2.6). Include additional tasks and responsibilities if you want.

2.10 SELECTED RESOURCES

Canter, Larry. (2015). *Cumulative Effects Assessment and Management*. Horseshoe Bay, TX: EIA Press.
Council on Environmental Quality (CEQ). (1970). National Environmental Policy Act. https://ceq.doe.gov/index.html.
Council on Environmental Quality (CEQ). (1999). Designation of Non-Federal Agencies to Be Cooperating Agencies in Implementing the Procedural Requirements of the National Environmental Policy Act. www.energy.gov/nepa/ceq-guidance-documents.

Council on Environmental Quality (CEQ). (2000). Identifying Non-Federal Cooperating Agencies in Implementing the Procedural Requirements of the National Environmental Policy Act. www.energy.gov/nepa/ceq-guidance-documents.

Council on Environmental Quality (CEQ). (2002). Cooperating Agencies in Implementing the Procedural Requirements of the National Environmental Policy Act. January 20, 2002. www.energy.gov/nepa/ceq-guidance-documents.

Council on Environmental Quality (CEQ). (2004). Reporting Cooperating Agencies in Implementing the Procedural Requirements of the National Environmental Policy Act. www.energy.gov/nepa/ceq-guidance-documents.

Council on Environmental Quality (CEQ). (2007). *A Citizen's Guide to the NEPA: Having Your Voice Heard*. Washington, DC: Office of the President, CEQ.

Environmental Protection Agency (EPA). www.epa.gov/nepa.

Hoskins, David E. (1988). Judicial Review of an Agency's Decision Not to Prepare an Environmental Statement. *Environmental Law Reporter*. 18 ELR 10331. https://elr.info/sites/default/files/articles/18.10331.htm.

Office of Management and Budget (OMB) and Council on Environmental Quality (CEQ). (2012). Memorandum on Environmental Collaboration and Conflict Resolution. Executive Office of the President, OMB and CEQ. www.energy.gov/nepa/ceq-guidance-documents.

2.11 TOPIC REFERENCES

Canter, Larry. (1996). *Environmental Impact Assessment*, 2nd ed. New York: McGraw-Hill.

Council on Environmental Quality (CEQ). (2020a). NEPA Practice: EIS Timelines. NEPA.Gov. https://ceq.doe.gov/nepa-practice/eis-timelines.html.

Council on Environmental Quality (CEQ). (2020b). Agency NEPA Implementing Procedures. NEPA.Gov. https://ceq.doe.gov/laws-regulations/agency_implementing_procedures.html.

Council on Environmental Quality (CEQ). (2021). State and Local Jurisdictions with NEPA-like Environmental Planning Requirements. NEPA.gov. https://ceq.doe.gov/laws-regulations/states.html.

Department of Energy. (1998). *Glossary of Terms Used in NEPA Documents*. Department of Energy, Office of NEPA Policy and Compliance. www.energy.gov/nepa/downloads/glossary-terms-used-doe-nepa-documents-doe-1998.

Eccleston, Charles. (2017). *Environmental Impact Assessment: A Guide to Best Professional Practices*. Boca Raton, FL: CRC.

Federal Register. www.federalregister.gov/agencies/environmental-protection-agency.

Glasson, John and Riki Therivel. (2019). *Introduction to Environmental Impact Assessment*, 5th ed. New York: Routledge.

Kershner, Jim. (2011). NEPA, The National Environmental Policy Act. HistoryLink Essay 9903. www.historylink.org/File/9903.

Morris, Peter and Riki Therivel (Eds.). (2009). *Methods of Environmental Impact Assessment*, 3rd ed. New York: Routledge.

Noble, Bram. (2015). *Introduction to Environmental Impact Assessment: A Guide to Principles and Practice*, 3rd ed. Don Mills, Ontario: Oxford University Press.

Rowell, Arden and Josephine Van Zeben. (2021). *A Guide to U.S. Environmental Law*. Oakland, CA: University of California Press.

Salzman, James and Barton H. Thompson, Jr. (2014). *Environmental Law and Policy*, 4th ed. St Paul: Foundation Press.

United Nations Environment Programme. (2018). *Annual Report 2018: Putting the Environment at the Heart of People's Lives*. New York: United Nations.

Weir, David and Mark Shapiro. (1981). *Circle of Poison: Pesticides and People in a Hungry World*. Oakland, CA: Food First Books.

Screening, scoping, and related aspects

3.1	NEPA agency guidelines and screening	46
3.2	The land development process as it relates to EIA	49
3.3	Roles of the players	51
3.4	Project scoping	54
3.5	Data collection and presentation	56
3.6	Public involvement	58
3.7	Concepts and terms	62
3.8	Selected resources	63
3.9	Topic references	64

3.1 NEPA AGENCY GUIDELINES AND SCREENING

The CEQ regulations charge each federal agency with developing its own procedures to carry out NEPA. The response to this mandate has been formalized and streamlined over the years and is a major role for some agencies. For example, the Department of Energy (DOE) has an Office of NEPA Policy and Compliance (www.energy.gov/nepa/office-nepa-policy-and-compliance) that provides, among other things, current lists and progress tracking of Categorical Exclusion Determinations, EAs, EISs, and Supplemental reports. The DOE NEPA office also maintains a list of NEPA-related reports and compliance procedures. The site is useful for the general public and practitioners using agency procedures.

Like their federal counterparts, state agencies list procedures and forms for compliance with state-level versions of NEPA. For example, New York State's Department of Environmental Conservation maintains forms useful for both screening and scoping under the New York State Quality Review Act (SEQR; www.dec.ny.gov/permits/6191.html). In California, the Governor's Office of Planning and Research or the California Environmental Quality Act (CEQA: https://opr.ca.gov/ceqa/) and the places to go. Other state-level versions include Connecticut, District of Columbia, Georgia, Guam, Hawaii, Indiana, Maryland, Massachusetts, Minnesota, Montana, Nevada (Tahoe Region), New Jersey, North Carolina, Puerto Rico, South Dakota,

DOI: 10.4324/9781003030713-3

Virginia, Washington, and Wisconsin. New York City is big enough and progressive enough to have its own version (CEQR), even though it is in a state with a state-level version.

Agency guidelines are particularly important for compiling, formatting and for other preparation tasks for the DEIS. The EPA Environmental Impact Statement Filing Guidance (www.epa.gov/nepa/environmental-impact-statement-filing-guidance) is a necessity for success in following instructions on determinations, filing, transmitting, reviewing, commenting on, and most other aspects of EIA/EIS. Subsequently, the EIS is among other "live" EISs for anyone using the internet (https://cdxnodengn.epa.gov/cdx-enepa-II/public/action/eis/search).

After decisions about which agencies may be involved and what kinds of documents may be published, a more detailed screening process begins. Screening is a part of the scoping process (see section 3.4) as it is applied to jurisdiction. Its questions address issues such as who has regulatory authority, under what type of authority, and what conditions are required for jurisdiction or project review. The outcome of screening determines what regulations apply to the review and potential approval of a project. This begins with the project description, which should originate from the proposer of the project. As part of screening we will understand project size, project type, project location, other project characteristics, the permits and licenses that might apply, potential stakeholders via the regulatory processes, and the overall review process. Many questions will arise to be addressed at this stage. Is it being undertaken by an agency or government entity? Perhaps the reviewer will be a member of the private sector that is being regulated by an agency. Other interested parties can also be involved. Will the proposed project require a detailed environmental impact assessment, and if so, at what level? Some agencies (state and federal) involve assessment forms that help with this screening along with the rest of the scoping process.

Canter (1996) described approaches to screening based on policies or environmental assessments. The main three approaches are individual case, threshold-based, and list-based. Some of these approaches are commonly combined in a hybrid approach. The individual case-by-case approach can be time-consuming. It typically involves an environmental assessment as a step in determining what regulations apply. This is particularly useful when the potential impacts themselves are a factor in evaluating jurisdiction. The threshold-based approach uses impact thresholds or project characteristic thresholds such as size (e.g., a parking lot of 1,000 or more cars requires an air quality permit). A potential problem is knowing if the thresholds are adequate to capture potential impacts. The list-based approach is similarly prescriptive. One asks whether the project on a list or occurring in a place or under a procedure on a list. Comparing undertakings to those on an agency's "categorical exclusion" list is one example of this application. A strength is that the determination can be made quickly, and a weakness is if there is a bias in the listed items that do not correctly match with the potential for impacts. A hybrid approach that combines other approaches is perhaps the best way to screen a project.

Assignment

The case-by case method is one approach to screening. What are other common approaches? Why might the hybrid approach be the best approach?

The Small Business Administration administers a site that helps people find out what Federal license and permits are required (www.sba.gov/starting-business/business-licenses-permits/

federal-licenses-permits). An EIS is undertaken on a much larger scale, involving many permits and many months of assessments. But the same idea of using categories of impact, jurisdiction, and geography usually applies.

Other countries have followed a similar approach. Egypt, typical of many countries, has a simple scale to screen projects. It uses a white list for minor impacts, a gray list for projects that "may result in substantial impacts" and black for projects that require a full-fledged EIA due to potential impacts. Other countries have more variation on what gets reviewed, who can participate, how to participate, and who pays the costs for the process.

Assignment

Examine the screening approach of another country that uses a sorting process for screening projects. Briefly describe the approach and categories. Has that approach worked well?

The United States by regulatory design leaves it to each agency to establish what gets reviewed and what does not. Projects deemed too minor to have potentially significant impacts are on a list of categorical exclusion (CX), and thus do not need an EA or a more in-depth EIS. Each agency maintains its own CX list. A list of exempt projects improves efficiency in the screening by concentrating effort into review to where it is most merited.

Assignment

Simulate a screening process by filling out the screening matrix below.

TABLE 3.1 Screening matrix for a project to construct an erosion control and stormwater barrier dike. Assume the dike is to be 100 feet long and 8 feet wide in a coastal project. Assume location is at your latitude and over to the closest coast (east or west), or your longitude north or south to the nearest coast if that is a more effective method.

Statute/Regulation	Applicable Phase of Project	Applicable Portion of Project	Integration into NEPA/EIA
Section 106 (National Register–eligible historic site)	Construction	Ground disturbance and archaeology/history	Coordinated review with NEPA
Army Corps 404 permit	Construction and operation	Wetlands disturbance	After NEPA, before construction

The screening matrix can track what laws and regulations apply to a project, when they apply, and to what portion of the project. Table 3.1 contains a simple matrix with examples for the first two laws we might screen for jurisdiction. We can also add cost, lead agency, assigned personnel, and other criteria for screening determination. Complete this matrix by screening four other laws/regulations that might apply.

3.2 THE LAND DEVELOPMENT PROCESS AS IT RELATES TO EIA

Screening begins with a description of the project. If it is a physical development of the land (or sea), what will be the change? The most frequent kind of projects that require environmental review are those that propose changes in the human use of land. An overarching conceptual approach to placing a project or the actions of a policy into the environment is to see if this intrusion is harmful or not. This is akin to the precautionary principle, which holds that the burden should be on the proponents to show their project does not do harm rather than on others prepared to show it is harmful. This should form the basis for proper screening and for the creation of rules/regulations that govern the screening process.

Figure 3.1 provides an overview of a private sector project as it travels through a land development review process. Proceeding through this process for a particular project will inform the screeners of what laws and regulations apply and bring out potential impacts. Much of this translates to the review of public sector projects as well.

Physical development projects begin as either a place in search of a development (e.g., "I have this site; what can I put on it?") or a development project in search of a place (e.g., "I want to build a post office distribution center; where can I put it?"). Some are a combination of these two approaches; it depends on the mission of the undertaking agency or developer and other factors. Knowing which approach or combination is present might prove very useful in developing or assessing impact mitigation strategies.

The EIA process should commence as soon as possible after most of the project proposal information is collected. Sometimes an early EIA process can identify potential significant impacts that can be dealt with by site planners. Ideally an agency will, like the EPA has done, adopt the "precautionary principle" for use in evaluating the impacts of projects (chemicals, in the case of the EPA) on the environment. The precautionary principle, defined in section 3.7, refers to the concept of having project and product proposers have the burden of first showing the project or product would not be harmful to the environment. It is an important concept with global implications. Other countries such as Canada have adopted the precautionary principle even more widely than the United States.

The phases of a project may overlap. They may also have separate categories of impact associated with them. And they may generate controversy. For example, a plan to drill offshore will need an EIS, but no developer would want to go through the long EIS process without some exploratory test wells to determine project viability. There are some boundary issues that arise. For example, at what point does putting in test wells cross over into being the project itself and trigger NEPA? Since NEPA is a planning law, it needs to occur within the first phase of a development project, prior to construction or other irrevocable commitment of environmental (and financial) resources.

FIGURE 3.1 Land development process for a private sector project.

We outlined the four phases of a project in the previous topic. Here we present them in slightly more depth as they inform the screening and scoping processes.

1. *Planning*: Defining the project is the first step in the project management process, and it is stated specifically in the EIA process. Going forward, the planners identify the need for and objectives of the project. Next, they clarify competing time and resource demands and watch for goals and boundaries that align from these demands. Organizing projects means

defining roles and responsibilities of all the players and developing high-level time and cost estimates. It also includes collection of data to be examined, which is sometimes called a case study. Information from agencies and outside experts can also be included. Screening of applicable laws and regulations begins early in the planning stage and be refined as the project is further developed.

The planning phase of a project continues with assembling the team and assigning tasks to individuals. Developing a budget for the project and creating a schedule comes next. While planning for the project, effective and clear estimates of the time and costs are forecast. If any changes need to be done in the estimates of time and costs, organization of the project may also be reviewed. This phase may include some preliminary physical site work in the form of testing or exploration as project or site feasibility is examined, and to develop information for an EA or DEIS.

2. *Development and construction of the project*: In executing the project, the team is launched. Team members monitor and control progress in terms of cost, time, and quality and manage risk. During the execution of the project, monitoring is continuous for timely completion of the project, as per schedules. All actions that affect environmental, social, and economic resources are carefully monitored, and all parties of interest are kept informed.

3. *Operation*: The bulk of the EIA should have dealt with anticipated impacts from construction (if a physical development) and operation of the project. If the project is a policy, then operation will constitute the majority of potential impacts. Monitoring of operations and environmental impacts should continue for the life of the project.

4. *Termination/closure/reclamation/long-term monitoring*: Once the execution of the project is completed, the project team evaluates performance, documents related to the project are achieved, lessons are learned, and experiences in the project are captured. The major impacts associated with this phase should have been anticipated by the EIA team. Ongoing monitoring and reporting should occur to see if modifications to the closure and reclamation plans are needed as the result of any necessary mitigation.

Assignment

1. Do you agree or disagree with the application of the precautionary principle to federal agencies in the United States? Support your reasoning.
2. The EIA process occurs within the planning phase of a land development project. Outline a subset of phases within the EIA process that concludes with the issuance of an FEIS.

3.3 ROLES OF THE PLAYERS

Participants in the EIA process will be designated by statute; there will be specific "parties" who have standing and some who request standing. Some parties may apply for "intervener status" due to their interest or expertise. Agencies and the public have a variety of roles: informational resource, archival, consultant, regulatory, and commenter. Perhaps the most useful way to organize the roles is by first looking at the steps in the EIA process and placing each stakeholder in the context of those roles.

3.3.1 Steps to go through in EIA preparation: a conceptual approach

These steps have been well articulated by Canter (1996) and others. They represent a logical sequence of events and are documented in an EA or EIS and Record of Decision.

1. Features of the project, including their relevance
2. Institutional information, including laws, regulations, and screening
3. Identify potential impacts
4. Describe affected environment (soil surveys, biological, geological, social)
5. Impact prediction (probably the most difficult step)
6. Impact assessment
7. Incorporated mitigation
8. Select the alternatives and deciding the appropriate action
9. Prepare written document
10. Plan and implement outcomes monitoring
11. Refine and improve the overall process, contributions to profession and to policy (essentially, input into how to do a better job).

The EPA has multiple roles, including those spelled out by the CEQ publications for handling EISs. The EPA comments on all parts of the EIS. It provides data, indices, and standards; stores and distributes EIA records; and even prepares its own EIS when it is preparing its own projects. For many years, the EPA used a set of criteria to share and comment on for draft EISs. On October 22, 2018, the EPA discontinued the rating system. However, it is worth summarizing here because the EPA still comments on EISs and because the framework is a useful system that may be reinstated or modeled by other agencies. The EPA's former system is instructive for anyone reviewing an EIA process or document.

3.3.2 EPA's former EIA/EIS results review system

Preparation of an EA/EIS is done with consideration of who will be reviewing it. The EPA reviews all federal EISs. It employed a rating system for many years that has now been abandoned in favor of individual comments. The former rating system, though superseded, forms a basis useful in determining what impacts merit further consideration and to evaluate the level of interest and concern that stakeholders have. The first review system below is of the adequacy of the draft Environmental Impact Statement. The second system is of the findings revealed in the EIS.

1. *Adequate*: The draft EIS adequately sets forth the environmental impact(s) of the preferred alternative and those of the alternatives reasonably available to the project or action. No further analysis or data collection is necessary, but the reviewer may suggest the addition of clarifying language or information.
2. *Insufficient information*: The draft EIS does not contain sufficient information to fully assess environmental impacts that should be avoided in order to fully protect the environment, or the reviewer has identified new reasonably available alternatives that are within the

SCREENING, SCOPING, AND RELATED ASPECTS 53

spectrum of alternatives analyzed in the draft EIS, which could reduce the environmental impacts of the proposal. The identified additional information, data, analyses, or discussion should be included in the final EIS.
3. *Inadequate*: The draft EIS does not adequately assess the potentially significant environmental impacts of the proposal, or the reviewer has identified new, reasonably available, alternatives, that are outside of the spectrum of alternatives analyzed in the draft EIS, which should be analyzed in order to reduce the potentially significant environmental impacts. The identified additional information, data, analyses, or discussions are of such a magnitude that they should have full public review at a draft stage. This rating indicates EPA's belief that the draft EIS does not meet the purposes of NEPA and/or the Section 309 review, and thus should be formally revised and made available for public comment in a supplemental or revised draft EIS.

The EPA had a system to classify the results of an EIA review. The system is based on the perceived depth or significance of environmental impacts, and is thus useful for comparative purposes.

- *LO (Lack of Objections)*: The review has not identified any potential environmental impacts requiring substantive changes to the preferred alternative. The review may have disclosed opportunities for application of mitigation measures that could be accomplished with no more than minor changes to the proposed action.
- *EC (Environmental Concerns)*: The review has identified environmental impacts that should be avoided in order to fully protect the environment. Corrective measures may require changes to the preferred alternative or application of mitigation measures that can reduce the environmental impact.
- *EO (Environmental Objections)*: The review has identified significant environmental impacts that should be avoided in order to adequately protect the environment. Corrective measures may require substantial changes to the preferred alternative or consideration of some other project alternative (including the no action alternative or a new alternative). The basis for environmental objections can include situations:
  1. Where an action might violate or be inconsistent with achievement or maintenance of a national environmental standard;
  2. Where the federal agency violates its own substantive environmental requirements that relate to EPA's areas of jurisdiction or expertise;
  3. Where there is a violation of an EPA policy declaration;
  4. Where there are no applicable standards or where applicable standards will not be violated but there is potential for significant environmental degradation that could be corrected by project modification or other feasible alternatives;
  5. Where proceeding with the proposed action would set a precedent for future actions that collectively could result in significant environmental impacts.
- *EU (Environmentally Unsatisfactory)*: The review has identified adverse environmental impacts that are of sufficient magnitude that EPA believes the proposed action must not proceed as proposed. The basis for an environmentally unsatisfactory determination

consists of identification of environmentally objectionable impacts as defined above and one or more of the following conditions:
1. The potential violation of or inconsistency with a national environmental standard is substantive and/or will occur on a long-term basis;
2. There are no applicable standards but the severity, duration, or geographical scope of the impacts associated with the proposed action warrant special attention;
3. The potential environmental impacts resulting from the proposed action are of national importance because of the threat to national environmental resources or to environmental policies.

Assignment

1. What might be a likely reason for why the EPA rating system was discontinued?
2. Should this system be reinstituted? Answer in terms of how its use might or might not improve the quality of EIAs.

Informational resources can be from almost any information delivery system, especially those found on the internet. These include university faculty or archived material. Professional journals, corporate records, and professional consultants are often consulted. Special interest groups such as environmental activists, social interests and development-related persons can also participate. Of course, regulatory agencies will have been sought out by the lead agency preparers of the EIS. As many of the sources as possible should be informed about the details of the project scoping session.

3.4 PROJECT SCOPING

After screening, we now have an idea of what laws and regulations apply and what persons, agencies and organizations may participate. The next step is to identify the major issues for analysis and recommendations in the assessment. If we do this correctly, we save time and money; efficiency and cooperation are the keys to success.

Scoping is a process that continues throughout preparation and early stages of preparation of an environmental impact statement. Scoping is required for an environmental impact statement and it also may be helpful during preparation of an environmental assessment. For an environmental impact statement, agencies must use scoping to engage state, local, and tribal governments and the public in the early identification of concerns, potential impacts, relevant effects of past actions, and possible alternative actions. Scoping is an opportunity to introduce and explain the interdisciplinary approach and solicit information as to additional disciplines that should be included. Scoping also provides an opportunity to bring agencies and applicants together to lay the groundwork for setting time limits, expediting reviews where possible, integrating other environmental reviews, and identifying any major obstacles that could delay the process. The lead agency shall determine whether, in some cases, the invitation requirement may be satisfied by including such an invitation in a notice of intent (NOI).

In scoping meetings, newsletters, internet connections, or by other appropriate communication methods, the lead agency must make it clear that it is ultimately responsible for determining the scope of an environmental impact statement and that suggestions obtained during scoping are options for the agency to consider.

Scoping also occurs in the examination of the environmental setting of a project. The project has a description by now. This description leads to a set of "portable" impacts that are associated with a type of project, like air emissions. The interaction of the project with its setting may lead to additional site-specific impacts that might occur from the project.

A checklist for describing the environmental setting can help insure sufficient information is collected (Canter, 1996). Canter's purposes in describing the environmental setting include:

1. Assessing the existing environment that could be affected
2. Looking for significant factors ("fatal flaws") in background
3. Obtaining information to bring back for further review
4. Determining or verifying basis for project need
5. Contributing to regional (or area) database
6. Establishing a basis to explore relevant internet-based resources.

For the scoping process and for the introductory parts of the EA or EIS, the environmental setting of the project needs to be described. This would include identity lists of environmental factors from participating agencies, professional consultants, prior baseline studies, recent EIA and EISs for similar projects and copies all previous studies of the site for the development plan. Agreement on methodologies to be used for each component should also be considered. With these tasks accomplished and appropriate graphics and other visual aids prepared, the scoping session will lead into the heart of the EIS document preparation.

With the scoping process underway, it important to choose what environmental factors to include and which ones merit detailed study. For each factor, the participants will carry out a review of written material; conduct field reconnaissance; consult with local community leaders, agencies, experts, political representatives; and then decide priorities for what to focus on. The scoping process also has some public participation elements. The public will want to find out what issues will be addressed and where they will be informed. The team may be able to pick up local knowledge from the public. To this end, the preparation team will publish notices of intent to prepare EA or EIS, notices for submitting comments about the proposed scope of the project, and directions for submitting comments on final scope, as well as the continued EIS/EIA development.

Assignment

Scoping is where you make your work efficient. You have to account for the major impacts of a certain category and determine how to assess them in the quickest way but without excluding anything that has the potential for a measurable impact. If you are preparing for fieldwork in which you assess archaeological/historical resources, you might use the following steps in scoping:
- Check archaeological/historical site files and records at SHPO/THPO (State/Tribal Historic Preservation Office)
- Review local historical documents and past reports that pertain to the project lands and general vicinity
- Determine regulatory requirements for site testing (what are protocols?)

- Identify the key resources you could encounter (e.g., Paleo-sites, historic mill, War of 1812 fortifications, 19th century African-American settlement)
- Determine portions of project land most likely to contain sites (sensitive areas)
- Develop a testing strategy based on time, money, requirements, and capabilities
- Assemble equipment, consultants, and other resources to carry out testing strategy.

Your project is 100,000 square feet of government buildings with 300 spaces for parking and numerous hiking trails, and five water access points. Project land involves 10 miles of a Class A river, with three streams feeding into it. The watershed drains to a coastal bay, located about 9 miles away from the downstream portion of your project. The area is moderately settled, and there are forestlands along much of the waterway. Discuss how you might plan to do the assessment. You have 6 months to do an EA. You have a budget of $100,000, not including salaries (which will cover three people for 6 months). Outline a scoping strategy for determining the potential range of impacts on one or more of the following: fisheries, water quality, air quality, archaeology, traffic, economic/fiscal, wildlife.

3.5 DATA COLLECTION AND PRESENTATION

We address characteristics of data in subsequent topics. Here we focus on general approaches and tools. Cantor (1996) describes five major methods for EIA data collection and impact forecasts:

1. Matrices (simple, stepped or cross-impact)
2. Checklists
3. Ad hoc (ask experts/Delphi)
4. Overlay map analysis (landscape architecture, GIS)
5. Networks (social, computer, infrastructural, financial, political).

Matrices and checklists are useful for a variety of reasons. The EIA process is not a cookbook recipe. Each one will entail handling significant amounts of complex information—a task for which matrices and checklists are often essential. Use of them reduces the chance of missing something and can lead to additional factors to consider, as part of the scoping of impacts and procedures.

Five uses for matrices and checklists are:

1. Identifying impacts
2. Showing background setting/baseline data
3. Predicting and assessing impacts
4. Selecting a proposed action
5. Managing the EIA project.

Checklists are commonly used by agencies for addressing impacts and determining procedures. Many checklists are complex and multidimensional, and some contain nested matrices. A matrix is particularly useful for finding a point of intersection that leads to an impact or item of concern that you might have otherwise missed. Table 3.2 shows how a matrix can indicate some potential impact intersections.

TABLE 3.2 Sample matrix for offshore wind turbine project impact assessment.

Phase	Visual	Energy	Benthic Ecology	Pelagic Ecology	Air Quality	Economics	Avian	Noise	Existing Hazards
Construction	Boat traffic	Fossil fuels and electricity for: manufacturing and transportation	Effects of cable installation, footings	Debris from jet plowing, risk of whale collisions	Increased emissions from fossil fuels	Manufacturing, distributing, and constructing jobs	Tall equipment may interfere with bird/bat paths	Pile driving WTG, ESP foundations may interfere with whale sonar	Vibrations and physical disturbance may dislodge harmful materials or unexploded ordnance
Operation	WTGs visible, lights should not be visible	Increases electricity supply, consumes energy in maintenance-related boat travel to site	Anchors provide substrate for benthic habitat	May provide habitat (as substrate or shading)	Maintenance boats exhaust, noise, vibration from the WTG	Jobs, power generation revenue	Turbines may interfere with birds	Boat noise interferes with whale sonar, vibrations may interfere with marine species	Vibrations may set off unexploded ordnance
Monitoring	Sight of boats	Fossil fuels used by boats to monitor the site	N/A	Boats may injure organisms	Boat fossil fuel use	Jobs	N/A	Boats may interfere with whale sonar	N/A
Decommissioning	Sight of boats going to and from site	Energy to recycle materials	Removing anchors and stirring up the bottom	Boat travel during decommissioning	Boat exhaust	Jobs, material to recycle, loss of power generation	Improve health of avian species	Noise effects on marine species	Vibration effect on unexploded ordnance

Source: Bureau of Ocean Energy Management (2017).

TABLE 3.3 Project and environment matrix.

	A	B	C	D	E	F	G
1							
2							
3							
4							
5							
6							
7							

Assignment

One way to think creatively about potential impacts is to make a list of environmental characteristics and a list of project characteristics, then find points of intersection between the two. One list becomes the X-axis and the other becomes the Y-axis and you thus create a matrix. This may lead to identification of potential environmental impacts that you might otherwise not have thought of. Make up an arbitrary list of seven types or categories of environmental impacts of your choice. Choose characteristics or aspects of the site where a project will be constructed. *Historic properties* might be one category; *agricultural soils* or *surface water* might be another—but choose your own, without worrying too much about it. It can be specific or general. List them as items 1 through 7 as the Y-axis of your grid.

Now list seven characteristics of a project. These can be from one or more of the following sets: location in the environmental setting (e.g., buildings, roads, sewer system, water supply, above-ground, airborne, topography); project phases (e.g., existing conditions, exploratory, site clearance, building construction, operation, reclamation, monitoring); time of year; number of employees; the project in 10 years. Alternatively, make your own—whatever you want in whatever combination. List them as items A through G. These will be in the X-axis of your grid.

The grid will form a matrix (Table 3.3). Enter a potential impact for each cell.

3.6 PUBLIC INVOLVEMENT

Active public involvement is a key to reducing antagonism, understanding the environmental values, improving decision-making, and incorporating new ideas. It sets the basis for ongoing relationships among all participants. Public involvement should play roles in the planning and EIA of a project and all other phases (build-out, operation, and closure).

NEPA builds in disclosure requirements, but agencies are already subject to public disclosure of information. Failure to do so impedes the process and can lead to Freedom of Information requests (FOI requests or FOIRs) for information "in the public record." Essentially, this information can include everything that is not part of a pending criminal investigation or personnel inquiry, although the results of both are subject to disclosure. Maine has a typical definition of public record:

> Pursuant to Title 1 M.R.S.A. Section 402(3), a public record includes any written, printed or graphic matter or any mechanical or electronic data in the possession or custody of an agency or public official that has been received or prepared for use in connection with the transaction of public or governmental business and contains information relating to the transaction of said business; therefore, the public is advised that any correspondence, whether by traditional method or email with Town offices or Town officials, with certain limited exceptions, is public record and is available for review by any interested party.

Most states and communities in the United States have similar processes and regulations.

NEPA conducts public meetings, but administrative law in agencies also allows for public hearings. There are three types of hearings on the formality spectrum: judicial, quasi-judicial, and informational. Knowing what type of hearing will help in knowing what you can expect from the participants. Hearings and other public meetings can be a continuum of fact-finding, fact-sharing, matters of law, procedures, and just getting acquainted.

Public meetings also vary by formality and purpose. One way to examine them is to look at the types of communication called for or specified for the meeting. Is the communication up, as in feeding information to the agency from the audience? Is it down, as in the agency communicating to the public? Is it a combination, with egalitarian give-and-take (lateral communication)? Nowadays, many public meetings allow for remote participation, usually via Zoom or a similar platform.

Arnstein (1969) described eight rungs on the ladder of citizen participation; we provide the first seven as ones most commonly used for EIA. Some researchers add more rungs or use fewer, but the concept remains useful in public policy no matter which model is used. Levels of participation may be affected by degrees of personal feelings of power, ranging from lowest to highest: (1) manipulative (propaganda), (2) feel-good (therapy), (3) consulting, (4) tokenism/representation, (5) partnership, (6) delegated power, and (7) citizen control. This is a standard hierarchy in the literature used to examine effectiveness of public involvement.

While meetings may be structured to improve public involvement, there are some potential roadblocks. Some of the biggest barriers to public involvement are cost, time frame, distraction from selected issues, and misunderstandings of meeting purposes. The major objectives of public participation should be to improve public relations (notably community-building and de-mythologizing), engaged information sharing, and conflict resolution.

The initial place for public involvement might likely be at the inception of the project, for it is to accomplish a public purpose and may be in response to a public demand. But public participation should occur within all stages of an EIA (Figure 3.2). Common steps or stages for public input can be any one or more of the following:

1. Project introduction and initial concerns/reactions;
2. Baseline studies that seek and describe public knowledge of local environmental conditions;
3. Prediction and evaluation of impacts ensuring an adequate range of alternatives, and developing project-specific criteria and feedback mechanisms;

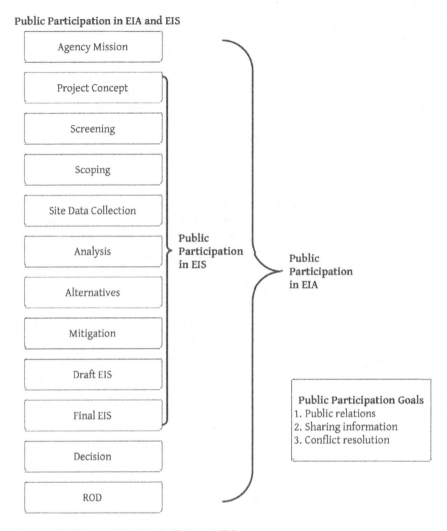

FIGURE 3.2 Public participation in EIA and EIS.

4. Mitigation planning with participant reactions, and negotiated/resolved acceptances;
5. Comparisons of impacts using weighting factors, methods for public input;
6. Decision-making that identifies preferred alternatives that resolve conflicts;
7. Working with EIA documents (EA, EIS, drafts and finals, supplements);
8. Monitoring: A plan for public involvement in monitoring may be helpful in dealing with concerns about the environmental consequences of project build-out or operation.

Assignment

1. Observe a public meeting and complete the following worksheet. This can be done in person or online (e.g., via Zoom or other format). Plan ahead so that you can locate an appropriate meeting to attend. Fill out the form below and Table 3.4.

TABLE 3.4 Stakeholder exercise. Enter up to six stakeholders and put a checkmark where you would assign each on the "degree of power" scale.

Name of Stakeholder	Description (private developer, federal agency, non-profit, advocacy)	Very low (1) Manipulative	Low (2) Feel-good	Somewhat Low (3) Consulting	Moderate (4) Tokenism	Somewhat High (5) Partnership	High (6) Delegated Power	Absolute (7)

A. Is the public meeting a hearing? If so, which type of hearing?
   __ judicial, __ quasi-judicial, __ informational
   If not a hearing, which form(s) of meeting? Public relations, informational, conflict resolution
B. Which public meetings types of communication do you see?
   Indicate all that apply: up, down, lateral
   Was the setting __ judicial, __ formal, __ informal? Neutral ground or stakeholder ground?
C. Stakeholder authority. For each stakeholder (other than the convener), indicate the apparent degree of power manifested in participation: 1 is low power, 7 is highest. Write the name next to the applicable level.
   (1) manipulative (propaganda), (2) feel-good, (3) consulting, (4) tokenism/representation, (5) partnership, (6) delegated power, (7) citizen control _____
   Where did most of the groups cluster?
D. Which of the following reasons for public participation apply?
   1. Regulatory requirement
   2. Public relations
   3. Obtain facts
   4. Reveal the decision-making process
   5. Identify local issues
E. Which of the following barriers to public involvement affected the meeting?
   1. Cost (materials, services, experts)
   2. Time
   3. Opposition from stakeholders
   4. Distraction from key issues
   5. "Popularity contest" of issues/people
F. What was accomplished? Check all that apply: __ information dissemination, __ information collection, __ planning/negotiations, __ responses to activities/policies, __ decision-making, __ scheduling, __ mutual trust.

2. The three main objectives for public participation are public relations, sharing information, and conflict resolution. Imagine a master plan for development of a new college campus, including residential dorms, a small teaching hotel and restaurant, and a recreational game center to be shared by residential students, hotel guests, and the public. Make a public involvement strategy for the master plan review in which you address basic principles of good public involvement, objectives for public involvement, stakeholders, and decision-making. Your plan should include filling out the matrix below (Table 3.5).

3.7 CONCEPTS AND TERMS

Categorical Exclusion (CX): A federal (or state) action or undertaking that is exempt from NEPA because the likelihood of significant environmental impact has a low threshold or does not exist.

Draft Environmental Impact Statement (DEIS): An initial formal statement documenting the potential environmental impacts of a proposed federal or state development or undertaking. It is circulated for comment before the final statement is issued.

TABLE 3.5 Master plan stakeholder matrix.

Stakeholder Name and Characteristics	Likely Concerns	When to Involve Them	How to Involve— What Activities and Objectives?
1			
2			
3			
4			
5			
6			

EIS Environmental Impact Statement (EIS): A formal document accounting for a range of environmental impacts of a federal or state action or undertaking in accordance with NEPA or state regulation.

Final Environmental Impact Statement (FEIS): A formal statement documenting the potential environmental impacts of a proposed federal or state development or undertaking.

Finding of No Significant Impact (FONSI): A declaration that a project will not have a significant impact and therefore no EIS will be prepared.

Matrix: A multidimensional array for arranging more than one type of information, such as characteristics of a project, and characteristics of a site. It is a useful analytical and management tool for complex information.

Precautionary principle: Policy of avoiding introduction of new chemicals or products into the environment until it is first shown that they will not cause harm to people or the environment. A variation on the principle of "first, do no harm."

Scoping: Process of narrowing down the focus in impact assessment to increase efficiency and effectiveness of reviews.

Screening: Process of determining what laws and regulations apply to a project.

Stakeholders: Someone or entity who may have an interest in a proceeding. Can include any potential party in an EIA or other process. Includes the proponent, regulators, adjoiners, experts, and the community.

3.8 SELECTED RESOURCES

Canter, Larry. (2015). *Cumulative Effects Assessment and Management*. Horseshoe Bay, TX: EIA Press.

Morrison-Saunders, Angus, Jos Arts, Jill Baker and Paula Caldwell. (2001). Roles and Stakes in Environmental Impact Assessment Follow-Up. *Impact Assessment and Project Appraisal*, 19(4), 289–296. doi: 10.3152/147154601781766871.

United Nations. (2002). *Environmental Impact Assessment Training Resource Manual*, 2nd ed. United Nations Environmental Programme. https://wedocs.unep.org/handle/20.500.11822/26503.

U.S. Environmental Protection Agency. National Environmental Policy Act Review Process. www.epa.gov/nepa/national-environmental-policy-act-review-process.

U.S. General Services Administration. National Environmental Policy Act. Includes NEPA Desk guide and NEPA Library. www.gsa.gov/real-estate/environmental-programs/national-environmental-policy-act.

3.9 TOPIC REFERENCES

Arnstein, Sherry R. (1969). A Ladder of Citizen Participation. *Journal of the American Planning Association*, 35(4), 216–274.

Bureau of Ocean Energy Management (IFC Research Planning, Pam Latham, Whitney Fiore, Michael Bauman and Jennifer Weaver). (2017), *Effects Matric for Evaluation Potential Impacts of Offshore Wind Energy Development on U.S. Atlantic Coastal Habitats*. US Department of the Interior, Bureau of Ocean Energy Management, Office of Renewable Energy Programs. OCS Study BOE 2017–014. Sterling, VA.

Canter, Larry. (1996). *Environmental Impact Assessment*, 2nd ed. New York: McGraw-Hill.

Department of Energy (DOE), Office of NEPA Policy and Compliance. www.energy.gov/nepa/office-nepa-policy-and-compliance.

Department of Environmental Conservation, New York State. State Environmental Quality Review Act (SEQR) forms. www.dec.ny.gov/permits/6191.html.

TOPIC 4
Assessing environmental impacts

4.1	Project planning and management	65
4.2	Data collection	67
4.3	Indices	69
4.4	Impact identification and prediction	70
4.5	Environmental Assessments (EA)	77
4.6	Concepts and terms	78
4.7	Selected resources	79
4.8	Topic references	79

Impact assessment begins with an overall strategy of project management within the context of public accountability for decision-making. This strategy includes developing a general framework for assessing impacts. This framework make come from agency protocols or be a project-specific one developed in an RFP. Many questions are asked at the beginning of the impact assessment process. An overarching question is, "What is the range of impact assessment methodologies useful for this project?" In evaluating methods we seek to determine what the specific strengths and weaknesses of each major method are. How much depth for each impact assessed? How much information will be needed for each impact that will be assessed?

4.1 PROJECT PLANNING AND MANAGEMENT

Section 102(A) of the National Environmental Policy Act (NEPA) calls for federal agencies to "utilize a systematic, interdisciplinary approach which will insure the integrated use of the natural and social sciences and the environmental design arts in planning and in decision-making which may have an impact on man's environment." This is the justification for standardizing approaches to data collection and analysis under federal compliance rules. It has promoted the use of planning and management in organizing impact assessment. Other benefits are improvement in the use of impact comparison and analysis, standardization of methods and reporting, development of environmental indexes, and the strategic employment of environmental assessments.

DOI: 10.4324/9781003030713-4

A project begins with the recognition of a need or obligation that will result in a change or intrusion on the landscape. This intrusion may be in the form of a specific development, such as constructing an airport, which has direct environmental impacts, or a policy such as disposing of hazardous waste, which leads to actions that have environmental impacts. A project description distills from these aspects. Once the project is defined and described in all of its potential ramifications, the next step, if not already done, is to secure acceptance by the project sponsors and agency leadership. If it is not already in place, a team forms to oversee the assessment.

An environmental assessment, like any major undertaking, has specific logistical steps from beginning to end. These arise from the field of project management and similar business approaches. While all these steps are familiar within the planning phase of a project, there is a specific start-to-finish process for producing the EIA report. These first seven steps are part of initial planning prior to undertaking the assessment, and they are the same steps one would take in undertaking any project, including the actual construction or building-out:

1. Secure acceptance by sponsors/managers
2. Provide clear statement of purpose/mission
3. Identify goals and milestones
4. Identify major tasks
5. Outline strategies
6. Establish organizational design
7. Secure support services.

Overall planning to accomplish an environmental assessment requires attention to detail and an overall strategy. Many environmental consulting firms will choose to specialize in certain types of projects in order to be even more efficient in providing project oversight. Modak and Biswas's (1999) five principles for managing the EIA process are scoping, participation, timing, mitigation, and information. Scoping can be effective in honing in on the main issues for the overall process, impacts, and the documents. Participation, as a management concept, refers to the EIA team, the consultants and contributors to the EIA process, the public, and the decision makers who comment on, approve, or act upon the EIA products. Effective timing for true planning/decision-making purposes involves multiple aspects. We must consider environmental issues at the beginning of process—what are the "portable" issues that arise from this type of project? We inquire into the environmental impacts known or likely to exist at the site. The intersection of project and site raises the issue of project feasibility. In addition to environmental concerns, what are the engineering concerns? The issue of timeliness of permit processes arises. Timeliness means certain protocols must be anticipated so sampling can be undertaken for the correct durations and conditions; water sampling during high flow and low flow is an obvious example. Monitoring, compliance corrections, and reporting need to be built into the project management.

Management of mitigation includes developing clear options for the reduction, off-setting, and compensation of impacts. A key task is to present clear choices and their consequences. This means planning for use of many issues: pollution control technology, waste disposal, social justice, identification of alternative sites, project size (final growth), local resource enhancement, government processes, monitoring and periodic reviews, contingency plans for regulatory action, and community/public involvement.

Modak and Biswas's final management item is to keep in mind the central function of providing useful information (1999). Information must be in a format that can be used by decision makers: hard facts; reliable, reasonable predictions; standard terminology and vocabulary; concise document; appendices; and visual clarity. Good, timely communication of useful information is the essence of an assessment.

The EIA report may reflect a project management technique called "decomposition." This approach separates a project into its components as a series of processes and activities within processes (Kerzner, 1997). The technique can be used for different scales ranging from detailed systems engineering to overarching cosmology (e.g., Kisak, 2015). As part of the approach, we can decompose (break down) the category of environment into components: receiving waters, labor pool, income, settlement (community) patterns, transportation routes, agriculture, forestry, and so forth. We can decompose other categories and components. Some agencies and impact assessment firms will have proprietary or commercial software to accomplish this.

The loop of decomposition can be illustrated by the following: *activity + component = impact, leading to assessment, mitigation, residual impact, modification, and back to activity + component*. The modifications include add-on systems and project redesigns.

Assignment

1. Select a recent EA report and make a table that applies the loop of decomposition to elements of the EA.
2. Assume you have a team of five people to do an EA for a 100-unit residential project in the Bayside neighborhood of Portland, Maine. Your first step is to see what is meant by "Bayside" in Portland. A quick search of Google tells you that there is an East Bayside and a West Bayside. You can also get an idea of the geographic boundaries, the socioeconomic demographics, the buildings, and the infrastructure. This is the area and context for this EIA project planning exercise.
   a. Describe Bayside.
   b. What skill set/expertise does the team need?
   c. Create a two-month timeline (Table 4.1) that produces an EIS. Assign roles and deadlines for tasks/responsibilities. Include additional tasks and responsibilities if you want.

4.2 DATA COLLECTION

There are many purposes for data collection. In EIA, data collection should contribute toward the essential objective of predicting future impacts. There are three steps in identifying potential impacts. The first and easiest is to link potential impacts to past, similar projects. The second step is to evaluate past impacts, and the third step is to create a model based on the results of steps 1 and 2.

In reviewing past assessments, older methods and/older information science may require more research. In collecting impacts, it is helpful to think about the categories of impacts. We

TABLE 4.1 Two-month timeline for EIA project management.

Task/Responsibility	Who in your group	When
Self-identification as consultants. Who you are, how you are structured, what you do.		
Respond to RFQ		
Respond to RFP (may include a plan to address some or all of the rows below)		
List of supplies and services		
Screening (determine what permits and laws apply)		
Site testing and evaluation		
Initial site visit		
Scoping of impacts		
Matrix/checklist of impact categories		
Literature review		
Write EA		

can conceive of them in terms of dimensions and descriptors: size, form, circumstances of occurrence, time, space, cost, interactive (synergistic), concatenating, direct, and indirect. An impact is an agent of change. We can describe change in many ways, including physical, economic, and institutional aspects.

Data collection is accomplished under protocols for sampling, analysis, and curation. These will be specific to the type of data or category of sampling, be it soils, water quality, air emissions, or most other forms or sources of data. Professional societies and government agencies are the dominant sources of protocols. EIA reports will typically need to use the data collection protocols that are commonly accepted industry standards. The protocols also include how the data and samples are handled, including curation and "chain of custody" or custodial "chain of command."

Assignment

Choose an environmental media category or component for sampling, such as soils, heavy metals, surface water, groundwater, or air quality, and describe an agency sampling protocol applicable to or required for EIA.

4.3 INDICES

EIA processes assemble complex information from all environmental aspects, distill it, bring in peer and public reviews, then submit it to analysis, all as part of environmental planning. Accordingly, it makes sense to use environmental indicators that are commonly recognized and translatable for comparing impacts. Indexes are one type of environmental indicator that has the job of distilling and simplifying a variety of environmental conditions for a particular media. They do not have to cover all pollutants in their categories, but they do need to capture the major ones. We get indices from a variety of sources including: academia, governmental agencies and professional societies (e.g., AASHTO, American Association of Civil Engineers, Certified Wildlife Biologists, American Institute of Architects). The regulated community (e.g., coal mining companies, manufacturer groups, energy groups) may also have valuable data.

There is always a need for more indices as new parameters arise and the field of EIA grows. We look for in an index to be understandable, encompassing major pollutants, relatable to ambient standards, relatable to "episodes," and consisting of simple calculations and reasonable assumptions. We also want an index that has scientific premise, is consistent with perceived levels, is spatially meaningful, and that exhibits (reflects) day-to-day variations. The index needs to be amenable to forecasting.

Thousands of different indices exist. They range from city/local to state, regional, national, and international. To show the variety, and for interest, we present a few here. The protocol for selecting indexes to use is to defer to the standard ones that facilitate the comparison of alternatives within and among different EIA/EIS documents. Indices are a major tool in evaluating past and current impacts and in predicting future ones and should be used whenever possible.

BEQI: The Bicycle Environmental Quality Index (BEQI) was developed by the State of California's San Francisco Department of Public Health. The mission of the Index is to assess the interaction of bicycles and roadways and evaluate improvements to promote bicycling in the city. The BEQI is based on five main categories: intersection safety, vehicle traffic, street design, general safety, and land use. The five factors are set to assess important environmental aspects for bicyclists (Program on Health, Equity and Sustainability, 2014).

BI: The Biotic Index (BI) is a score for a specific surface water site in New York, based on macroinvertebrate biodiversity.

CPI: Corruption Perception Index (CPI) from Transparency International (a non-governmental organization). It uses data from the World Bank, UN Economic Commission, World Economic Forum, and other sources to provide a combination from nine separate measures. It correlates with HDI.

EPI: Environmental Performance Index (EPI) is another composite index. It uses 32 indicators in 11 issue categories and ranks 180 countries. It can be considered as a scorecard for policy makers (https://epi.envirocenter.yale.edu/).

ESI: Environmental Sustainability Index (ESI), World Economic Forum, Yale and Columbia Universities. The index uses 76 environmental variables distilled through 22 indicators. The index covers ambient environmental conditions, current emissions, and predicted future emissions based on development rates and regulations. There are many other factors, notably mortality, education, income, health, and crime. There are approximately 100 sustainability indexes in use (Gan et al., 2017), with more in development. These are considered composite indexes because they assess a variety of factors, including economic, social, and ecological.

HDI: Human Development Index (HDI), UN Development Programme. Computed from life expectancy (a proxy for healthcare and living conditions), adult literacy, and gross domestic product (GDP) per capita (a proxy for disposable income).

QLI: The Economist Intelligence Unit, of the Economist Group, established the Quality of Life Index (QLI). This index is set to quantify the overall satisfaction with life of an individual living in a certain place. The QLI uses five scores in four domains: health and functioning, psychological/spiritual domain, social and economic domain, and family (The Economist, 2005).

SQI: The Soil Quality Index (SQI) was established to assessment the management-induced changes in soil quality, reflected in crop yields. Relative soil properties and overall crop yields are the main factors considered in an SQI. Soil layer composition, minerals, and organic materials are contributing factors of the SQI (Mukherjee, 2014).

WQI: The Water Quality Index (WQI) is a tool to summarize and report the State of Washington's Ecology Freshwater Monitoring Unit routine stream monitoring data. The WQI is a unitless number ranging from 1 to 100; a higher number is indicative of better water quality. Scores are determined for temperature, pH, fecal coliform bacteria, dissolved oxygen, total suspended sediment, turbidity, total phosphorus, and total nitrogen. Constituent scores are combined and results aggregated over time to produce a single yearly score for each sample station.

Assignment

1. Define "environmental media."
2. Identify media likely to have indexes of some sort at the international level and suggest why this is so.
3. What are some media most likely to have indexes at the state or local level. Why is this so?
4. What are two of the more common indexes used for water quality in the United States?
5. Locate and describe a national index not provided in the examples above.

4.4 IMPACT IDENTIFICATION AND PREDICTION

Successful impact identification starts with a framework for collecting data. It may be helpful to recall the meaning of accuracy and precision in referring to data. Consider a target (Figure 4.1). Inaccurate and imprecise data is analogous to strike points that are scattered all over the target. Inaccurate but precise data are clustered but off the target. Accurate and imprecise data is represented by scattered points that average out to being on target even if no one data point is actually in the exact center. Accurate and precise means clustered and on target.

Data used to understand and identify impacts reminds us that data collection is about understanding impacts (Canter, 1996). To this end, data collection about impacts should address all of the following: how the impacts are displayed, understood, and incorporated into a project. We must specify each potential impact, estimate probability of occurrence, and identify the likely conditions of occurrence. They should have a quantitative set of dimensions that create a

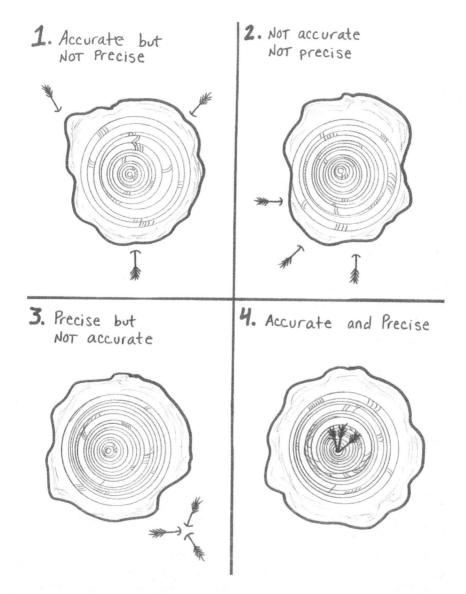

FIGURE 4.1 Accuracy and precision illustrated.

basis for quantification. Unquantified data should made quantifiable if possible. Finally, remediation costs for mitigation measures should be included.

Descriptions of variables can be used to help describe impacts: proximity, magnitude, importance, probability, quantifiable, and stability are examples. These attributes are just for detailing the impacts, not whether we can mitigate them or manage them—that comes later as a result of description and analysis of the impacts.

The context for the impact is within the environmental setting of the project. The EIA may be compared against baseline data from the setting component. An issue that may arise from this action is that the baseline may have changed due to normal or induced fluctuations. For example, Chesapeake Bay has had significant intrusions into productivity of its fisheries, and people need to make different assumptions on what to compare change to (Kennedy, 2018).

4.4.1 Types of impact

It is helpful to think of the environmental effects of a proposed action as variables with myriad potential attributes and outcomes. Canter (1996) points out some of the aspects or sources of variation. Consideration of the multiple characteristics of impacts will be helpful in documenting the impacts and in developing mitigation. Some of the ways impacts may vary (or be described) are:

- Predictable or unpredictable
- Direct or indirect
- Positive (beneficial) or negative (harmful)
- Temporary or permanent
- Short-term, medium-term, or long-term
- One-off, intermittent, or continuous
- Immediate or delayed
- Certain or uncertain
- Avoidable or unavoidable
- Reversible or irreversible
- Localized or widespread
- Small or large
- Individual or cumulative
- Significant or of no consequence.

Assignment

1. Do any other types of impacts exist in addition to the ones above commonly described by Canter (1996)?
2. Regarding data collection and categories of impact matrix: look online for a project that involves a proposed housing development in your municipality or another municipality nearby. Fill out the cells in Table 4.2 in a manner in which the data might address the variable characteristic in the column heading.

An EIA uses a variety of models for data collection, project management, impact prediction, and other aspects of the process. Models can address portable impacts that come with a particular impact and place-bound impacts associated with the intersection of space and project. In either case, the characteristics we seek in an impact model are that it must be general enough to allow for repeatability and must be realistic so that it models actual natural systems. It also must be precise (level of exactness) and be accurate (measure what it is supposed to measure).

Predictive modeling (predictive analytics) attempts to analyze patterns and trends to predict future outcomes. It is not about getting the most data, it is about getting the best data in

TABLE 4.2 Categories of impact matrix.

Impact	Stakeholder	Proximity	Magnitude	Importance	Probability	How Quantified	Stability	Duration
Endangered Species	State F&W, Audubon	On/ Adjacent	Breeding Populations, # of Species	Ecological, Social, Legal— if on ESL List	Habitat Models	# of Individuals	Predictable Based on Specific Species	Permanent
Deer habitat/ wintering area disturbance								
Project partly in floodway								
Aesthetics of lighting								
Agricultural land								
Water supply loss								
Groundwater contamination								

a cost-effective way that is statistically reliable for predictions. A model is by definition not a complete replica. It must leave many things out. Models are divided into full effect and partial effect, depending on how complete they are. A full effect model is typically used for disaster planning and management. It is a relatively complex model that leaves out as little as possible. Most models are partial effect. The process of deciding what to include is decided by "scoping" for efficient use of significant variables. In evaluating EIA reports and modeling, keep in mind the general principle that if the model is wrong, more data does not necessarily make it better.

There is an old saying that "All models are wrong but some are useful." A great many impact prediction models exist, such as methods of scenario analysis, data overlays, simulations, expert judgments, checklist and matrix tools, and use of analogues (Canter, 1996). Many models are a combination of these methods. This summary of model types is informed by Schmidt et al. (2008).

4.4.2 Mechanistic model

Mechanistic models are cause-and-effect relationships, flow charts or mathematical models. A model can be qualitative or quantitative, which means it will be either deterministic or stochastic.

4.4.3 Deterministic model

A deterministic model seeks a single solution from fixed relationships. One example is the gravity model used in economic and spatial patterning. This model is not concerned with uncertainty or variability because it is based on averaging tendencies.

4.4.4 Stochastic model

The stochastic model is a probabilistic approach. It considers randomness, changing environmental conditions, and probability flows. A common use is in dealing with flooding and river control systems.

4.4.5 Physical model

Physical models are direct scaled reproductions and may allow for experimentation. An example would be an architectural model of a building on a site.

4.4.6 Scenario

Another approach to modeling is the scenario. Scenario analysis focuses on identification of effects on fauna, flora, landscape, and cultural heritage (including archaeological and historical information). A scenario enables us to describe possible states of a system with the help of factual knowledge or other techniques of prediction, even though the basic knowledge might be incomplete. Scenarios have the advantage of taking different aspects into account at the same time. They are especially suitable for situations lacking easily quantifiable elements, as is often the case in estimations of environmental impact concerning fauna, flora, or landscape.

The purpose of a scenario is not to make an accurate and precise forecast of future developments, but rather to find out if there is only one, virtually inevitable development or if other developments are conceivable and possible. The variety of possible future states of the system is supposed to make citizens, politicians, and planners think and avoid false feelings of security.

Schmidt et al. (2008) point out that scenarios are usually backed up by statistical evidence, but at their core, they are nevertheless primarily verbal. They are a blend of intuition and creativity, systematical scientific work, and experience. Scenarios are built from a few premises that lead to a logical sequence of hypotheses of the change from an actual to a future state.

The three main elements of a scenario are general description of the examined object and its behavior, premises for the examined object, and a description of the expected changes, consequences, and impacts.

4.4.7 Overlay technique

The overlay technique, which typically uses geographical information systems (GIS), is a general method for identification of effects on the environment. The overlay of maps with environmental sensitive areas and impact factors and their spatial range has become a standard technique for all environmental components. Overlays allow calculations about direct loss of environmentally important areas or indirect effects on special ecological functions.

4.4.8 Simulation

Simulation models are used for many types of impacts, particularly identification of effects on soil, water, terrestrial systems, air, and climate. Most simulation models are computer-based manifestations of stochastic or deterministic (or other) models. Data is input from professionals, usually consultants who specialize in the subject. Like any model, although large and complex, computer simulations are still an approximation of a larger system. They describe the examined system with a specific set of parameters that can be adjusted. Simulations can address any resource issue, notably air pollutants, noise emission, aquifer recharge, resource extraction, surface water flow and contamination, transport of pollutants in groundwater, climate change, stormwater runoff, traffic growth, demographics, economic systems, and other systems.

4.4.9 Expert judgment

Professional judgment always has a role in impact assessment but it can be considered a modeling technique for unique cases. Environmental effects may be uncertain and not predictable with normal methods. In such cases, experts can give their opinion and expectations from their own experiences. This method is known as the Delphi method, after the Oracle of Delphi from Greek mythology. Expert judgment can be used in any area, particularly identification of potential impacts and effects on fauna, flora, or interactions with other environmental resources

4.4.10 Checklists and matrices

Checklists and matrices are a common tool or method for impact assessment and related processes. We address them in multiple places throughout this text. Most agencies and consultants use them in some form for identifying and handling complex information and processes. Matrices are a more advanced version of checklists. At their most basic level, they can be adequate to gain first impressions about the main effects likely to be expected. In large, complex EIAs, they are just one of many tools and approaches needed.

4.4.11 Analogues

Analogues are a general method based on similarities of projects or conditions with other sites. Often, this approach is a referred to as a case study comparison. Experiences from case studies with similar environmental conditions and projects can be helpful in making predictions about a new project or site. Since analogues can be replication of another system elsewhere that is similar and has already yielded data and been tested, it can be a simple physical model. It is often linked with threshold-based predictions. At this level, the model supports a conclusion such as "this next step is needed" or "that reaction will occur." These are often simple physical models.

Predicting impacts with a model usually requires three steps, as mentioned in section 4.2:

1. Connect the potential impacts to past or similar projects.
2. Evaluate the impacts from past projects.
3. Incorporate past effects and construct a model.

Sources for identifying similar projects from the past this include the "gray literature," meaning EISs and other government and consulting group documents and reports that do not appear as formally published literature typically found in academia or popular media (i.e., journals, books). A big caution in using these sources is that past assessments used older methods and did not necessarily address all impacts. In describing impacts that arose from those projects, the assessors will use scalar or vector quantitative terms, like size, form, circumstances of occurrence, time, space, cost, interactive (synergistic, antagonistic), concatenating, proximity, magnitude, importance, probability, and stability.

In doing EIA, we evaluate and determine the meaning (significance) of past impacts. We consider a wide area of change that can occur from the past to the present: biological change, chemical change, physical change, infrastructural services change, demographic change, human health/epidemiological change, institutional change, social/cultural change, economic change, stability change, sustainability change, regulatory change, and ecosystem change (some combination of other changes). Further, we also have to anticipate changes for the future, within the lifespan of the project being assessed.

In step 3 we construct an impact model informed by steps 1 and 2. The impact model should be general enough to allow for repeatability. It must be realistic enough to model actual natural systems. It must be precise (level of exactness), and it must be accurate (measure what it is supposed to measure).

Assignment

Develop an outline of a model to predict impact change for impacts that might result from actions on your university campus. Use the above information about predicting change to help you construct a model that a team could use for an EIA. What type of model(s) will you use? What impact categories will the model(s) address? Choose one of the following scenarios for the conditions under which the model will operate.
1. Scenario A: No residential halls and reduced enrollments on campus.
2. Scenario B: Increased residential halls; student population doubles.

4.5 ENVIRONMENTAL ASSESSMENTS (EA)

In lowercase letters, environmental assessment refers to the general process or action of assessing environmental impacts. In uppercase letters, an Environmental Assessment (EA) is a formal document. Also called an Initial Environmental Examination (IEE) in Europe, in the United States it can be called an EA, an EIA, or a Phase I Report. It can be preparatory to or in place of a full EIS. It contains:

1. A description of the project
2. A brief description of the predicted or expected environmental changes
3. Measures, steps, and procedures to avoid/reduce impacts
4. Additional study requirements/recommendations.

Canter (1996) describes the components of an ideal EIA/EIS. Some agencies use checklists to review and evaluate the EIA/EIS based on these and related factors.

An ideal EIA report:

- Promotes timely participation
- Catches significant projects
- Addresses all reasonable impacts
- Properly considers and compares alternatives
- Is prepared by qualified, capable personnel
- Involves the public
- Is clear and understandable
- Provides for monitoring
- Allows feedback into policy
- Facilitates proper decision-making.

Assignment

1. Fill out this timeline (Table 4.3) for preparation of an EIA report for a project assigned to you by the instructor. Be able to support your choice of dates and personnel.
2. Make a flow chart that addresses the items in your timeline.

TABLE 4.3 EIA report timeline.

Due Date	Component	Description	Lead Personnel
	Screening	Laws and regulations	
	Site visit		
	Write job plan	Who does what when	
	Scoping	Environmental history	
	Public meetings	PR/data collection	
	Scoping/data	Aesthetics, traffic	
	Scoping	Biological, water	
	Scoping	Fiscal/infrastructure	
	Mapping	Lithosphere/landscape	
	Impact evaluation		
	Impact mitigation		
	Interagency review		
	Residual impacts		
	Write draft EIA report		
	Circulate draft		
	Final EIA		
	Debriefing		

4.6 CONCEPTS AND TERMS

Amelioration: Improvement of something. It can be considered a form of mitigation that, rather than compensating for an impact, actually provides a benefit.

Environmental Assessment: Either a *process* of assessing environmental impacts or a *document* resulting from such a process, depending on agency nomenclature.

Finding of No Significant Impact (FONSI): A specific legal conclusion in the NEPA process issued as an Environmental Notice.

Index: A type of environmental indicator, often a ratio, which provides a numerical or descriptive categorization of a large quantity of environmental data or information. It summarizes and simplifies data and information for use by various stakeholders and decision makers (Canter, 1996). Is the plural form "indexes" or "indices"? Both are correct, but *indices* tends to be used more outside of North America and to refer more to mathematical or statistical use. *Indexes* tends to be used more in the United States and Canada than in other countries. Regardless of what they are called, they are growing in popularity because they are becoming more robust and they facilitate comparisons.

Indicator: A variable or aggregation of variables that symbolizes or captures a measurement of environmental quality.

Mitigation: reduces the effects of an environmental impact by offsetting or compensating for the impact.

Strategic Environmental Assessment (SEA): An assessment of programs and policies. Important because programs and policies lead to projects and an EA is a planning tool. Essentially, an

SEA is done in advance of any environmental action or undertaking, so a pathway for proper resource commitment is created.

4.7 SELECTED RESOURCES

Canter, Larry. (2015). *Cumulative Effects Assessment and Management*. Horseshoe Bay, TX: EIA Press.
Council on Environmental Quality (CEQ). (2007). NEPA & Environmental management Systems (EMS). NEPA.Gov. www.energy.gov/nepa/ceq-guidance-documents.
Council on Environmental Quality (CEQ). (2010). Establishing, Applying, and Revising Categorical Exclusions under the National Environmental Policy Act. Executive Office of the President, CEQ. www.energy.gov/nepa/ceq-guidance-documents.
Council on Environmental Quality (CEQ). (2011). Appropriate use of Mitigation and Monitoring and Clarifying the Appropriate Use of Mitigated Findings of No Significance. www.energy.gov/nepa/ceq-guidance-documents.
Environmental Protection Agency. (2014). *Environmental Quality Index Overview Report*. Office of Research and Development.
Environmental Protection Agency. (2021). Environmental Quality Index (EQI). www.epa.gov/healthresearch/environmental-quality-index-eqi.
Environmental Protection Agency. Environmental Dataset Gateway. https://edg.epa.gov/metadata/catalog/main/home.page.
Wendling, Zachary A., John W. Emerson, Alex de Sherbinin and Daniel C. Esty. (2020). *2020 Environmental Performance Index*. New Haven, CT: Yale Center for Environmental Law & Policy. epi.yale.edu.

4.8 TOPIC REFERENCES

Canter, Larry W. (1996). *Environmental Impact Assessment*, 2nd ed. New York: McGraw-Hill.
The Economist. (2005). The Economist Intelligence Unit's quality-of-life Index. *The World in 2005*. The Economist.com.
Gan, Xiaoxu, Ignacio C. Fernandez, Jie Guoc, Maxwell Wilson, Yuanyuan Zhaoe, Bingbing Zhoub and Jianguo Wu. (2017). When to Use What: Methods for Weighting and Aggregating Sustainability Indicators. *Ecological Indicators*, 81, 491–501.
Kennedy, Victor S. (2018). *Shifting Baselines in the Chesapeake Bay: An Environmental History*. Baltimore: Johns Hopkins University Press.
Kerzner, H. (1997). *Project Management: A Systems Approach to Planning, Scheduling, and Controlling*, 6th ed. New York: John Wiley & Sons.
Kisak, Paul F. (Ed.). (2015). *Functional Decomposition: "Breaking Things Down into Their Fundamental Components*. CreateSpace Independent Publishing Platform.
Modak, Prasad and Asit K. Biswas. (1999). *Conducting Environmental Impact Assessment for Developing Countries*. Tokyo: The United Nations University.
Morse, Stephen and Ernest Tollner. (2007). Development Indicators and Indices. In Cutler J. Cleveland (Ed.), *Encyclopedia of Earth*. Washington, DC: Environmental Information Coalition, National Council for Science and the Environment. [Published in the *Encyclopedia of Earth* March 12, 2007; Retrieved January 4, 2022]. www.eoearth.org/article/Development_indicators_and_indices.
Mukherjee, A. (2014). Comparison of Soil Quality Index Using Three Methods. *PLoS One*, 9(8), e105981. doi: 10.1371/journal.pone.0105981. Retrieved January 4, 2022, from http://journals.plos.org/plosone/article?id=10.1371/journal.pone.0105981.
Program on Health, Equity and Sustainability, San Francisco Department of Public Health. (2014). *Bicycle Environmental Quality Index (BEQI): Draft Report*. San Francisco: San Francisco Department of Public Health.
Schmidt, Michael, John Glasson, Lars Emmelin and Hendrike Helbron (Eds.). (2008). *Standards and Thresholds for Impact Assessment*. Series 3 Environmental Protection in the European Union. Berlin: Springer.

Geology, topography, and earth resources

5.1	Environmental setting	81
5.2	Project description	84
5.3	Potential or predicted impacts	85
5.4	Mitigation and monitoring	88
5.5	Concepts and terms	89
5.6	Selected resources	90
5.7	Topic references	91

This earth-based topic addresses how various components of an environmental study are compared and evaluated within categories of impact. A first step is an assembly of background information about the project location and setting. Next is a discussion of the project plan as described by its proponents; this is associated with a "portable" set of impacts that will manifest in the setting. Then comes a detailed analysis and evaluation of the project discussing potential or predicted impacts. Following that is a general range of suggested mitigation measures common to geological, topographical, and earth resource aspects of a project. Finally, there are descriptions of how the residual mitigations can or will be monitored. This format is recommended for writing sections on individual categories of impact in an EIA report. The impacts and mitigation will be site specific and project specific in the EIA, flowing from the impact analysis.

Pioneering environmental impact assessment specialist Larry Canter (1996) advised a six-step approach to EIA (identify potential impacts, describe existing conditions, obtain relevant standards, predict impacts, determine impact significance, and identify and incorporate mitigation). This has become the standard practice and is particularly suitable to earth resources (geological, soils, topography, and land stability) because they are generally fixed in place on a site.

As with other impact topics, earth resource factors can comprise subchapters for most impact categories in an EA. The identification of potential ground-related impacts arises from a project description intersecting with the general project area. For example, extracting oil from below the ocean floor in the Gulf of Mexico may have wide-reaching impacts. The second topic, description of existing conditions, gets much more specific. The use of relevant standards

DOI: 10.4324/9781003030713-5

refers to soil quality standards, reclamation procedures, and specific environmental resource laws that govern the nature of the project, its location, and its likely potential impacts. This sets the stage for the next step, the prediction of impacts, through qualitative and quantitative approaches. The relevant standards provide the lens through which impact significance is determined, although the scope of the assessment need not confine itself just to the legal significance, since some impacts may be predicted that are not yet fully regulated. The EIA component will conclude with recommended mitigation and monitoring for the impacts. The residual impacts are those that are not fully mitigated, and against which the decision to go forward or not on the project is made.

5.1 ENVIRONMENTAL SETTING

The environmental setting of the proposed project includes the topography, the geology, environmental and cultural features, and the earth resources value of a site. Earth resources are important as a structural support for a physical undertaking, such as construction of a building, or an environmental policy, such as timber harvesting or other resource management that could affect soils and land surfaces. Another aspect of earth resources is as a product of excavation or removal. Earth extraction can occur in all settings, even including the ocean bottom.

The properties of the soil to support a project relate to its ability to support suitable on-site septic systems, withstand erosion, regulate and support biological activity, and support buildings and other structures. How a soil measures up is described as either soil *suitability*, soil *quality*, or soil *health*. The US Department of Agriculture (2001) defines soil quality as the fitness of a specific kind of soil to maintain air and water quality, support plant and animal productivity, support human habitation, and preserve human health.

The important basic properties of soil are texture structure, color, redoximorphic features, horizons (its profile), and effective depth (Figure 5.1). From these directly measured properties, information can be learned about other properties, such as permeability, leaching potential, heavy metal immobilization, infiltration, and expansion. The assessment of soil utility for forestry, agriculture, buildings, on-site septic systems, and other purposes depends on soil properties. These properties should be clearly described in the EIA and identified on site plans at an appropriate scale.

Soil-related information is usually presented as a general description using previously developed regional soil maps. If the proposed development contemplates specific structures or buildings, more detailed soil information, including with erosion control information, will be provided as part of the assessment package. This may also be in the form of local applications for subsurface wastewater disposal, septic design, or connection approvals. The qualifications for soils expertise include training as an environmental or civil engineer, geologist, licensed site technician, and soil scientist. Site evaluators are qualified to make judgments on where to place a septic system, and engineers are qualified to do on-site septic systems and erosion and sediment control plans. Licensed or certified soil scientists do all other soil evaluations. Some states license locally, but there is also a national certification available through the Soil Science Society of America. The soil information should be communicated in clear, understandable terms so that non-experts are able to understand the environmental assessment report.

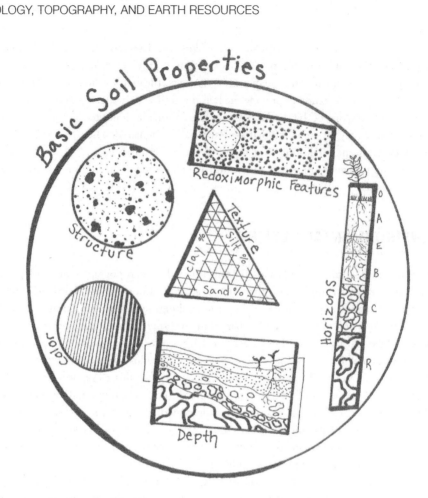

FIGURE 5.1 Basic soil properties.

Characteristics of the environmental setting that involve topography, geology, or that may otherwise affect earth resources extraction are shown in Table 5.1.

Each of the features in Table 5.1, if existing in the project area, should be described in the EIA in terms of what could happen under worst case conditions in the project area. This is an analysis of current conditions in the context of what could happen in the near future.

Assignment

You are advising a developer on the feasibility of developing a subdivision in the foothills to the east of Springfield, Oregon, in the southern Willamette Valley (other sites may be selected). Research the topology, geology, and environmental characteristics online.
1. What specific considerations should be given to geology, topography, soils, groundwater, and natural hazards?
2. Are there any apparent restriction on the size and location of the developable site?

TABLE 5.1 Earth resource components of an environmental setting.

Feature	Issue
Landforms and topography	The most important factor about slope is the steepness of slopes expressed in degrees or percent. (A 90-degree angle is a 100% slope.) The implications of slope may include potential runoff, potential for land sliding, and difficulty of construction.
Faults and seismic history	Knowing the location of possible active earthquake faults is obviously important.
Fault movement	Knowing the history of fault movement (earthquakes) is equally important. This kind of research is almost always done by geologists or geological engineers.
Creep	Tectonic creep refers to the slow movement of the earth along a fault. Field observations are likely to include ground rupture.
Ground rupture	Cracking or breaking open as a result of faults, landslides, or other movement.
Ground shaking	Shaking from earth movements largely depends on the nature of the earth material and how saturated it is.
Flooding due to structural failure	Examples include the Johnstown flood ("The Great Flood of 1889") that killed over 2,200 people in Pennsylvania when the South Fork Dam broke. A more recent example is the Oroville, California, dam failure, where no lives were lost.
Volcanic or geothermal characteristics	Consult records of volcanic events. A particularly important factor in the Pacific Northwest and Hawaii in the United States. Remember the Big Island in 2018.
Slope stability	Risk of slides, slumps, and debris flows. As we continue through an era of climate change, this possibility becomes an increasingly significant issue for EIA.
Foundation support	Soils engineering is required in identified zones.
Soil classification	For agriculture, silviculture, and landscaping.
Engineering properties	Studied with recommendations from a geological engineer.
Permeability	How fast does the soil drain? This information is also needed in the hydrology section.
Differential compaction	This is bad for foundations. Engineering services are required.
Erodibility	Measurement of erosion and downstream deposition. This information is also used in a hydrology assessment.
Corrosivity	Effects of soils and earth materials and groundwater on structures and structural features such as metal or concrete pipes and buried fixtures.
Subsidence	Land sinking due to groundwater extraction, such as wells or shale fracking.
Earth resources	Value of and feasibility of extraction such as gravel, minerals, or topsoil.
Unique features	A catch-all for anything notable and not otherwise addressed: natural disasters, landfills, mines, quarries, and other earth extraction sites. Wetlands, agricultural uses, coastlines, cultural sacred sites, historic sites, and educational features all fall into this category.

5.2 PROJECT DESCRIPTION

The project description should include and fully describe all four phases of development that could affect earth resources: planning and development, construction, operation, and project closure/reclamation. Earth resources are involved either in the project itself, as in the case of a mining operation, or in a phase of the project, as in building access roads for the development of a construction project, or clearing and excavating for land development. If clearing and excavating for construction, the usual approach is to save any available excavated topsoil for landscaping and to thereby practice a balanced cut-and-fill as much as possible. The latter is particularly true for road construction, where time and cost savings are maximized by using excavated material as fill to level out hills and valleys (the "cut-and-fill method"). Earth resources may also be stockpiled for eventual use in construction or in reclamation. Such piles may be seeded, mulched, and revegetated if the reclamation is not to occur within the season. In any case, the project description and site plans should address issues associated with earth resources.

Canter (1996: 250–252) illustrates the range of impacted projects associated with significant earth resources development processes and issues (Table 5.2). The EIA should anticipate impacts, evaluate impacts, and address mitigation.

The list of earth-specific projects is lengthy. Many other types of project may also generate impacts under this criterion. Each type of project will have a range of impacts associated with particular stages or phases. For mining projects, the exploration and development of the

TABLE 5.2 Projects associated with significant earth resource issues.

Land subsidence from over-pumping groundwater, oil, gas, or other materials (including the use of fracking)
Excavation of earth materials for construction
Erosion from construction grading or surface exposure
Landslides from slope instability or construction in steep or seismically unstable areas
Strip-mining of any form
Construction of jetties
Military training activities that lead to soil compaction and erosion
Projects that contribute to acid rain projects that involve earth excavation for waste materials disposal
Coastal developments
Most reservoir projects
Grazing leases and other wide-area agricultural pursuits that affect soils
Buried pipelines
Brownfield development of sites that may have hazardous waste materials
Land disturbances from off-road vehicles on delicate soils and topography
Mining operations

Source: Based on Canter (1996: 250–252).

site will contain potential impacts to be addressed, including constructing access roads and the actual preparation of the site itself. Active mining operations include open pit mining, placer mining, underground mining, and the reworking of inactive or abandoned mines. Reworking usually occurs due to the development of more efficient extracting and processing methods or increased value of the earth material itself.

5.3 POTENTIAL OR PREDICTED IMPACTS

Major potential environmental problems that can occur from earth resources extraction and mining include erosion, acid rock discharge (ARD), aesthetic losses, water pollution, increased traffic, noise, and vibration. Mining operations often raise issues pertaining to environmental justice, social impacts, economic impacts, water resources, air quality, wildlife, soil quality, energy, traffic, public health, cultural resources, and aesthetic resources. In summary, almost anything can happen from improper surface and sub-surface development.

In evaluating these impacts, the EIA should compare present and potential conditions of land uses as well as other site features and conditions that are considerations for construction impact considerations. The environmental hazards prediction and assessment should be very specific about how events are likely to occur, how they are measured, and what the effects on the project will be. The effects need to be evaluated in terms of risk assessment. Significance of impacts is usually identified with reference to set engineering standards. The range of hazardous effects is fairly large: earthquakes and ground movement, volcanism, land sliding, liquefaction,

TABLE 5.3 Sample checklist for a soil engineering report.

√	Factor
	Data regarding the nature, distribution, strength, and erodibility of existing soils
	If soil is to be brought to the site, data regarding the nature, distribution, strength, and erodibility of that soil
	Grading specifications and procedures
	Temporary soil stabilization/erosion control devices
	Designs and details for permanent soil stabilization
	Design criteria for corrective measures when necessary
	Opinions and recommendations covering the stability of the site
	Subsurface conditions, including soil profile
	Soil boring results
	Summary of surficial and bedrock geology of site
	Seismic stability
	Test units data from borings, trenches, and other excavations
	Source of soils report and qualifications of preparers

Source: Based on Sanford (2018: 99).

TABLE 5.4 Sample checklist for erosion control plans and details.

√	Factor
	Identify qualified individuals or firms
	Show site-specific (not generic) erosion control measures
	Show the limits of construction disturbance
	Minimize area of construction disturbance
	Stabilize swales and streams
	Refer to agency/state/municipal erosion control standards, details, and objectives
	Phase construction to reduce amount of exposed land
	Mark and describe buffers (e.g., "undisturbed, naturally vegetated")
	Show all staging areas for soil, equipment, temporary parking, and construction
	Designate topsoil stockpiled and surrounded by stone check erosion control dams
	Describe how buffers, staging areas, critical areas, utilities, and other features will be flagged in the field
	Accommodate and protect steep slopes
	Depict typical erosion control devices, including silt fence, stone check dams, inlet protection, stabilized construction entrance, berm, swale, and mulch
	Provide a timetable and schedule for the application of erosion control devices
	Include winter erosion control plans if site work continues after October 1 (or other date)
	Describe monitoring, especially during and after storm events
	Provide routine maintenance schedule for erosion control devices
	Provide system for reporting problems, inspections, and compliance with erosion control requirements
	Potential effects of seismic stability
	Include contingency plan in the event an erosion control device or treatment fails
	Propose performance bond, irrevocable letter of credit or escrow agreement for large projects, or for where erosion control is a crucial aspect of the project
	Address removal of obsolete erosion control devices
	Coincidence with local plans and ordinances
	Provide plan and details for final site treatment, including rapid seeding, mulching, permanent stormwater drainage, and long-term erosion control/site stability
	Language for covenants or other agreements to maintain erosion control devices, stormwater control systems, and other utilities

Source: Based on Sanford (2018: 101).

soil expansion and/or contraction, soil subsidence, unstable cut or fill slopes, cut-fill balance, and overall slope stability.

A project assessment should include impacts on public infrastructure and public works, and conformance to local comprehensive plans and zoning ordinance with regard to environmental hazards. The following tables and checklists reflect the categories of impact and mitigation that would be expected in an EIA report. The specific range and details will be governed by site and project conditions, agency guidelines, and professional standards.

TABLE 5.5 Sample California checklist for sites in unstable areas.

Is there exposure of people or structures to potential substantial adverse effects, including the risk of loss, injury, or death, involving the following?

    Rupture of a known earthquake fault, as delineated on the most recent Alquist-Priolo Earthquake Fault Zoning Map issued by the state geologist for the area or based on other substantial evidence of a known fault

    Strong seismic ground shaking

    Seismic-related ground failure, including liquefaction

    Landslides

Is there a potential result in substantial soil erosion or the loss of topsoil?

Is project located on a geologic unit or soil that is unstable, or that would become unstable as a result of the project, and potentially result in on- or off-site landslide, lateral spreading, subsidence, liquefaction, or collapse?

Is project located on expansive soil, as defined in Table 18-1-B of the California Uniform Building Code (1994), creating substantial risks to life or property?

    Are soils incapable of adequately supporting the use of septic tanks or alternative wastewater disposal systems where sewers are not available for the disposal of wastewater?

Assignment

1. Minnesota Department of Natural Resources and co-lead agencies US Army Corps of Engineers and US Forest Service issued an Environmental Impact Statement for the North New Mining Project and Land Exchange. These agencies issued a mining alternative assessment on September 27, 2013, and attached it as Appendix B in the FEIS (www.dnr.state.mn.us/input/environmentalreview/polymet/feis-toc.html). Read this alternative assessment and the project description in the FEIA, and then describe why you think the agencies did not accept the alternative.

2. The Callahan Mining Corp. is a Superfund site in coastal Maine. Zinc-copper sulfide ore deposits were extracted from the late 19th century until closure in 1972. It needs an EIA Report on the plan to clean up the site. You are to do a brief summary for use in the EIA report. Applicable laws for these actions. Include the Resource Conservation and Recovery Act, state and federal mining laws, the Environmental Conservation Act, and the Coastal Zone Management Act. The mitigation is to stabilize the site, lower the dam, remove 130,000 cubic yards of tailings, add a drainage swale, install horizontal and vertical drains, and add easements to facilitate installation of protective measures.
   a. What are the major contaminants?
   b. What are the conditions of the site that appear to have resulted from mine operations?
   c. Create an impact assessment matrix for an EA that addresses general categories for clean-up and reclamation of the Callahan Mine located on the coast of Hancock County, Maine (Figure 5.2). The matrix should address protection of the bay, protection of surface waters in the watershed, protection of groundwater, erosion, and site security/safety. The matrix should include monitoring of principal impacts, what is to be done about exceedances (corrective measures and reporting), and timing.

88 GEOLOGY, TOPOGRAPHY, AND EARTH RESOURCES

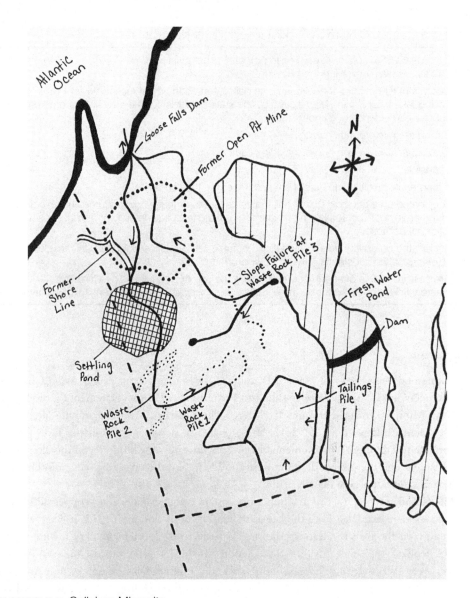

FIGURE 5.2 Callahan Mine site.

5.4 MITIGATION AND MONITORING

Some mitigation measures for an earth modification project might include required disclosure statements, use of preventative engineering and design, impact/site avoidance, off-site compensation, regulatory changes, monitoring, warnings, education, and public information. The residual impacts expected after mitigation should be provided in the assessment. The monitoring should be appropriate to the issues of site stability, public concern, and agency mission.

TABLE 5.6 Factors to consider in control of erosion through vegetation.

√	Factor
	Design minimizes stripping or clearing for construction, shows stockpiles of resources for reclamation.
	Designs and plans show routes, materials, and procedures for temporary diversion of stormwater runoff.
	Plans address ground preparation, such as tilling, application of topsoil, and fertilizer.
	Plans address planting scheme with placement, species, and maintenance documented. Usually, fast-growing indigenous species are preferred.
	Plans address use of mulch or other stabilizing measure to protect and nurture.
	Plans address timing and seasonality.
	Plans address mitigation compliance, monitoring, and repair scheme.
	Plan show final appearance after operations cease and/or reclamation occurs.

Source: Based on Sanford (2018: 103).

Assignment

Look up the Oso Landslide (USGS, 2017; www.usgs.gov/news/revisiting-oso-landslide).
1. Could the damage and loss of life from the 2014 Oso Landslide in Washington State have been prevented?
2. How could have environmental planning and environmental assessment lessened the impacts of the event?

Stability is a function of earth material and slope. However, it is also greatly affected by vegetative cover, impaction, climate, human and animal disturbance, and other environmental factors. The EIA report should address erosion impacts and mitigation. Table 5.6 represents some vegetation factors that might be considered in a report. The EIA should discuss vegetative conditions and other environmental factors and the role they play in potential environmental impacts from the project.

5.5 CONCEPTS AND TERMS

Acid rock drainage (ARD): Acidic water discharge from mines. Oxidation of metal sulfides (commonly iron sulfide or pyrite) found at the surface from existing material or recently deposited overburden leads to acidity in the runoff.
Borehole: Vertical shaft to collect sample cores or groundwater samples.
Cyanidation: Use of a weak cyanide solution to dissolve ore and reveal gold or silver grains.
Earth resources: Physical materials residing in and extractable from the earth. Includes sand, gravel, dimensional stone, minerals, topsoil, fill material, ore, oil, gas, and coal.
Fracking: Hydraulic fracturing to free oil and gas from rock through application of a high-pressure water and chemical mixture.

Leaching: Use of cyanide solutions on crushed ore to dissolve and extract gold, copper, or other precious metals.

Overburden: Earth materials to be removed to gain access to an ore deposit.

Redoximorphic features (mottles): Spots of one color within a soil layer of a different color, resulting from fluctuations of the water table.

Soil horizon: Discernable stratigraphic layer in the soil profile. The top layer might be the O layer (for organic). Below that is the A horizon (subsoil), then the B horizon (substratum). An E horizon, light colored "master horizon" of clay, some organics, and some depleted minerals might exist. Parent rock material (source of rock for some of the above layers) is the C horizon. The bedrock layer designation is R.

Structure: Groupings ("peds" when referring to soil) of earth particles by size and type, which affects the movement of water.

Tailings: Residual material from the extraction of recoverable minerals or other earth resources.

Texture: Refers to relative proportions of particle size in soils or other earth materials. Soils have particle size <2 mm.

5.6 SELECTED RESOURCES

Association of State Dam Safety Officials. Lessons Learned from Dam Incidents and Failures. https://damfailures.org/.

Brady, Nyle C. and Ray R. Weil. (2015). *The Nature and Properties of Soils*, 15th ed. Boston, MA: Pearson.

Brown, Whitney E. and Deborah S. Caraco. (1995). Muddy Water in—Muddy Water Out? A Critique of Erosion and Sediment Control Plans. *Watershed Protection Techniques*, 2(3), 55: 68. February 1995.

Canter, Larry. (2015). *Cumulative Effects Assessment and Management*. Horseshoe Bay, TX: EIA Press.

Dewberry, Sidney O. (2008). *Land Development Handbook: Planning, Engineering and Surveying*, 3rd ed. New York: McGraw-Hill.

Environmental Law Alliance Worldwide. (2010). *Guidebook for Evaluating Mining Project EIAs*. Eugene, OR: Environmental Law Alliance Worldwide.

Hodson, Martin J., C. Stapleton and R. Emberton. (2001). Soils, Geology and Geomorphology. In P. Morris and R. Therivel (Eds.), *Methods of Environmental Impact Analysis*, 2nd ed. (pp. 170–196). London: Spon Press.

International Atomic Energy Commission (IAEA). (2005). *Guidebook on Environmental Impact Assessment for in Situ Leach Mining Projects*. Vienna: IAEA.

Marsh, William M. (2010). *Landscape Planning: Environmental Applications*. Hoboken: Wiley.

Soil Science Society of America. http://soilslab.cfr.washington.edu/S-5/.

Spitz, Karlheinz and John Trudinger. (2019). *Mining and the Environment: From Ore to Metal*, 2nd ed. London: CRC Press.

United States Consortium of Soil Science Associations. http://soilsassociation.org/.

USDA Natural Resource Conservation Service. www.nrcs.usda.gov/wps/portal/nrcs/soilsurvey/soils/survey/state/.

5.7 TOPIC REFERENCES

Canter, Larry W. (1996). *Environmental Impact Assessment*, 2nd ed. New York: McGraw-Hill.

Minnesota Department of Natural Resources, U.S. Army Corps of Engineers, and US Forest Service. (2015). NorthMet Mining Project and Land Exchange Final Environmental Impact Statement. November 2015. www.dnr.state.mn.us/input/environmentalreview/polymet/feis-toc.html.

Sanford, Robert M. (2018). *Environmental Site Plans and Development Review*. New York: Routledge.

United States Geological Survey ((USGS). (2017). Revisiting the Oso Landslide, November 30, 2017. www.usgs.gov/news/revisiting-oso-landslide).

U.S. Department of Agriculture (USDA). (2001). *Soil Quality – Introduction*. Washington DC: Natural Resource Conservation Service.

Hydrology, water quality, and water supply

6.1	Environmental setting	93
6.2	Project description	95
6.3	Potential or predicted impacts	96
6.4	Mitigations and monitoring	100
6.5	Concepts and terms	100
6.6	Selected resources	101
6.7	Topic references	102

It can be argued that fresh water is the most important resource on earth. Here we are restricting our views to ways that fresh water is part of a development project under environmental review. We suggest the following approach, based on Canter (1996):

1. Impacts of project ("portable impacts" that come from the particular type of project being assessed);
2. Existing conditions of the site;
3. Regulations/standards (include NEPA, Clean Water Act, and other federal, state, and local laws);
4. Impact prediction (a function of the project plus existing conditions);
5. Impact significance (a function of impact prediction and regulations/standards);
6. Mitigation of all identified impacts;
7. Residual impact calculation (impact minus mitigation = residual impact);
8. Monitoring of residuals;
9. Evaluation and feedback into the assessment process (this might not necessarily be in the EIA itself, but the kernels of it might be in the final recommendations/conclusions section).

The successful review of impacts will have addressed all four phases of a project: planning and exploration/testing, construction, operation, and closure/post-closure.

DOI: 10.4324/9781003030713-6

6.1 ENVIRONMENTAL SETTING

An environmental setting for the EIA project may contain surface waters and subsurface waters. These waters can be flowing or still. There is, of course, water in the atmosphere, and projects have the potential, to varying degrees, of affecting the hydrological (water) cycle. Water itself may also be the main subject or aspect of a project, as in the case of relicensing a dam. Water may have interior uses in a project under one or more of four categories: cooling, sanitation, cleaning, and processing. Water topics may be addressed in many sections of an environmental assessment report, including environmental hazards, soils, erosion, air quality, habitat, biology, utilities, public services, recreation, boat traffic, and aesthetics.

Water in the atmosphere is "in residence" there for a brief time as part of the hydrological cycle of precipitation, evaporation, transpiration, runoff, and infiltration; all aspects can be associated with project-related imposed impacts. Residence time in various aspects of the cycle is an important factor in the assessment of pollution or treatment of water.

The product setting can contain point source and non-point source water discharges from the project or from its environs. Non-point sources will likely be diffuse, intermittent, and tied to meteorological events. Non-point sources will have an area of land surface rather than a specific origin such as a spring or lake. It will require site-specific treatment and land/water management techniques. Water conservation practices prior to discharging off-site will also be important.

Hydrologic simulation models use meteorological information (particularly precipitation), ground cover, surface features, landforms, soils, historical records, and subsurface attributes to describe flow rate, volume, and behavior in a watershed. The environmental scale can be large; working at a watershed level is a particularly important factor in all development projects. The Environmental Protection Agency (2020a) BASINS program provides GIS modeling for EIA that can range from small watersheds to entire states. The EPA site also has links for groundwater models, surface water models, other models, and modeling products.

Settings that contain surface waters, flowing or standing, will require some special attention such as the use of buffering strips and other forms of set-back protective areas, management practices, specific types of reviews, and other impact assessment and mitigation schema as promulgated by federal and state agencies. The impact assessment and mitigation analysis will mesh with how the surface water is classified under the Clean Water Act:

Class AA: scenic or otherwise significant natural resources and Class A quality water for drinking.
Class A: suitable for public water supply but may need some filtering.
Class B: Suitable for swimming, recreation, irrigation habitat, aesthetics, and water supply if filtered and disinfected.
Class C: Suitable for recreational boating, irrigation of crops that will not be eaten raw, wildlife and fish habitat, and industrial uses.

Wetlands are one form of surface water in the project setting that regularly requires attention to detail in the assessment. In non-desert environments, wetland analysis is frequently addressed in the habitat and biological sections of an EA or EIS. Identifying and delineating wetlands is a major component of describing the setting in non-arid environments. Soil, water, and vegetation are the three main factors considered by wetlands evaluators in determining the presence of a wetland.

If a wetland is present, the next step is to delineate its boundaries. This requires coordination with the US Army Corps of Engineers. After that task is complete, the wetland values and functions are assessed (Figure 6.1). Most states will have a wetlands checklist for this assessment, as in the example below, based on that of Maine's Department of Environmental Protection. *Function* and *value* can sometimes be used interchangeably. The value is the weight or significance attached to a function. The overall wetlands quality will be reported and will be useful in determining impacts and in evaluating mitigation.

The environmental setting may also include historical flooding information, maps of existing wetlands, water rights and ownership of harvested water, and of course, the role of water in creating and maintaining biological habitats. In addition, if not addressed in another section of the EA or EIS, the matter of flooding will need to be addressed in a section on water or water resources. Increasing runoff or flooding on an adjacent property without a permit can also be described as a criminal event. Preventative design is the best solution.

TABLE 6.1 Sample wetland functions and values checklist for an EIA on a site that contains wetlands.

Function/Value	Suitable	Not Suitable	√ if a Primary Function	Comments, Rationale, Degree of Suitability
Groundwater recharge/discharge				
Surface water quality/interaction				
Flood flow alteration				
Fish and shellfish habitat				
Sediment/toxicant retention				
Nutrient removal				
Production export				
Sediment/shoreland stabilization				
Wildlife habitat				
Recreation				
Educational/scientific				
Uniqueness/heritage				
Visual quality/aesthetics				
Endangered species habitat (include vernal pools)				
Other:				

Assignment

1. A tiny droplet can contain more than 300 trillion water molecules. How many zeros are to the right of the three in 300 trillion? Now express this number using scientific notation.
2. The average discharge of the Mississippi River is about 450,000 cubic feet per second. What is this amount in gallons? (Be sure to use scientific notation.)
3. The average fishing pond is 7.1 hectares (17.5 acres). If water weighs 1 tonne (1 tonne is 1 metric ton, equivalent to 1,000 kilograms) per cubic meter and the average depth of a catfish pond is 3 m, how many tonnes of water are in the average fishing pond? (Hint: assume 1 acre = 4047 m^2.)

6.2 PROJECT DESCRIPTION

A project is a "federal action" or a "state action": something at the federal or state level that requires review for an environmental impact assessment. There are many ways water can be involved. Questions need to be asked and answered, such as "What is the water use for the project? Where does it come from? Where does it go?" If the project involves any significant amount of land, as in a timber harvest plan, or range management, it will involve water resources. If the project involves coastal stabilization or wetlands, as in many Army Corps projects, or dam relicensing by FERC, the focus will be water related. Or the project may be for a building or other structure that has water use and development of a specific site. Water is used in material processes and as such, it may be discharged into the environment. We look at what is done to it prior to discharge, and then at its behavior on the ground or into receiving water, and within or below ground. From these comes an assessment of potential impact and liability, and from that come management practices in response. We can take on measures that include cleanup and emergency responses, stormwater management, supply development, channeling, source reduction, treatment, filtering, pumping, and other mitigation.

We use water in many ways that are not readily seen. The largest amount of water withdrawal in the United States is for thermoelectric power (45%), with irrigation coming in second (32%), and public water supply a distant third (12%) in 2010 (Maupin et al., 2014). The EPA is among the major sources for estimating water use. For example, it has tools that can help commercial and institutional facilities estimate water use and identify best management practices (BMPs; Environmental Protection Agency, 2021). Remember that flowing water can be used for several purposes at once or in succession, which can be an issue, particularly in arid areas and in these times of rapid environmental change. Consider the volumes and services of the Colorado River as an example of such variation, with myriad impact issues.

States and municipalities generally assume a rate of water use for generating permits, typically between 50 and 150 gallons per day (GPD) per bedroom, for residential units, with a certain minimum flow. Maine requires a minimum design flow of 240 GPD. There may be other alternatives to the GPD per bedroom system. For non-residential construction, there are estimates based on square footage, or visitor accommodation capacity.

What is the water supply for the project? It can be a private water supply or a public water supply, the determination of which will be by state or federal regulations. The Environmental Protection Agency defines a public water system as having at least 15 service connections or serving at least 25 people or 60 or more days per year (Environmental Protection Agency, 2020b). A public water system may be privately owned, which is worthy of note in evaluating risk, BMPs, and mitigation.

The ability of a project to affect flooding is addressed in an environmental impact assessment or report and can be anticipated to a certain degree by the type of project considered.

Some projects commonly associated with increased flooding are:

- Removal of stabilizing vegetation adjacent to stream and riverbanks;
- Construction of structures that deflect or inhibit the flow of floodwaters, such as levees, which modify flow paths and can spread flooding problems and increase erosion;
- Construction of buildings, bridges, and culverts on floodplains, which reduce the storage area available for floodwaters;
- Construction of impervious surfaces or drainage systems that generate more stormwater runoff and channel it into a receiving water body;
- Straightening of meandering watercourses to increase drainage or improve shipping, which transfers flooding problems downstream.

6.3 POTENTIAL OR PREDICTED IMPACTS

Surface water is the most visible arena for perceiving an environmental impact related to water. Four main areas exist in which environmental impacts are assessed and reviewed in the context of surface waters: aesthetics, land use, water supply, and water pollution.

If a project requires water from a well, then the potential impacts of that well on the surrounding environment need to be documented. The physical installation and maintenance of the well may generate impacts, as will the effects of the well on area groundwater. The well would need the capacity for the sustained yields required, or drawdown could cause pressure that could deplete local aquifers or cause the movement of contaminants into vulnerable areas. Similarly, the well itself could introduce contaminants if it is not properly secured and maintained (Figure 6.1).

TABLE 6.2 Major pollutants affecting water management.

Pollutant Category	Effects	Common Project Sources
Biological oxygen demand (BOD)	Organic pollutants that consume oxygen during degradation	Food processing, paper manufacturing, agricultural runoff, aquaculture
Inorganic chemicals	Acids, salts, mercury, lead, and other heavy metals	Manufacturing, stormwater runoff
Nutrients	Nutrients, especially nitrates and phosphates, fertilize the growth of algae and aquatic plants, which lead to oxygen depletion	Agriculture, concentrated feedlot operations, lawn fertilizer runoff

HYDROLOGY, WATER QUALITY, AND WATER SUPPLY

Pollutant Category	Effects	Common Project Sources
Sediments	Sediments from erosion form the bulk of water pollutants	Agriculture, feedlot operations, construction sites, disturbed areas, meteorological events
Radioactive material	Water-soluble radioactive isotopes	Storage and disposal of radioactive materials
Thermal pollution (heat)	Thermal pollution lowers dissolved oxygen (DO) and can increase temperature beyond the range of tolerance of some species	Heat discharge from manufacturing and power generation
Organic chemicals	Oil, gasoline, plastics, pesticides, cleaning solvents, detergents, medications; some of these chemicals are endocrine disruptors, causing organisms to mutate and change gender	Manufacturing, pharmaceuticals, feedlot operations, household chemicals
Pathogenic (disease-causing agents)	Viruses, bacteria, protozoa, parasitic worms	Sewage, animal waste (agriculture and feedlot operations)
Biological (invasives)	New species that invade an ecosystem such as Asian carp, northern snakehead, sea lampreys, zebra mussels, green crab; also terrestrial species, and indirect effects like cheatgrass, pepperweed, gypsy moth, hemlock, woolly adelgid	Accidental or intentional introduction by humans

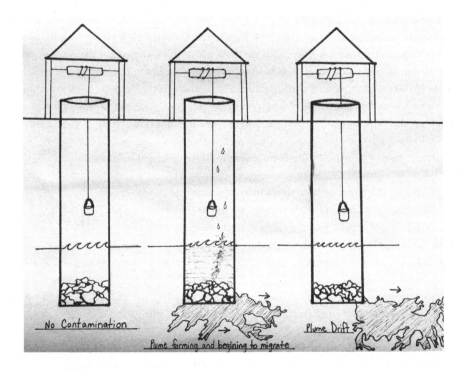

FIGURE 6.1 Well plume.

TABLE 6.3 Water impact categories.

Sort these terms into the appropriate column below: Color, BOD (biological oxygen demand), coliform bacteria, solids, thermal discharge (temperature), TOC (total organic carbon), COD (chemical oxygen demand), pathogens, TOD (total oxygen demand), oil, grease, taste, viruses, plastic litter, salinity, *Cryptosporidium*, Eurasian milfoil, hardness, pH, acidity, heavy metals, odor, turbidity, alkalinity, dissolved nitrogen, H_2SO_4 (dissolved sulfuric acid from oxidized sulfides), dissolved organic material, dissolved calcium, endocrine disruptors, phosphates, sediment, sulfuric acid.

Physical	Chemical	Biological

Assignment

1. Environmental scientists deal with water quality based on its physical, chemical, and biological characteristics. Examine the following terms associated with or describing water quality and place them in the most appropriate category of characteristics (Table 6.3). Note that not everything may fit nicely into a specific category, and you may need to make a judgment call.
2. To prepare for flooding, we calculate the peak flow expected during a particular flood event. Peak formulas derive from regression equations that vary based on the *n* year event, the total drainage area, and the total amount of impervious surface in the drainage area. Although computer programs are used by the US Geological Service and others to calculate more accurate peak flows, it is good to get a quick sense of what might occur for a 100-year event, a common threshold in environmental planning, using the simple formula $Q100 = 364 \times A^{.772}$. In this formula, Q_{100} is the 100-year peak flow (in cubic feet per second), and A is the contributing drainage area (in square miles). Use this formula to calculate the peak flow of a 100-year flood where the drainage area is 24 square miles.

The characteristics of non-point water pollution indicate potential impacts:

- Are diffused, intermittent, tied to meteorological events
- Involve an area of land rather than a single pipe or "point"
- No one point of origin, making them hard to monitor
- Require site-specific treatment
- Are linked to land management techniques and conservation practices.

These characteristics suggest variation in impacts depending on the phase of the project (Table 6.4). They also become categories for development of mitigation.

Category I: Discharge from septic systems, injection wells, and land applications.
Category II: Unplanned releases from storage, treatment, and disposal sites.

TABLE 6.4 Checklist of water resource categories of impact for project phases in an EIA.
Groundwater Contamination

√	Category	Planning	Construction	Operation	Closure
	Groundwater				
	Aquifers				
	Surface water, flowing (rivers, streams)				
	Surface water, standing (lakes, reservoirs, ponds, wetlands)				
	Ocean/coastal				
	Water supplies				
	Erosion				
	Stormwater				
	Water as habitat (breeding, migratory stop, threatened and endangered species)				
	Scenic/recreational				
	Hydrologic cycle (how does the project intrude)				

Category III: Contamination from transport or transmission—shipping, pipelines, vehicles.
Category IV: Planned discharges such as salt on road, fertilizer, urban runoff, irrigation, mine drainage.
Category V: From construction and other alterations of flow patterns.
Category VI: Natural occurrences (leaching, exchanges between groundwater and surface water).

Aquifer classes

Class I: Sole source water supply.
Class II: The majority of aquifers, with moderate protection.
Class III: Limited use potential.

Surface water classes are AA, A, B, and C. These are aspirational for management in addition to actual reflections of water quality. Mitigation of water use can be reduced by using less water, changing design, changing use process, changing treatment after use, changing management, and utilizing off-site actions.

Assignment

Briefly explain how the following terms pertain to EIA of water resources.
1. Recharge area
2. Wild and Scenic Rivers Act
3. Zone of saturation
4. Section 404 General Permit

5. Dry well
6. Fish and Wildlife Coordination Act
7. National Water Use Information Program
8. Drawdown
9. Impervious surface
10. Wellhead Protection Area
11. Category II Discharge
12. Phase I requirement from the National Pollutant Discharge Eliminating System (NPDES).

6.4 MITIGATIONS AND MONITORING

Mitigation and monitoring of water resources should address all four phases of a project: planning, construction, operation, and closure/post-closure. Common mitigation measures include decreasing surface water usage, decreasing wastewater generation, reducing erosion during construction, using integrated pest management (IPM), and use best management practices (BMPs) to control non-point source pollution. Many projects today propose constructed or enhanced wetlands as a way to reduce off-site impacts and improve the aesthetic and wildlife characteristics of a site.

Mitigation of water from a site can be thought of in terms of improving quality and managing quantity. Mitigation can include reducing the amount of water, treatment to improve quality of the water, and controlling of the flow of water. Flow is controlled by volume and rate. Basins and other storage can reduce volume of discharges, as can transport via canals or pipes, which also can be used to control the rate of discharge. Water rates can be controlled by barriers that spread the amount over time and space. A good mitigation plan will address multiple aspects, including the feasibility of the most viable methods, and how they can be used in conjunction with an environmental management plan of some sort. Most federal agencies and states will have guidance for best management practices and other standardized procedures.

Assignment

Watch *Last Call at the Oasis* (2011), directed by Jessica Yu, about the world's water crisis. This production documents the role of water in our lives and includes solutions for effective management of this resource. Make a suggestion for the role of EIA to address a major water problem illustrated in the video.

6.5 CONCEPTS AND TERMS

Accretion: A gradual increase in land area adjacent to a river.
Acre-foot: The amount of water required to cover 1 acre to a depth of 1 foot. An acre-foot equals 325,851 gallons, or 43,560 cubic feet. A flow of 1 cubic feet per second produces 1.98 acre-feet per day.

Alluvium: Sediments deposited by erosional processes, usually by streams.
Aquifer: A geologic formation that will yield water to a well in sufficient quantities to make the production of water from this formation feasible for beneficial use.
Best Management Practice (BMP): Methods or measures designed and selected to reduce or eliminate the discharge of pollutants from point and nonpoint source discharges.
Biological impacts: Pathogens, coliforms, viruses, parasites, and invasives.
Chemical impacts: Water quality monitoring and assessment uses a variety of common measurements, notably, BOD (biochemical or biological oxygen demand, an indicator of the organics present in the water), COD (chemical oxygen demand), TOC (total organic carbon), and TOD (total organic demand).
Clean Water Act (CWA): Federal legislation enacted in 1972 to restore and maintain the chemical, physical, and biological integrity of the surface waters of the United States. The stated goals of the Act are that all waters be fishable and swimmable.
Contaminant plume migration and drift: Movement of contaminated water or other contaminant material affected by hydraulic pressure, soil matrix, volume, and flow patterns.
Deposition: The laying down of material by erosion or transport by water or air.
DO: Dissolved oxygen, a common measurement in water quality assessment.
Floodplain: Land next to a river that inundated by water when the river overflows its banks.
Impact prediction: Hydrological simulation models. Flow: volume (stored/transported) and rate moved, spread over time and space.
Inorganic chemical parameters: Salinity, hardness, pH (acidity, alkaline), iron, magnesium, other heavy metals, and dissolved inorganics of sodium, calcium, sulfate.
Irrigation water: Refers to water applied to assist crops in dry areas or during times where rainfall is inadequate.
Marsh: An area periodically inundated and treeless and often characterized by grasses and cattails.
Physical impact variables: Color, odor, temperature, solids, turbidity, oil and grease, taste, dissolved, organics, inorganics, and suspended.
Reclaimed water: Domestic wastewater that is under the direct control of a treatment plant owner/operator and has been treated to a quality suitable for a beneficial use.
Stormwater discharge: Precipitation that does not infiltrate into the ground or evaporate due to impervious land surfaces but instead flows onto adjacent land or water areas and is routed into drain/sewer systems.
Water quality: Chemical, physical, biological, radiological, and thermal condition of water.

6.6 SELECTED RESOURCES

Bedient, Philip B., W. Huber and B. Vieux. (2019). *Hydrology and Floodplain Analysis*, 6th ed. New York: Pearson.
Canter, Larry. (2015). *Cumulative Effects Assessment and Management*. Horseshoe Bay, TX: EIA Press.
Fennessy, M. Siobham, Amy D. Jacobs and Mary E. Kentula. (2004). *Review of Rapid Assessment Methods for Assessing Wetland Condition*. EPA/620/R-04/009. Washington, DC: Environmental Protection Agency. https://training.fws.gov/courses/csp/csp3112/resources/Wetland_Assessment_Methodologies/RapidAssessmentMethods.pdf.

Gordon, Nancy D., T. A. McMahon, B. L. Finlayson, C. J. Gippel and R. J. Nathan. (2004). *Stream Hydrology: An Introduction for Ecologists*, 2nd ed. New York: John Wiley & Sons.
Lyon, John Grimson and L. K. Lyon. (2011). *Practical Handbook for Wetland Identification and Delineation*, 2nd ed. Boca Raton: CRC Press.
Mitsch, William J. and James G. Gosselink. (2015). *Wetlands*, 5th ed. New York: John Wiley & Sons.
Morris, Peter, J. Biggs and A. Brookes. (2001). Water. In P. Morris and R. Therivel (Eds.), *Methods of Environmental Impact Analysis*, 2nd ed. (pp. 197–242). London: Spon Press.
Newson, Malcolm. (1994). *Hydrology and the River Environment*. London: Clarendon Press.
U.S. Environmental Protection Agency (EPA). (2017). *Water Quality Standards Handbook: Chapter 3: Water Quality Criteria*. EPA-823-B-17–001. EPA Office of Water, Office of Science and Technology, Washington, DC: US Government Printing Office.
U.S. Environmental Protection Agency. (2021). Water Sense. www.epa.gov/watersense.

6.7 TOPIC REFERENCES

Canter, Larry. (1996). *Environmental Impact Assessment*, 2nd ed. New York: McGraw-Hill.
Environmental Protection Agency. (2020a). BASINS Framework and Features. www.epa.gov/ceam/basins-framework-and-features.
Environmental Protection Agency. (2020b). Information about Public Water Systems. www.epa.gov/dwreginfo/information-about-public-water-systems#:~:text=A%20public%20water%20system%20provides,be%20publicly%20or%20privately%20owned.
Environmental Protection Agency. (2021). Tools for CI Facilities. www.epa.gov/watersense/tools-ci-facilities.
Maupin, M. A., J. F. Kenny, S. S. Hutson, J. K. Lovelace, N. L. Barber and K. S. Linsey. (2014). Estimated Use of Water in the United States in 2010: U.S. *Geological Survey Circular*, 1405, 56 p.
Yu, Jessica (Director). (2011). *Last Call at the Oasis* [Film]. Documentary. Participant Productions.

TOPIC 7
Biological Species and habitats

7.1	Environmental setting	104
7.2	Project description	106
7.3	Potential or predicted impacts	106
7.4	Mitigations and monitoring	109
7.5	Concepts and terms	111
7.6	Selected resources	113
7.7	Topic references	114

Biological resources are often thought of in terms of wildlife, particularly threatened and endangered species. Wildlife as a resource includes the monitory value of activities such as hunting, zoo visiting, camping, tourism, and such but also many non-monitory values such as education, ecosystem services, medical advances, and biological diversity. Wildlife environments are as small as one's backyard and as great as whole continents. The most serious threats to wildlife resources involve the potential for plant and animal decline and extinction. These threats include climate change, habitat loss, hunting/fishing/poaching, non-native invasions, and pollution. Thus, wildlife is understood and managed in the context of habitat and ecosystems. Biological resources as a concept include much more than just wildlife. It includes the aforementioned habitat and ecosystems. It includes biogeochemical cycling and ecological functions. It includes plants, interactions among different species and communities, interactions with humans, and so much more.

In preparing to do a biological assessment, one should begin with some questions such as: What is the jurisdiction of the decision makers? Which agencies issue permits or otherwise manage wildlife or fisheries resources? Are there current problems with wild plants or animals at or near the project site? Is there any controversy? What data sources are available, such as watershed inventories? Is there any GIS mapping? Are there any local organizations or agencies that have information about the area? These questions are all relevant to the reviewers of a biological assessment, as well as the preparers of them. It is useful to frame these questions in

DOI: 10.4324/9781003030713-7

the context of the typical steps used in the environmental assessment process. These steps are particularly appropriate for biological impact assessment (Canter, 1996):

1. *Biological impacts of proposal project*: These are the "portable" impacts that come with the project. The actual assessment of these impacts will be informed by the nature of the project, including the type of environment it occurs in, how much land alteration, and other characteristics from the project description.
2. *Existing biological conditions*: This is the site itself, including background setting in which the project occurs.
3. *Relevant laws and standards*: This forms the legal basis for the government determination of what impacts are meaningful. The screening process in EIA deals with what laws, regulations, and rules apply.
4. *Impact prediction*: This uses a variety of methods, including expert opinion, assessment "in the field," and the use of existing inventories, forecast models, and other tools.
5. *Impact significance*: This is the interpretation of item 4, impact prediction, through the lens of item 3, relevant laws and standards.
6. *Impact mitigation*: This includes everything that can reasonable be done on-site and off-site. Some mitigation may be in the form of alternatives to the project. Others might be modifications to the project, including construction, operation, and reclamation.
7. *Residual impact calculation*: What remains after you have mitigated what you can?
8. *Monitoring*: This may include build-out, operation of the project, and post-operation/closure.
9. *Evaluation*: This involves examining the overall issues and effectiveness of the assessment process. It may include feedback into improving the process for the future.

7.1 ENVIRONMENTAL SETTING

The EIA process may start with a description of a project as it relates to biological resources or a description of the project setting. This is in keeping with the two common approaches: that of a project searching for a place, or of a place searching for a project. Some approaches are a combination of these two, as in the case where an agency has a range of settings or a range of projects. In keeping with our topic format, we begin with the setting.

The environmental setting usually includes the tract of land on which a project will occur. But it may include surrounding areas or even an entire watershed if necessary. It would include the direct footprint of any construction and the area of potential effect (APE), which might involve off-site land or impacts. Any research into a setting should include the mapped biological resources known to exist in the geographic area. If the site is large enough or well known, it may contain documented resources such as habitat maps available at state or local levels, if not directly from the US Fish and Wildlife Service. These resources may include individual species or habitats.

The setting gets examined in the EIA from three perspectives. One is mapped resources of what is known at local, state, and federal levels. The second is any federal or state lists of threatened and endangered species of animals and plants. These lists might also include species that are neither threatened nor endangered but are of special concern. There may be local and citizen-made lists as well. The third is any relevant habitat descriptions that may be available. These would be evaluated from previous documentation and through on-site investigation. Consultation with local professional experts is usually a good idea. Find out how well the habitat has been documented? How does the habitat relate to overall biodiversity of the area? Is the habitat being actively managed?

The degree to which these three perspectives are addressed will provide insight into the eventual assessment of environmental impacts. It may be that the project's impacts can be compared easily due to existing documentation, or there may need to be an extensive amount of fieldwork for on-site evaluation.

This section of the EA is most often prepared by a biologist or ecologist. Below is a checklist of various things that could be included in the Environmental Setting or Site portion of the draft EA document depending upon the nature of the project and its location (Table 7.1).

The lead agency that will prepare the EA or EIS will probably have a list similar to the one above. Most of these factors will or should appear in several different sections of the assessment

TABLE 7.1 Checklist of items for assessing an environmental setting.

√	Item
	Definition and description of a regional setting
	Maps and documents that show relationships with other environmental resources
	Species lists with information about important species
	Local field surveys
	Land use inventory (usually a map and database)
	Ownership information on project related properties
	Disturbance level by humans and human-related sources (e.g., cattle)
	Habitat evaluation (can be a formal or informal document)
	Uniqueness as a measure of significance
	Endangered species listings (federal or state listing)
	Biomes or other ecologic communities
	Species evaluation (e.g., diversity, extent, density, productivity, stability)
	Indicator species, keystone species
	Carrying capacity of area for critical species
	On-site or nearby territorialism (usually birds or large mammals)
	Heritage trees or other kinds of plants or animals having cultural value
	Wildfire potential
	Commodity resource potential (e.g., forests, grazing land, hunting)
	Visual significance
	Educational use/value
	Flood potential
	Potential mitigation use/value
	Erosion control importance
	Recreational use/value (e.g., hunting, fishing, wildlife viewing)

document. Commenting or reviewing agencies may have their own criteria to apply in assessing the project and its setting. Ideally, these are anticipated and addressed in the EA/DEIS and should be addressed in the FEIS.

7.2 PROJECT DESCRIPTION

An understanding of the environmental impacts of a project will stem from the project description. The project should be viewed in terms of four phases—planning, construction, operation, and closure—because each phase may have associated or specific impacts to the biotic community. Generally, it can be assumed that the planning phase will not have high impacts, unless it involves something like testing, well drilling, or related disturbances in critical habitat areas. Even something like introduced noises during nesting times of threatened or endangered birds could be quite significant before construction of a project has even begun. Thus, a thorough project description will help establish a baseline of criteria for determining potential impacts.

A project description should include contingencies, such as the capacity for overflow parking at special events of a proposed facility. There is a range of reasonableness in determining what contingencies are probable. Experience will help determine what is reasonable. A full, thorough description of the project contributes to making a suitable range of potential impacts—a potentially contentious issue when it comes to biological impacts.

According to the IAIA (International Association for Impact Assessment, 2021), a thorough project description should include:

- The results of an aggressive exploration into the project details.
- A search for and use of guidance documents.
- Comparison with other similar project descriptions.
- Documentation of who owns, builds, operates, and monitors the project.
- The flow of energy and materials into and through the project.
- Suitable databases, maps, and charts.
- A detailed schedule for project phases.

Assignment

Go to the US Fish and Wildlife Service (FWS), Federal Register of Documents, and choose a text or PDF of an environmental notice (www.fws.gov). The FWS should have a link near the bottom for notices. The notice will have a description of the project. Compare the notice with the IAIA recommendations above. How much are you able to tell about the project's description based on the notice or the documents referenced in the notice? Does the notice seem to address all the IAIA recommendations?

7.3 POTENTIAL OR PREDICTED IMPACTS

Impacts are usually predicted by using standard models in conjunction with on-site field assessments. Numerous methods exist to quantify habitat attributes and importance. A Biological

Assessment is an analytical document for use in applying the Endangered Species Act (ESA) and focuses on threatened or endangered species. An early step in evaluating potential biological impacts is to check the lists of threatened and endangered species. This can be done before even setting foot on the project site, and it gives some insight into what is already known. Under Section 7 of the ESA, agencies consult with the Fish and Wildlife Service (FWS) and the National Marine Fisheries Service (NMFS) to see if federal actions could affect listed species. The Biological Assessment will help these services determine if they need to get more involved in the federal action. The Biological Assessment is thus a sort of specialized Environmental Assessment, and the two can be integrated depending on the authoring agency, or they can be separate documents. Methods used to evaluate species impacts and habitats can appear in Biological Assessments, Environmental Assessments, and EISs.

Two popular methods are the Habitat Evaluation System (HES) developed by the Army Corps of Engineers and the Habitat Evaluation Procedure (HEP) developed by the US Fish and Wildlife Service. Both of these two methods use an index to quantify the habitat in a specific area, and both address mitigation. Table 7.2 compares these methods.

Impacts to biodiversity should be expressed in more than just species diversity. Genetic diversity within species should be assessed, as should ecosystem diversity. Consumptive use of biological resources should be addressed, as should non-consumptive and indirect effects. The mitigation efforts should be tailored to the range of impacts identified in the assessment.

Assignment

1. What is the primary difference between Endangered and Threatened Species? How could this difference affect the biological mitigation plans for an EIA report?
2. Select a bird on your state/territory/reservation Endangered Species list, and explain what the major environmental impact restrictions or considerations would be for this species based on habitat and behavior.
3. Select an insect on your state/territory/reservation Threatened Species list. Provide the major environmental impact restrictions or considerations for this species based on habitat and behavior.

TABLE 7.2 Brief comparison of Habitat Evaluation System and Habitat Evaluation Procedure.

Army Corps Habitat Evaluation System (HES)	F&W Habitat Evaluation Procedure (HEP)
Focus on rivers and water resources	Water and terrestrial habitat
Habitat Quality Index (HQI)	Habitat Suitability Index (HSI), assumes a linear relationship between HIS values and carrying capacities
Habitat Unit Values (HUV)	Habitat Units for evaluation species (HU)
Smaller sites	Most frequently used for large projects/sites
Lesser cost and time	Greater cost and time
Focus on habitat quality in general	Focus on one or a few species, often four to six overall; may not imply suitability for other species
Not as detailed or accurate as HEP	Highly detailed and accurate

4. How are rare plants protected from development in your state/territory/reservation?
5. What are the differences between your state/territory/reservation Endangered Species list and the federal Endangered Species list? How does this affect EIA planning?
6. Choose a reptile or mammal threatened or endangered in your state/territory/reservation that might be found at your campus or another designated site. What wildlife buffer dimensions and characteristics would be sufficient to protect this species from adjacent development? Include conditions of maintenance of the buffer you would recommend and conditions or restrictions of operation of development/project usage outside the buffer. Consider such factors as noise, vibration, air pollution, disturbance, pets, human traffic, off-road cycles, and so on.

Birds are an important and highly visible part of urban and rural ecosystems. The birdwatching industry generates more than $17 billion a year, and the ecosystem benefits of birds are much greater. EIAs of projects such as wind power generation stations, power line corridors, and forest management require particular attention to potential bird impacts. Observations of bird behavior and flights should be throughout all seasons and times of the day. Pre- and post-construction monitoring of bird flights in a project area include corridor observations, circular scans, surveillance radar, and dead bird searches. These should all be mapped and documented in the EIA.

Aviation projects are another potential impact to bird flyways, nesting areas, and habitat. Projects in sensitive, vulnerable areas such as wetlands and coastal zones also receive special attention to bird impact assessment. Table 7.3 shows a potential checklist for birds. An authoring or reviewing agency will likely have its own checklist, as might the state or US Fish and Wildlife agencies.

TABLE 7.3 A checklist for avian impact assessment.

√	Item
	Bird habitat and species mapped
	Levels of significance (protected species or habitat listings are always significant)
	Stream or lake or other aquatic feature alteration
	Species removal (take) intentional or not
	Food web alteration resulting from the project?
	Introductions (invasive plants or animals)
	Maintenance needs for species
	Nutrient cycling and role of bird species
	Natural successions of plants or animals and interaction with birds
	Primary and secondary (direct and indirect) impacts to birds
	Flyways, breeding grounds, or other specific use habitat
	Avian effects and interactions from adjacent and previous projects in area
	Has a HEP (habitat evaluation procedure) or HES or similar assessment been done?

TABLE 7.4 Checklist of biological impacts from the California Environmental Quality Act for your state.

√	Impact
	Would the project have a substantial adverse effect, either directly or through habitat modifications, on any species identified as a candidate, sensitive, or special status species in local or regional plans, policies, or regulations, or by the California Department of Fish and Game or the US Fish and Wildlife Service?
	Would the project have a substantial adverse effect on any riparian habitat or other sensitive natural community identified in local or regional plans, policies, or regulations, or by the California Department of Fish and Game or the US Fish and Wildlife Service?
	Would the project have a substantial adverse effect on federally protected wetlands, as defined by Section 404 of the Clean Water Act (including, but not limited to, marsh, vernal pool, coastal wetlands), through direct removal, filling, hydrological interruption, or other means?
	Would the project interfere substantially with the movement of any native resident or migratory fish or wildlife species or with established native resident or migratory wildlife corridors, or impede the use of native wildlife nursery sites?
	Would the project conflict with any local policies or ordinances protecting biological resources, such as a tree preservation policy or ordinance?
	Would the project conflict with the provisions of an adopted Habitat Conservation Plan, Natural Community Conservation Plan, or other approved local, regional, or state habitat conservation plan?

What would you suggest be added to this checklist?
  Your suggestion:

  Your suggestion:

Assignment

What would you add or change to this checklist of impacts questions from the California Environmental Quality Act for your state (Table 7.4)? Make changes you deem appropriate and fill in the impact blanks by adding a response for the last two items.

7.4 MITIGATIONS AND MONITORING

Effective mitigation is predicated upon an adequate assessment of biological impacts, and it is part of an overall process. In 1993, the CEQ identified several weaknesses in NEPA practices of the time (CEQ, 1993: 18). The first was an inadequate consideration of species that were not federally listed as threatened or endangered. This raised the issue of proper management because waiting for species to be threatened or endangered before properly conserving or managing them is not an effective practice for ecosystem health. A second concern was inadequate consideration of non-protected land areas. The obvious weakness here is identical to that of non-listed species. A third area was inadequate consideration of "non-economically important" species. An example of this problem is the adverse impact of abundantly managed game species such as rainbow trout (*Oncorhynchus mykiss*) of native brook trout (*Salvelinus fontinalis*), chubs (*Cyprinidae*), and other species. These weaknesses can lead to the fourth identified category, inadequate consideration of cumulative impacts. The nature of NEPA is project specific, making cumulative impacts an implicitly difficult consideration. NEPA has continued to improve since the 1993 issuance of these concerns but they still remain good points to consider in current project reviews.

The US Fish and Wildlife Service may issue a "biological opinion." This opinion states whether a federal action is likely to jeopardize the continued existence of listed species or result in the destruction or adverse modification of critical habitat. It includes "reasonable and prudent" alternatives and "reasonable and prudent" measures. It also addresses irreversible or irretrievable commitment of resources, "action areas," and "incidental takes," and is based on "best available scientific and commercial information."

Mitigation may be project specific, or it may be for multiple projects, especially if for programmatic EIAs or multiple EIAs in a particular area. Mitigation "banking" is increasingly common, as is "in lieu fee mitigation" and conservation banking. Relocating or redesigning a project are forms of mitigation.

The Council on Environmental Quality (CEQ) regulations (40 CFR § 1508.20) define mitigation as:

1. Avoiding an impact altogether by not taking a certain action or parts of an action;
2. Minimizing impacts by limiting the degree or magnitude of the action and its implementation;
3. Rectifying the impact by repairing, rehabilitating, or restoring the affected environment;
4. Reducing the impact over time by preservation and maintenance operations during the life of the action;
5. Compensating for the impact by replacing or providing substitute resources or environments.

When mitigation occurs away from the site of impact, it may be required in greater amounts. For example, off-site wetlands mitigation might be at least a 3:1 ratio. The ratio may increase if the off-site quality is not equal to or greater than the on-site habitat or if there are time delays or other factors. Agencies may have formulae they use for ratio selection.

Off-site enhancements may be particularly useful for small projects and may allow multiple project combinations to increase the effectiveness of particular mitigation efforts. The connectivity of habitat is an important factor, and the relationship of habitat to particular functions, such as breeding, nesting, migration, and flyways in evaluating mitigation.

Mitigation banks and conservation trusts require monitoring and perpetual management. Other forms of mitigation may be quite simple, such as controlling the timing of construction and other disturbances to protect selected sites and seasons (Table 7.5).

Assignment

1. The late renowned biologist E. O. Wilson proposed the concept of "biophilia," which includes the notion that we humans prefer species that resemble our own babies, with rounded, soft features and big eyes (1984). How might we compensate for this bias in our plans to mitigate wildlife impacts?
2. Should off-site mitigation acreage ever be less than a 2:1 ratio? Why or why not?

TABLE 7.5 Common types of biological impact mitigation measures.

√	Mitigative Measure
	Conditions for approval in state or federal permits
	Incorporation of other land use controls: permits, conservation easements, and land use designations
	Consultation inputs from US Fish and Wildlife and other agencies
	Law enforcement at local, state, and federal levels
	Habitat Conservation Plan (per Endangered Species Act)
	Species Recovery Plan (per Endangered Species Act)
	Monitoring program
	Relocation program
	Project redesign
	Phasing and timing of project build-out or implementation
	Enhancement: site or habitat reconstruction or restoration
	New acquisitions or dedications of habitat
	Mitigation banking
	Best on-site management practices (BMPs)
	Education and public information
	Warning signage
	Collaborative agreements with local institutions
	Reserved regulatory jurisdiction for authority to revisit issues if findings change

7.5 CONCEPTS AND TERMS

Abiotic factors: Nonliving or physical factors (temperature, light) in the environment. A change in these or in biotic factors can change the **carrying capacity** for various species in a habitat.

Biodiversity (also **biological diversity**): Variation in and among species and life-sustaining processes and ecological elements. Includes ecosystem diversity, species diversity, and genetic diversity.

Biogeochemical cycles: Nutrient cycles of carbon, oxygen, nitrogen, phosphorous, sulfur, and metals. Water also has a cycle, in addition to its elements. Every cycle has a reservoir pool and a faster exchange pool.

Carrying capacity: Maximum abundance of a population that can be maintained by a habitat or ecosystem without degrading the habitat or the ecosystem. Achieving carrying capacity means a future collapse or some other negative feedback to drop the population. Some species increase the number of their offspring but reduce the age and size of maturation just before collapse, giving the illusion based on mass that the population is well.

Charismatic megafauna: Comparatively large animals that have popular appeal or symbolic value that may color the public perception of their role or value in an ecosystem. An EA report will need to pay attention to these species without neglecting other species.

Climax community: Final stage (self-perpetuating) of plant community in ecological succession. Some scientists consider this an overly deterministic model, but it does suggest a stable, mature condition of an ecosystem. It gets complicated when some communities thrive on disturbance (see **resilience**).

Clime: Change in a trait of a given species over a geographic region or range.

Community: Multiple species in a common habitat.

Conservation bank: Parcel of land containing natural resource values. These parcels are conserved and managed in perpetuity for listed species as an offset for environmental impacts elsewhere. Conservation banks are a somewhat broader category than mitigation banks, which tend to be for aquatic resources.

Conservation easement: Legal restriction that, to protect or enhance biological or other resources, limits the use of land.

Disturbance: Any force disrupting the pattern of species diversity and abundance, community structure, and community properties. Measured in terms of frequency, scale, intensity, and similar factors. Includes natural disturbances and human-caused (anthropocentric) disturbances. Some species thrive on disturbance (likely to be pioneers, especially invasive species).

Diversity: Measurement of variation. There are many different kinds. We usually think of this in terms of number of different species per area, but we can include diversity within species, genetic diversity, phylogenetic diversity, habitat or ecosystem diversity, and other types to measure variation. Diversity implies stability.

Dominant (plant) species: Plant species with greatest canopy cover.

Ecological diversity: Variety of **habitats**.

Ecosystem: A concept of shared characteristics of a **habitat**.

Ecosystem engineer: A species that can significantly change a habitat, usually a **keystone** species with an unusually strong effect on other organisms.

Ecotone: Transition zone between adjacent **communities**.

Endangered Species Act: Federal legislation enacted in 1973 to protect endangered and threatened (likely to become endangered) species and their habitats. Declares federal policy and calls for cooperation with states.

Genetic diversity: Variation within species.

Habitat: Type of environment in which a species is usually found.

Habitat Evaluation Procedure (HEP): US Fish and Wildlife Service method for assessing biological impacts.

Habitat Evaluation System (HES): US Army Corps of Engineers method for assessing biological impacts.

Habitat Suitability Index: US Fish and Wildlife Service tool for Habitat Evaluation Procedure (HEP) using carrying capacity of select species and GIS.

Indicator species: A species to watch in determining the health of an ecosystem because it is an early warner of potential harm.

Keystone species: Role of a species in an ecosystem is critical in proportion to its numbers or biomass. Not necessarily a top predator. Some examples: sea star, honeybee, frog-eating salamander, alligator, and sea otter. Relationships can be complex, and it can be hard to tell which species is a keystone.

Lacey Act: 16 USC §§ 3371–3378. Federal ban on trafficking in illegal wildlife. Enacted in 1900, amended in 2008 to add plants and plant products.

Lethal concentration: Amount of a chemical that causes death to organisms; most commonly measured at LC_{50}—the concentration that kills half of the test organisms. The lower the LC_{50} value, the greater the toxicity.

Natural capital: An approach in which biodiversity and other natural resources are treated analogous to economic capital as depletable and subject to related qualitative and quantitative characteristics.

Phase I Site Assessment: The first phase of an environmental site inspection, including the basic observation of the environmental health of a designated area. It asks the question, "Is something there?"

Phase II Site Assessment: Seeks to verify the environmental observations identified in a Phase I Site Assessment and delineates the problems through sampling and analysis. It asks the question, "We know something is there; Where is it specifically located?"

Phase III Site Assessment: Generally refers to remediation (clean-up) of delineated contamination sites. In archaeology, the salvage or excavation of a site. Addresses the question, "The site is contaminated or at risk; what can we do about it?"

Potential natural community (PNC): The biotic community that would likely be established and maintained under the present environmental conditions if no further human-caused disturbances occur.

Primary succession: Ecological progressive change of plant community through gradual weathering and actions of time. Reflects a deterministic perspective.

Range of tolerance: The varying range of environmental conditions that a species can tolerate; individuals within a species may also have slightly different ranges of tolerance.

Resilience: Ability of a system or component to "bounce back." Usually due to possessing alternate pathways and abilities to absorb or process stress and energy. Resilience does not necessarily mean no disturbance. Grassland ecosystems systems, for example, are more likely than not to depend on disturbances. In such cases, fire itself might be a stabilizing disturbance.

Secondary succession: Ecological progressive change of plant community through gradual weathering and actions of time after an initial disturbance. Reflects a deterministic perspective.

Serial status (or **serial stage**): One or more stages of secondary successional development before the PNC is reached.

Species diversity: Variation among species, including microorganisms, fungi, plants, insects, fish, reptiles, birds, mammals, and other animals.

Threatened and endangered species lists: Threatened species are those that are likely to become endangered. Endangered species are already in trouble and in danger of becoming extinct. The federal government and each state maintain lists of threatened and endangered species.

Umbrella species: A species particularly representative of their ecosystem and that has a disproportionate management benefit. Protecting these species includes protection of other species below them. Similar to a **keystone species** but more likely to be keyed in to a particular geographic species range.

7.6 SELECTED RESOURCES

Burke-Copes, Kelly A., Antisa C. Webb, Michael F. Passmore and Sheila D. McGee-Rosser. (2012). HEAT—Habitat Evaluation and Assessment Tools for Effective Environmental Evaluations: User's Guide. U.S. Army Corps of Engineers. https://usace.contentdm.oclc.org/digital/collection/p266001coll1/id/6536/.

Canter, Larry. (2015). *Cumulative Effects Assessment and Management*. Horseshoe Bay, TX: EIA Press.

Council on Environmental Quality (CEQ). (1993). Incorporating Biodiversity into Environmental Impact Analysis Under the National Environmental Policy Act. January 1993. NEPA.GOV. https://ceq.doe.gov/publications/incorporating_biodiversity.html.

Federal Highway Administration. Environmental Review Toolkit. Eco-Logical: An Ecosystem Approach to Developing Infrastructure Projects. www.environment.fhwa.dot.gov/env_initiatives/eco-logical/report/eco_index.aspx.

Hall, Frederick C., Larry Bryant, Rod Clausnitzer, Kathy Geier-Hayes, Robert Keane, Jane Kertis, Ayn Shlisky and Robert Steele. (1995). Definitions and Codes for Serial Status and Structure of Vegetation. US Department of Agriculture, Forest Service. General Technical Report PNW-GTR-363.

Mitsch, William J. and James G. Gosselink. (2015). *Wetlands*, 5th ed. New York: John Wiley & Sons. This Is a Standard Reference for College Courses and Professional Work Pertaining to Wetlands.

Morris, Peter and R. Therivel (Eds.). (2001). *Methods of Environmental Impact Analysis*, 2nd ed. London: Spon Press.

United Nations. (2002). *Environmental Impact Assessment Training Resource Manual*, 2nd ed. New York: United Nations Environmental Programme. https://wedocs.unep.org/handle/20.500.11822/26503.

U.S. Environmental Protection Agency (EPA). (2002). *Methods for Evaluation Wetlands Conditions: Biological Assessment Methods for Birds. Office of Water*. Washington, DC: US EPA. EPA-822-R-02-023.

U.S. Fish & Wildlife Service. Endangered Species. www.fws.gov/endangered/. Searchable by state or county and by species.

U.S. Fish & Wildlife Service. www.fws.gov/.

U.S. Fish & Wildlife Service. (1981). Habitat Evaluation Procedures Handbook. www.fws.gov/policy/esmindex.html.

7.7 TOPIC REFERENCES

Canter, Larry. (1996). *Environmental Impact Assessment*, 2nd ed. New York: McGraw-Hill.

Council on Environmental Quality (CEQ). (1993). Incorporating Biodiversity into Environmental Impact Analysis Under the National Environmental Policy Act. January 1993. NEPA.GOV. https://ceq.doe.gov/publications/incorporating_biodiversity.html.

Doub, J. Peyton. (2012). *The Endangered Species Act: History, Implementation, Successes, and Controversies*. Boca Raton: CRC Press.

IAIA. (2021). International Association for Impact Assessment. Project Description. www.iaia.org/wiki-details.php?ID=20.

Johnson, K. Norman, F. Swanson, M. Herring and S. Greene (Eds.). (1999). *Bioregional Assessments*. Washington, DC: Island Press.

Liebesman, Lawrence and Rafe Peterson. (2010). *Endangered Species Deskbook*. Washington, DC: Environmental Law Institute.

Lyon, John G. and Lynn K. Lyon. (2011). *Practical Handbook for Wetland Identification and Delineation*. Boca Raton, FL: CRC Press.

Martin-Ortega, Julia, Robert C. Ferrier, Ian J. Gordon and Shahbaz Khan. (2015). *Water Ecosystem Services*. Cambridge: Cambridge University Press.

Meffe, Gary K., Larry A. Nielsen, Richard L. Knight and Dennis A. Schenborn. (2002). *Ecosystem Management: Adaptive, Community-Based Conservation*. Washington, DC: Island Press.

Tourbier, J. Toby and Richard N. Westmacott. (1981). *Water Resources Technology: A Handbook of Measures to Protect Water Resources in Land Development*. Washington, DC: Urban Land Institute.

Treweek, Jo. (1999). *Ecological Impact Assessment*. Malden, MA: Blackwell.

Wilson, Edward O. (1984). *Biophilia*. Cambridge, MA: Harvard University Press.

Air quality and climate change

8.1	Environmental setting	116
8.2	Project description	118
8.3	Potential impacts	118
8.4	Mitigation and monitoring	121
8.5	Concepts and terms	125
8.6	Selected resources	125
8.7	Topic references	126

The Clean Air Act (CAA) is the major guide for evaluating existing or predicted air and climate impacts from projects subject to NEPA review. Air pollution is a human-defined and assigned attribute, though we further define its affects in terms of science and ecosystem functioning. The common approach for evaluating air quality impacts in an environmental assessment is similar to other components of an EIA. The startup tasks are a description of existing air quality conditions. Then, using current air quality standards and guidelines, the team predicts impacts under various project alternatives. Next, the preparers assess impact significance by comparing predicted impacts with the standards and regulations. Last, the report will identify and incorporate mitigation measures as necessary in order to make the project compliant with all existing standards and regulations. In addition, the results of the environmental assessment should provide feedback about assessment methods used to comply with EIA policies. The results of individual assessments should contribute to improving the overall efficiency and accuracy of EIA and NEPA and to understanding cumulative impacts.

The categories of air quality impacts are quite broad: noise, odor, light, particulate matter, radiation, vibration, thermal, and chemical vapors. They may include many sources such as vehicle traffic, mining, industrial processes, wildfires, blasting for construction projects, vapors emitted from industrial facilites, smoke from residential woodstoves, and other sources. Most impacts under these categories are treated by local and state review boards as aesthetic concerns until they reach a level at which they become pollutants or health hazards. Some of these sources are specifically regulated under the Clean Air Act. All are subject to review in an EIA under NEPA.

DOI: 10.4324/9781003030713-8

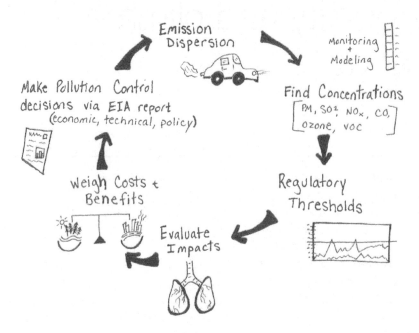

FIGURE 8.1 Air quality assessment cycle.

An air impact assessment is an iterative process (Figure 8.1). The air impact assessment may be contracted out as a separate workpiece or undertaken as an impact category in NEPA compliance work. If done as the latter, it may follow the impact steps as outlined in previous topics of environmental setting, project description, potnetial impacts, mitigation, and monitoring.

8.1 ENVIRONMENTAL SETTING

Since air covers the entire planet, researchers must select the boundaries and other limits of what is to be addressed in the assessment. In dealing with air quality, it is not just the biological and physical environment, but also the human effects from air and climate. The scale can be extensive, and there are potentially infinite attributes of a setting. These attributes can be largely aggregated in terms of potential impact variables, including exhausts, noise, radiation, radon, thermal, chemical vapors, odor, vibration, particulate matter, and toxins. The cumulative effects of these impact variables contribute to changes in climate.

Most EIAs involve specific projects that directly or indirectly alter an environmental setting. The air component of this setting is the "air basin." The existing air quality in this basin is an attribute of the project environmental setting. It is documented through comparison with the EPA's National Ambient Air Quality Standards (NAAQS) for the six criteria air pollutants of atmospheric particulate matter (PM_{10} and PM_2), carbon monoxide (CO), lead (Pb), sulfur oxides (SO_x), nitrogen oxides (NO_x), and ozone (O_3).

Part of understanding the environmental setting involves researching the historic air quality trends up to the current conditions. The EPA maintains a color-coded Air Quality Index (AQI), as do some states. Revisions to the AQI by the EPA strengthened fine particle pollution standards (EPA, 2012). The AQI descriptors, levels of concern, and ranges are shown in Table 8.1. Environmental change and increasingly dynamic weather conditions means it is risky to forecast the future based on the past, but historic air quality trends should be documented for comparison and for understanding the consequences of environmental change.

Characteristics of the physical setting (topography, weather, land use) affect the potential impacts of a project. The setting has characteristics that affect the very flow of air. For example, cold air flows downslope into valleys toward the sea at night like a river, as the cold makes the air heavier. In the morning, it warms and rises back up the valleys. Conditions such as fog, wind speed, and precipitation affect local air conditions. As part of the assessment, the setting is described in accordance with NAAQS as "attainment area" or "non-attainment area." "Hotspots" such as corridors and highway intersections may be identified in the setting for additional analysis.

Assignment

Locate the AQI reports for the nearest city.
1. How has air quality been for the past year?
2. What are the implications for the review of NEPA projects within or adjacent to this city?
3. Does the city have a plan for climate change? What predictions for air quality are addressed in the plan?

TABLE 8.1 AQI for ozone and particle pollution effects on humans.

Color	Level of Concern	Index Range	Description
Green	Good	0–50	Satisfactory, little or no health risk
Yellow	Moderate	51–100	Acceptable, but a risk for very sensitive people
Orange	Unhealthy for sensitive groups	101–150	Risk for sensitive groups
Red	Unhealthy	151–200	Affects some members of general population and sensitive groups.
Purple	Very unhealthy	201–300	Health alert; risk for everyone
Maroon	Hazardous	301+	Emergency conditions

Source: EPA (2012).

8.2 PROJECT DESCRIPTION

The project description sets up the basic criteria for potential air quality impacts. These are the impacts that come from the project itself and from the project's interaction with the site. The EIA should thoroughly describe the project. This reduces the likelihood of leaving out potential impacts. For example, the project description should include the parking areas and vehicle traffic generated by the project because these will result in air emissions and temperature "islands." If the project engenders parking because it includes large or continuous events and its description omits parking lots, then the EIA is forced to deduce the implicit parking needs and treat those as if they had been included as part of the project description.

The project may include fixed or portable sources of air pollution. In the United States, mobile sources are the biggest source of air pollution. Sources may be categorized as "point" or "area." The project may include an "emission inventory" with potential source categories of fuel usage, manufacturing and processing, solid waste, transportation, and natural. Facilities such as power plants, refineries, and chemical producers should have existing histories of all emissions related to the location.

8.3 POTENTIAL IMPACTS

An environmental specialist usually performs the air quality impact prediction, interpretation, and mitigation in an Air Quality Impact Assessment (AQIA). This person is often an engineer, using proprietary computer models and programs, with standardized monitoring techniques. The reviewer's job is to see if the relevant information is provided in the EIA and to ask questions about the EIA air quality component. Air quality impacts can include odors, which we have chosen to address in the unit on aesthetics, but odors can indicate air quality problems or an immediate health hazard, as in the case of the smell of hydrocarbons (propane is an obvious example). The EPA provides a Reference Guide to Odor Thresholds for Hazardous Air Pollutants Listed in the Clean Air Act Amendments of 1990.

The anticipation of potential impacts generally follows the conventional understanding of air pollution as greenhouse gases (GHGs) and related emissions. The EIA will identify the relevant (usually nearest) monitoring stations that collect this information. The gases include SO_2 or SO_x, NO_x, CO, O_3, volatile organic compounds (VOCs), hydrogen sulfide (H_2S), and hydrogen fluoride (HF). A very simplified list of common air emissions follows.

8.3.1 Carbon dioxide

Over half of the anthropogenic greenhouse effect is attributed to CO_2. Atmospheric concentrations of CO_2 today are much higher than the preindustrial level and exceed the highest natural levels of the past 160,000 to 400,000 years. Though poorly understood natural processes in the carbon cycle are removing much of the anthropogenic CO_2 from the atmosphere, it is reasonable and prudent to assume that CO_2 concentrations and resulting global warming will increase in the future due to human activities.

8.3.2 Methane

Natural sources of methane include termite activity, cows, and freshwater wetlands; anthropogenic sources include biomass burning, coal and natural gas production, landfills, and certain agricultural and forestry practices. Agriculture is the largest source. Massive amounts of methane exist in the seafloor. Some is making its way into the ocean and then to the atmosphere as a result of warming.

8.3.3 Chlorofluorocarbons

Chlorofluorocarbons (CFCs), used as aerosol propellants, blowing agents, and refrigerants, are powerful and persistent GHGs; they may remain in the atmosphere up to 100 years. Other gases such as hydrochlorofluorocarbons and halons are powerful, too, but are emitted in smaller amounts.

8.3.4 Nitrous oxide

Nitrous oxide, commonly released as a result of artificial fertilizer use and fossil fuel combustion, is a minor but persistent GHG. Anthropogenic trace gases contributing to the greenhouse effect accumulate in the atmosphere at much faster rates than CO_2.

8.3.5 Particulates

Dust, smoke, fumes, and particles emitted into the air fall under the category of particulate matter. Particulate matter may be described as a quality, such as soot or smoky haze, or in quantitative terms, such as micrograms per cubic meter. There are many sources for particulate matter: burning things, traffic, construction, demolition, manufacturing, processing materials, and quarrying.

Particulates are solids or liquids classified according to size (e.g., PM_{10} is a particulate less than 10 μm in diameter).

8.3.6 Radon gas

Radon (Rn) is a naturally occurring radioactive gas that is colorless, odorless, and tasteless. It is produced from the decay of ^{238}U. Radon is a risk factor for lung cancer. Indoor radon gas poses risks that are hundreds of times greater than those from outdoor pollutants in the air.

The production of radon gas varies with the local geology and, thus, varies spatially. It is a particular problem in the northeastern United States. Radon enters buildings by migrating up from the soil into basements and lower floors. It can come from groundwater that sometimes is pumped into structures, or it can result from radon-contaminated building materials such as building blocks. There are simple test kits available through commercial testing laboratories. When it is identified as a problem, it can be controlled for new construction by design with preventative and treatment systems, and for existing structures by sealing the entry points and improving the ventilation.

8.3.7 Air quality models

The EIA report should identify the air quality model used and describe its appropriateness for the type of project. Typically, models include the project sources and potential influences to air quality external to project sources. These factors include settlement patterns and density, solar radiation, ground contours, weather, and climate. There may be a Regional Air Quality Management Plan (AQMP) and a State Implementation Plan (SIP) for air quality that sets the context for what model to use for interpreting air impacts and for planning the mitigation.

Models consider sources, which they categorize into "emission factors," background conditions, receptors, and other factors. *Dispersion modeling* is a mathematical approach that estimates concentrations of pollutants and specified ground-level receptors. It determines compliance with NAAQS, New Source Review (NSR) and Prevention of Significant Deterioration (PSD). The EPA Support Center for Regulatory Atmospheric Modeling (SCRAM) addresses preferred air quality dispersion systems and models. Conceptually, a box model for dispersion will assign a constant emission rate as a function of mass divided by time (Figure 8.2). This rate is further divided by (wind speed × area) to obtain the concentration. This method works for lots of small sources within an area.

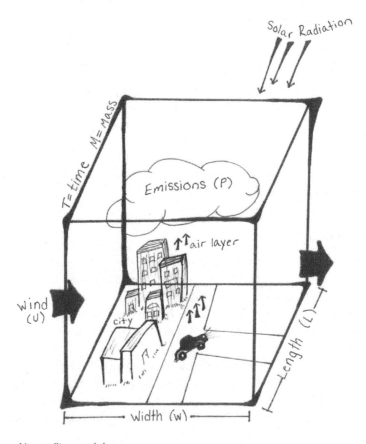

FIGURE 8.2 Air quality model.

Photochemical modeling is another modeling approach addressed by SCRAM. It simulates changes of pollutant concentrations with mathematical equations. It can be used at a wide range of scales from local to global. The two most common examples for photochemical modeling are the Lagrangian trajectory model (uses a moving frame of references) and the Eulerian grid model (uses a fixed coordinate system).

Unlike dispersion models and photochemical models, *receptor models* use the characteristics of gases and particles as measured at sources and receptors. This observational data is inputted into mathematical and statistical programs. The EPA versions of these models include Chemical Mass Balance (CMB), UNMIX, and Positive Matrix Factorization (PMF). Receptor models are often used with State Implementation Plans (SIPs) for identifying sources contributing to AQ problems.

The rollback model, also called proportional scaling, is used for short-term variations in dispersal. It, embedded in proprietary software, assumes an average value at a point $c_i = (k) \times$ (mass/time) $+ b$, where c is in $\mu g/m^3$, b is the background concentration of the pollutant in $\mu g/m^3$, and k is a proportionality factor (De Nevers and Morris, 1975). It determines the percent of time concentrations are exceeded.

A Gaussian model is used for point sources, as in the case of a particular factory. It relies on assumptions of constant rate of pollutant emission, wind speed, and direction. It is best for stable emission and flat land. However, computers can incorporate variation and make reasonable predictions.

The State Implementation Plan (SIP) for air quality may incorporate the recommendation of specific models for comparison. The EIA should address how the project complies with or "fits" the SIP and any applicable agency implementation plans.

Assignment

1. Does your state have a State Implementation Plan (SIP) for air quality? If so, what are its key provisions?
2. How are the air quality and transportation sections of an EA connected?
3. Identify and describe a predictive computer model used in EIA air quality work.
4. What is the role of the EPA in the air quality impact assessment process?
5. Should carbon dioxide (CO_2) be classified as a pollutant under the Clean Air Act? Explain why it should or should not be listed.

8.4 MITIGATION AND MONITORING

Air quality assessment needs to incorporate Worst Condition Analysis—determining potential impacts and mitigation for when air pollution tends to be the highest, such as during hot days in August. Monitoring needs to include sensitive "receptors," such as hospitals and nursing homes.

Industrial particulate matter (PM) can be controlled with electrostatic precipitators, scrubbers, and filters. Catalytic converters are used on autos to reduce VOCs and carbon monoxide (CO). Auto emissions of NO_x are reduced by recirculating exhaust gas and adjusting the air:fuel ratio. Sulfur dioxide (SO_2) can be controlled by using scrubbers on smokestacks and by using gasified coal as a fuel source.

Short-term and long-term emissions are different perspectives for mitigating a single pollutant. Short-term emissions are usually mitigated at the point where it enters the atmosphere. Long-term emissions are those that enter the atmosphere over time and accumulate regional characteristics.

Direct emissions are directly injected into the atmosphere. Examples are from combustion, welding, grinding and abrasion, evaporation, and spills. Direct emissions can be accidental or intentional. Indirect emissions are those that undergo a chemical change, sometimes creating a new chemical substance. Examples are sulfur oxides, hydrocarbons, and VOCs that interact with sunlight.

One example of a serious impact category calling for long-term mitigation is vehicle traffic. Mitigation for traffic means use of Transportation Control Measures (TCMs) and Transportation System Management (TSM). Common TSM tools are promotion of ride-sharing and carpooling, public transportation, energy efficient vehicles, work-hour/work-from-home management, park-and-ride lots, high-occupancy lanes, alternative transportation, and incentive programs.

Not surprisingly, every air pollution source from ongoing and new projects calls for professional analysis. Large, complex projects call for systematic coordinated impact documentation and management. Organizing categories of impact and their sources will prove useful in developing monitoring and mitigation measures.

Assignment

Assume a sand pit in your town generates 100 trip loads per day (one-way 10 yd^3). The pit has loaders and other excavation and transport equipment. The project operates for 12 hours a day, seven days a week. The site is on 50 acres in the middle of a square 300-acre tract. A dirt road connects the project to a main road. The project owners propose to double in size and operation, triggering a review that involves an environmental impact assessment.
1. List the possible air pollutants and air-related impacts likely from this project.
2. Complete a table (Table 8.2) that shows the types of pollutants you identified in item 1, a potential source, and one or more mitigation measures for each type.

Plume discharges to air can be monitored through visible stack emissions. The pattern of dispersal from stack emissions will vary according to seasonality, operating conditions, the

TABLE 8.2 Types of air pollutants for a gravel pit.

Pollution Type	Pollution Source	Mitigation Measure(s)

weather, and other factors. Patterns should be documented and incorporated into models in the EIA. A simple glance at a stack emission can reveal information about atmospheric conditions. For example, if the plume is looping, it indicates cold temperature. A plume that forms a cone ("coneing") indicates a warm day, whereas a plume that "fumigates" spreads out all over and indicates a hot day. A plume that fans out reflects an external force or pressure acting on it. Prevailing winds may direct the plume and will be part of air quality modeling. The mitigative measures will be derived from the modeling.

8.4.1 Climate change

Climate change is an environmental effect that should be considered within NEPA compliance documents. Considerations may include both potential effects of a proposed action on climate change and the implications of climate change for the environmental effects of a proposed action. Which effects to analyze and the depth of analysis will vary by the nature of the project proposal and the needs of decision makers. On January 2, 2021, the president signed Executive Order 14007, Tackling the Climate Crisis at Home and Abroad, which "establishes a government-wide approach to climate crisis and ensures Federal permitting decisions consider the effects of greenhouse gas emissions and climate change" (Council on Environmental Quality, 2022).

A federal or state agency considering a proposed action will likely have procedures that require them to consider the potential effects of climate change as they develop their projects. However, some projects may have characteristics that could contribute to climate change but still be necessary. Once the project proposal is developed, much of the climate change analysis will be in the form of determining and evaluating impacts in the EIA process, which will result in an EA or EIS.

Climate change and projects intersect in multiple categories of impact. Essentially, there are the contributions a project makes *to* climate change, there are the effects of climate change *on* a project, and there are the synergistic implications of climate change and the environmental effects or impacts of a project. Brant and Schultz (2016: 3–4) summarize the three ways to consider climate change at various stages in the NEPA process:

1. The first is the effects of a proposed project on climate change through greenhouse gas. (GHG) emissions and sequestration. Examples include short-term GHG emissions and alteration to the carbon cycle caused by hazardous fuels reduction projects; GHG emissions from the extraction of fossil fuels and minerals; or avoiding large GHG emissions pulses and effects to the carbon cycle by thinning overstocked stands to increase forest resilience and decrease the potential for large scale wildfire.
2. The effects of climate change on a proposed project . . . will climate change influence the affected environment in such a way that it will affect the purpose and need of a project? Examples could include current or projected influences of climate change on habitat suitability for target species or ecosystems in restoration projects; effects of increased flooding on site selection for recreation areas; or effects of decreased snowfall on a ski area expansion proposal at a marginal geographic location, such as a southern aspect or low elevation.

3. The implications of climate change for the environmental effects of a proposed action should also be considered. In addition to consideration of emissions and sequestration caused by the project, it may be necessary to consider the effects of a project on a particular resource in combination with those caused by climate change. Will the action and climate change combine to create increased impacts on a resource? Will other reasonably foreseeable actions add further impacts creating cumulative effects? Examples include the potential for climate change and habitat fragmentation caused by the project and outside the project area to lead to jeopardy or listing under the Endangered Species Act, or the potential for climate change and project activities to foster the spread of non-native invasive species. Some projects may not require detailed analysis of all or any of these effects. Which effects to analyze, and the depth of analysis, will vary by the nature of the proposal, the needs of the decision maker, the intensity of the effects, scientific uncertainty or controversy, and public interest as determined from scoping.

Integrating climate change considerations into the scoping of a project and its impacts is still fairly new—but it is necessary. Brant and Schultz (2016) remind us that climate change is properly within the scope of an analysis of cause-and-effect relationships. It is difficult to quantify consequences, but we can set into place data collection and monitoring schemes. Thus a project could incrementally and cumulatively contribute to future understandings about GHGs and other effects on climate change.

Assignment

Your state is rebuilding a pier on the coast of a lake and installing waterfront commercial and recreational facilities. The project will take 5 years to complete and will then be good for at least 50 more years. At the end of that time, the piers still need to be viable. What climate change considerations might you take into account? Incorporate your answers into Table 8.3.

TABLE 8.3 Climate change and waterfront development.

Type or Category of Change	Environmental Consequence (Impact)	Mitigation Measures
Sea level rise	Water depth increases ecological change	Anticipate and elevate pier and structures
Warmer weather		
Increase in storms and other weather events		
Trend in local human population change accelerated		
Warmer water		
Increased runoff and impervious surface		
Other		

8.4.1.1 Indoor air quality impacts

Most NEPA EIAs focus on external air quality. Projects that serve as indoor worksites or living areas, such as federally funded manufacturing incubators, can have indoor air quality impacts. Federal housing projects can also have indoor air impacts stemming from cooking, heating and cooling, and improperly regulated construction materials that off-gas. Super-tight buildings also may fail to have sufficient air exchanges. Climate change can worsen indoor air quality issues. The EPA and the Occupational Safety and Health Administration (OSHA) are the primary federal agencies that promulgate indoor air quality regulation and evaluation.

There are many potential sources of air quality issues. Tobacco smoke is no longer encountered in US public buildings, although it can be in residential or commercial buildings. Other sources include *Legionella pneumophila* (bacteria), mold spores, radon gas from soil/rock under the building, other pathogens, pesticides used in the building, asbestos from insulation and tile, formaldehyde gas from decomposing particle board, cleaning materials and other stored/used chemicals, and dust mites.

Indoor source issues may be exacerbated by improper ventilation design and function. COVID-19 has greatly increased public awareness of the importance of air exchanges, filtering, and air flow. Responses need to be coordinated. The two best ways to conserve energy in buildings (increase insulation and eliminate air leaks) tend to worsen the problem of indoor air pollution by reducing volumetric room exchange rates.

8.5 CONCEPTS AND TERMS

Air toxics: Air pollution typically present in low concentration but with toxic characteristics to human or ecological health.
Climate Impact Assessment (CIA): An assessment of the changes of greenhouse gas (GHG) emissions that may result from a project or policy proposal.
Criteria pollutant: Any of six of the most common air pollutants defined under the Clean Air Act for NAAQS as particulate matter, lead (Pb), ground-level ozone (O_3), nitrogen oxides (NO_x), carbon monoxide (CO), and sulfur dioxide (SO_2).
Hazardous air pollutant: Any of 189 carcinogens and other toxic pollutants.
National Ambient Air Quality Standards (NAAQS): EPA standards for criteria pollutants.
Residence time: How long a pollutant or other material remains suspended in the air, water, soil, or other media.

8.6 SELECTED RESOURCES

California Air Resources Board. (2003). Air Quality Handbook on Land Use. www.arb.ca.gov/ch/landuse.htm.
Canter, Larry. (2015). *Cumulative Effects Assessment and Management*. Horseshoe Bay, TX: EIA Press.
Council on Environmental Quality (CEQ). (2021). Guidance on Consideration of Greenhouse Gases. NEPA.GOV. Contains a Grouping of Documents and Resources. https://ceq.doe.gov/guidance/ceq_guidance_nepa-ghg.html.

DiGiovanni, Franco and Miguel Coutinho. (2017). Guiding Principles for Air Quality Assessment Components of Environmental Impact Assessments. *International Association for Impact Assessment*. Special interest publication, February 2017.

Elsom, Derek M. (2001). Air Quality and Climate. In P. Morris and R. Therivel (Eds.), *Methods of Environmental Impact Analysis*, 2nd ed. (pp. 145–169). London: Spon Press.

Environmental Protection Agency. www.epa.gov/airnow/where/. A Radon Zone Map Can Be Found at www.epa.gov/iaq/radon/zonemap.html. EPA's. www.epa.gov/oar/oaqps/supplies information on air quality standards and air quality planning.

Environmental Protection Agency (EPA). (2020). Air Quality Guidance for National Environmental Policy Act Reviews. www.epa.gov/nepa/air-quality-guidance-national-environmental-policy-act-reviews.

EPA State Inventory Tool. Interactive Spreadsheet Model to Develop Greenhouse Gas Emission Inventories. www.epa.gov/statelocalenergy/state-inventory-and-projection-tool.

IPCC (Intergovernmental Panel on Climate Change). (2022). *Climate Change 2022: Impacts, Adaptation and Vulnerability*. Cambridge University Press. www.ipcc.ch/report/sixth-assessment-report-working-group-ii/.

Mid-Atlantic Regional Air Management Association, Inc. www.marama.org/about-us/introduction-to-marama/82-resources/resources. Links to Federal, Government, Regional, and Other Resources Pertaining to Air Quality and Air Resources.

National Academies of Sciences, Engineering and Medicine. (2022). *Wildland Fires: Toward Improved Understanding and Forecasting of Air Quality Impacts: Proceedings of a Workshop*. Washington, DC: The National Academies Press.

National Park Service, Natural Resource Program Center. Technical Guidance on Assessing Impacts of Air Quality in NEPA and Planning Documents. January 2011. Natural Resource Report NPS/NRPC/ARD/NRR-2011/289. https://irma.nps.gov/DataStore/DownloadFile/423676.

Northeast States for Coordinated Air Use Management NESCAUM). www.nescaum.org/about.html. This consortium of Connecticut. Maine, Massachusetts, New Hampshire, NJ, New York, Rhode Island, and Vermont Provides News Items, Resources, and Links.

Provides Links to Arizona, California, Colorado, Idaho, Montana, Nevada, New Mexico, North Dakota, Oregon, South Dakota, Utah, Washington, Wyoming, Tribal, Federal, and non-Government Organization Web Sites.

Western Regional Air Partnership. www.wrapair.org/wraplinks.htm.

8.7 TOPIC REFERENCES

Brant, Leslie and Courtney Schultz. (2016). Climate Change Considerations in National Environmental Policy Act Analysis. U.S. Department of Agriculture, Forest Service, Climate Change Resource Center. www.fs.usda.gov/ccrc/topics/nepa.

Canter, Larry W. (1996). *Environmental Impact Assessment*, 2nd ed. New York: McGraw-Hill.

Council on Environmental Quality. (2022). National Environmental Policy Act Implementing Regulations Revisions. Federal Register. 87 FR 23453.

De Nevers, N. and J. R. Morris. (1975). Rollback Modeling: Basic and Modified. *Journal of the Air Pollution Control Association*, 25(9), 943–947.

Environmental Protection Agency. The Plain English Guide to the Clean Air Act. 2007. www.epa.gov/clean-air-act-overview/plain-english-guide-clean-air-act.

Environmental Protection Agency (EPA). (1992). *Reference Guide to Odor Thresholds for Hazardous Air Pollutants Listed in the Clean Air Act Amendments of 1990*. Washington DC: Office of Research and Development. EPA/600/R-92/047.

Environmental Protection Agency (EPA). (2012). *Revised Air Quality Standards for Particle Pollution and Updates to the Air Quality Index (AQI). Fact Sheet*. Washington DC: EPA. www.epa.gov/sites/default/files/2016-04/documents/2012_aqi_factsheet.pdf.

Archaeology and historic preservation

9.1	Environmental setting	127
9.2	Project description	128
9.3	Potential and predicted impacts	129
9.4	Mitigations and monitoring	132
9.5	Concepts and terms	133
9.6	Selected resources	136
9.7	Topic references	136

The comprehensive nature of environmental impact assessment enfolds archaeological and historical resource assessment into that of other environmental resources, including economic, social, natural, and infrastructural resources. Archaeology performed for EIA represents a form of public archaeology, with attendant expectations in terms of outreach to the public, documentation of resources, addressing research questions, and in essence, finding out what really happened in the past on a particular site or tract. It also represents addressing the custodial or managerial aspects of cultural resources at various levels from federal and state to local.

9.1 ENVIRONMENTAL SETTING

The EIA defines an area of potential effect (APE) for the project; this is incorporated into a description of the setting along with the geographical and environmental descriptors. If the setting is in a sensitive area for archaeological resources, additional assessments and mitigations may be expected. In some regulatory schemes (especially federal), the resource must be evaluated by an agency and determined to be "significant" to be officially designated a "cultural resource." Operationally, cultural resources are physical features and sites. They can include anything related to humans. Examples include cemeteries, canals, fortifications, wagon trails, mills, stone quarries, houses, campsites, and old shopping streets, without regard to regulatory or official designation.

DOI: 10.4324/9781003030713-9

Sometimes sites and sensitive areas are known to locals yet not documented officially or in state records. In other instances, sensitive sites can be found if only people look and, of course, talk to the traditional community involved. For example, a Traditional Cultural Practice (TCP) of Native Americans might require a patch of sweetgrass, so knowing where that occurs might tell you where a sensitive area is. In such cases, the site review process might be proactive in identifying areas important to people and taking action to mitigate impacts. The cultural practices, beliefs, and values are what acts to give an area or property significance. A TCP is a property that is eligible for inclusion on the National Register of Historic Places (Parker and King, 1992). Identification and evaluation of TCPs form a large part of cultural resource management in EIA.

9.2 PROJECT DESCRIPTION

The project description includes direct aspects of the project where construction or land management or other aspects of the "undertaking" will occur. It includes aspects of the project that are beyond just the "footprint" of construction or land management, such as infrastructure for things like overflow or emergency parking, stormwater discharge, and timber staging areas. These aspects should be addressed for all four phases of a project: planning, construction, operation, and closure/post-operation. The planning of projects needs to address whether they will be located in areas known to contain or likely to contain (be sensitive to) cultural resources. Construction represents the actual physical disturbance of land in the footprints of structures and features and in areas disturbed in the creation of these footprints. The operation of a project can raise potential impacts through the bringing of people onto a site or through the expansive consumption of a project as in quarrying or mining. Closure usually represents minor threats to cultural resources but can have potential impacts if the site represents an "attractive nuisance" or is not stable or secure.

Some projects falling under NEPA may also require assessments under other federal laws (Table 9.1). The EIA should address the coordinating or intersection of such laws. Federal highway improvements may require an environmental assessment to take into account the historic significance of sites or public lands. This accounting, known as a "Section 4(f) evaluation," is "a major preoccupation of DOT agencies" (King, 2000: 11).

Dams, federally funded housing, and other large projects may require archaeological assessments under various state and federal laws. Wetlands alterations can fall under Army Corps jurisdiction, with some of the review conducted at the state level. The federal Coastal Zone Management Act of 1972 (CZMA) includes scenic, cultural, and historic values in addition to natural and economic resources. The CZMA requires evaluation of coastal archaeological sites for projects under its jurisdiction. All 23 coastal states except Alaska regulate their coastal areas under the federal CZMA (Alaska has local district coastal management plans). Projects that involve federal land with cultural resources may be subject to the Archaeological Resources Protection Act of 1979 (ARPA), which requires permits for the excavation of archaeological resources on federal or Indian lands. Notably, the 1906 Antiquities Act, the Historic Sites Act of 1935, the National Historic Preservation Act of 1966, and other legislation assign federal responsibilities for managing heritage assets. Other cultural resource federal laws and executive

TABLE 9.1 NEPA and three intersecting federal laws involving cultural resources.

Federal law	NEPA	Department of Transportation (DOT) Act	National Historic Preservation Act	Land and Water Conservation Fund (LWCF) Act
Historic and cultural resource portion	Sections 101(4), 202, 1502.16(g)	Section 4(f)	Section 106	Section 6(f)
Jurisdiction	Federal permit, license, funding, undertaking	US DOT projects. ROW or easements on public property of cultural, historic, or recreational value	Federal projects that contain potential sites or structures eligible for listing on the National Register (significant)	Federal projects; anything that affects recreational lands subject to purchase or improvement with LWCF funds
Action	Address impacts in EIA, including alternatives and mitigation	Address impacts	Assess impacts, determine significance	Address impacts, coordinate with Secretary of Interior
Coordination	Coordinate with Sections 106, 4(f) and 6(f)	Coordinate with Sections 106 and 6(f)	Coordinate with NEPA	

orders may apply to federal undertakings, depending on the project description and intended action. Federal agencies manage lands that contain significant historical resources, and they often have huge collections of cultural artifacts. In 2013, the US Army Corps of Engineers estimated it had over 49,000 cubic feet of artifacts in collections from over 500 federal projects stored in 165 repositories.

9.3 POTENTIAL AND PREDICTED IMPACTS

The EIA should present the likely or potential impacts for all phases of the project, with particular emphasis on the construction and operation of the project. The first step is the documentation of known or likely resources in the project area (APE) through a site assessment.

A Phase I site assessment is when a project or APE is examined for the presence of historical or archaeological resources. A Phase II involves the delineation and extent to archaeological or other resources. Phase III refers to salvage or recovery of the resources. In the case of archaeological resources, Phase III means a destruction of the resource although it may be documented, with artifacts recovered.

In calculating impacts, the resources encountered are compared with lists of known historic and archaeological properties, and any cultural resource plans at the tribal, state, or federal level. Some plans may ascribe values to resources based on research questions and other factors not immediately apparent in the resources themselves. Consequently, an assessment should reference these plans as an aspect of evaluating potential impacts.

The cultural resource assessment report should provide documentation of the site or APE investigation. It should provide the history and prehistory from a literature review done before the fieldwork began, the results from the fieldwork and laboratory analyses, and recommendations for impact reduction. Table 9.2 summarizes a typical format for the report content.

TABLE 9.2 Contents of an archaeological or historical site assessment or survey report.

Cover, title page, documentation page. In addition to the title, author, firm or organization, client, and date, the cover of the compliance report sometimes requires specific information in accordance with the sponsoring agency. The principal investigator may also be required to sign the report. Cover page protocols may be specified in a contract or by state regulation or agency rules. The report title page should contain the site number, county, and state, and the title itself should be descriptive.
Table of Contents
List of Tables and List of Figures
Abstract/Summary. The abstract is a paragraph or two orienting the reader to the report for organizational purposed. The summary is more detailed than the abstract. The summary may be in the form of a "letter from the field" or "executive summary" and is likely the only portion of the report that many people will read. Therefore, it needs to be complete enough for the reader to understand what the report contains.
Acknowledgments (may be placed at the end of the document).
Chapter/Section 1: Introduction and Statement of Problem. This section identifies the nature of the project, including where and when it was done; the laws and regulations mandating the cultural resource assessment; the people who directed or supervised the work; the history of the undertaking and reason for the field investigation; a summary of project impacts; research design and objectives; and the contents of the report.
Chapter 2/Section 2: Environmental Background. This section describes the past structure and history of the ecological system, so that the cultures represented by the archaeological remains can be understood in their larger ecological context. It includes a description of the landforms, soils, geology, and notable environmental features and conditions.
Chapter 3/Section 3: Prehistoric and Historic Background. This section summarizes what is known about existing pre-contact and historic cultural resources within the vicinity of the proposed undertaking and provides a cultural context for understanding any existing or potential cultural resources. Table, figures, maps, and historical photographs may be included.
Chapter 4/Section 4: Field and Analytical Methods. The field method includes the number and location of units or transects, method and location of any visual surveillance of the surface ("surface collection") and excavations, including depth. Also included are mapping and sampling protocols, such as flotation (how much from what contexts, how stored, and what is screened first), screen size, and any kind of in-field discard policy (for example, deciding not to keep thousands of shards of fairly modern window glass). The analytical method treats how artifacts and other materials from the site were processed, catalogued, labeled, and prepared for curation; how individual artifacts were analyzed; and if the information is contained in a computer file in addition to hard-copy storage. The methods chapter also must state where the materials will be curated and under what conditions.
Chapter 5/Section 5: Results of Investigations. This portion of the report covers the analysis and interpretation of the archaeological data and applies this information to basic compliance issues. The archaeology is used in Phase I and Phase II analyses to support recommendations concerning the existence of sites and possible adverse effects on them. The analyses must be sufficient to answer the questions implied by the particular stage of the compliance process, as well as to satisfy any professional obligation arising from the archaeological investigation. All reports have a site or project map showing where excavations or surface collections were done and in relation to proposed construction and to the distribution of known or discovered archaeological deposits. All features and diagnostic artifacts are drawn or photographed.

Chapter 6/Section 6: Summary and Recommendations. The summary and recommendations should be quite short. They should address the next steps for the client in regard to the resources. The summary and recommendations may also be in the form of a cover letter.

References Cited. Complete references are given in a standard format for all sources of information, including books, internet sites, interviews, manuscripts, maps, and artifact collections.

Appendices. The appendices generally include the artifact inventory from the site and, for Phase I surveys, the site forms for any new sites discovered. Additional materials may include the Scope of Work (SOW), which sets out the framework for what the investigation is supposed to accomplish); the Request for Proposals (RFP), a solicitation for contract services); and various correspondence and communications, including summaries of all interviews.

Sources: Neumann et al. (2022); Sanford (2018: 187–188).

Assignment

1. The National Environmental Policy Act (NEPA), 42 USC §§ 4321–4370h, requires an environmental impact assessment for federal actions. Environmental impacts assessed are comprehensive and include historic, architectural, archaeological, and cultural resource impacts. What are three other prominent federal laws dealing with archaeological and/or historical resources that might intersect with NEPA? Provide each law's title and legal citation, and explain how they might interact with a NEPA project.
2. A museum has three Native American or indigenous people skeletons from the Archaic Period in storage. Which federal law is most likely to apply and why?
3. What are three major changes in the emphasis on professional (non-academic) archaeology now compared with 50 years ago?
4. What is "Part 800"? How does it relate to Section 106?
5. What role(s) could untrained workers or volunteers take in an archaeological assessment undertaken for NEPA compliance? Explain.
6. A historical site is on the State Historic Register but not on the Federal Register. Your project, a new power plant, is proposed for on top of that site. Does the National Historic Preservation Act (NHPA) apply? If so, how?
7. Who is the single largest employer of archaeologists in the United States? Why does this entity hire the most archaeologists?
8. Ideally, how should Section 106 of the NHPA relate to or coordinate with NEPA?
9. How is public archaeology different from academic archaeology and compliance archaeology?
10. Write a memorandum of recommendations for a project that involves a transmission line crossing the archaeologically sensitive area in Figure 9.1.

Archaeology uses physical science methods to answer social science questions. Archaeological information can be important for more than just curating the nation's past. It can be useful in determining the environmental and cultural history of a project area. It can inform matters of environmental and social justice. Most professional archaeology is done outside of academia as part of environmental legal compliance. In a world of increasing change, archaeology is growing in importance for showing how people coped with change in the past. Sea level rise, climate change, and development pressures are imperiling archaeological and other cultural resources,

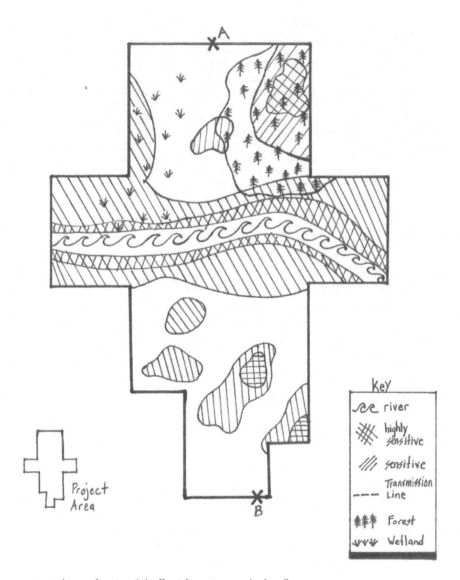

FIGURE 9.1 Area of potential effect for a transmission line.

making avoidance, preservation, conservation, and mitigation ever more important. Some successful development projects have included educational components as art or design features. Some states, such as Vermont, require a public education or awareness component as part of archaeological assessment work.

9.4 MITIGATIONS AND MONITORING

During project planning, impacts to cultural resource sites can be mitigated through redesign of the project, designation of "not-to-be-disturbed" buffer zones or green space, or recovery through an excavation or other documentation process. Mitigation can also include purchase of land, acquisition of development rights and other conservation efforts. The EIA may recommend a number of these strategies to compensate for impacts. These can include off-site actions such as protecting other historical or archaeological properties as part of an impact trade-off. Some mitigative efforts have include monitoring of earth disturbances and public education presentations. Imperiled resources may be excavated in a salvage operation that, while destructive by nature, recovers archaeological information in the form of artifacts and contextual analysis of them and features of the site.

The EIA document may have a public version with archaeological site locations concealed. Information about archaeological sites can be kept from the general public under specific conditions deemed necessary to protect the resource for the public good. For example, Section 304 of the National Historic Preservation Act (16 USC § 470w-3) restricts disclosure on the basis of significant invasion of privacy, risk of harm to the resource, or impediment on the use of a traditional religious site by practitioners. Each state has laws that exempt the location of archaeological sites from the public's "right to know." This exemption is necessary to protect sites from looting and damage.

Since there are so many potential points of intersection with the Section 106 process, the trend now is to do NEPA and Section 106 in tandem rather than sequentially as had been done in the past. The Council on Environmental Quality (2013) issued a report addressing this collaboration. Figure 9.2 shows the relationship between these two laws.

9.5 CONCEPTS AND TERMS

Archaeological site: "A location that contains the physical evidence of past human behavior that allows for its interpretation," whether listed on the National Register or not (Little et al., 2000).

Area of potential effect (APE): The "geographic area or areas within which an undertaking may directly or indirectly cause alterations in the character or use of historic properties, if any such properties exist" (36 CFR § 800.16(d)). The term is commonly used in reference to archaeological and historical resources, particularly for federally regulated projects. It implies that the environmental impacts of a project may include a larger area than just the tract of land in which a project is located.

Building: A built structure that is intended to contain people. Examples include houses, barns, outhouses, businesses, factories, churches, meeting halls, taverns, and government administrative offices. A "compound" (like a farm compound) is considered to be a building, provided all associated structures are essentially unchanged and part of the original group that functioned as a unit.

Certified Local Government (CLG): The National Historic Preservation Act of 1966 established a nationwide program of financial and technical assistance to local governments through the Certified Local Government Program. If a municipality has its own historic preservation commission and an ordinance meeting federal and state standards, including inventories of historic structures, surveys of prehistoric and historic archaeological sites, programs for public education, and activities related to comprehensive community planning (including things like the development of design guidelines), then that municipality is eligible for designation as a CLG. This greatly increases its efficiency in an EIA process.

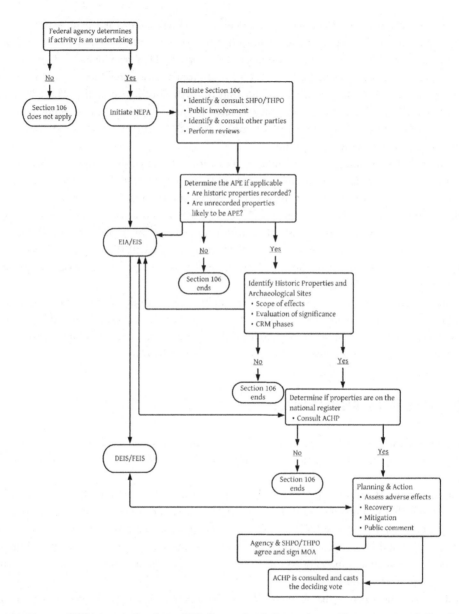

FIGURE 9.2 NEPA and Section 106 flow chart (based on Neumann and Sanford, 2010: 43).

Cultural resources: These are not clearly defined in statute but are generally considered to include architectural, historical, and archaeological resources. They also include areas valued by particular cultural groups that might not necessarily be historical, such as a place where ceremonies are currently held. Thus, a cultural resource can be a landscape in addition to an archaeological site, a building, a foundation, or other feature.

District: A collection of "buildings, sites, structures," or even "objects" unified by time, architecture, aesthetics, historical association, or cultural affiliation. Districts may be continuous (close boundary) or discontinuous (interspersed).

Effect: "Alteration to the characteristics of a historic property qualifying it for inclusion in or eligibility for the National Register" (36 CFR § 800.16(i)).

Historic property (or **historic resource**): Any "prehistoric or historic district, site, building, structure, or object included in, or eligible for inclusion on, the National Register of Historic Places, including artifacts, records, and material remains related to such a property or resource" (NHPA, 54 USC § 300308).

Integrity: How intact an archaeological or historical site is, compared to the information it contains. In effect, integrity is the ability of the property to convey its significance through its physical features and context.

Mitigation: "A way to remedy or offset an adverse effect or a change in a historic property's qualifying characteristics in such a way as to diminish its **integrity**" (ACHP, 2022). Mitigation likely emphasizes data recovery and other treatment plans.

Objects: Things like monuments, mileposts, statues, fountains, and similar location-specific items whose significance is related both to where they were placed and the purpose they served. Objects that have been moved may have their integrity and significance lowered unless they are by nature portable, such as a ship.

Significance: A key concept in the application of law to cultural resource assessments. In the Section 106 process, significance means being eligible for listing on the National Register of Historic Places. In 36 CFR § 60.4, "the quality of significance" means having **integrity** while also being associated with events, people, or information considered "important." More broadly and in somewhat looser usage, significance has come to mean being eligible for listing on the National Register, since a cultural resource eligible for such listing has the "quality of significance."

Site: A location of significant events, prehistoric or historic occupations or activities, buildings, or structures. Sites can include areas with petroglyphs, battlefields, settlements, burial mounds, and any other physical remnant of human activity. Things in the site, such as buildings or structures might be intact, in ruins, or exist only as archaeological traces.

State Historic Preservation Office (SHPO, pronounced "ship-oh"): Represents the state in identifying, reviewing, and commenting on historic and archaeological properties under the National Historic Preservation Act, NEPA, and other legislation.

Structures: Elements of the built environment that are not buildings. A bridge would be a structure, as would a highway, a railroad tunnel, a Civil War fortification, an aqueduct, or a canal, among others.

Traditional Cultural Property (TCP): An area, place, or feature of importance to Native peoples, such as where basketmaking materials were gathered, where ceremonies were held, or where other cultural functions occurred. A TCP may have importance even though there may be no actual archaeological site or building or discernable feature.

Tribal Historic Preservation Office (THPO): This is the tribal equivalent of SHPO. It represents the tribe in identifying, reviewing, and commenting on historic and archaeological properties under the National Historic Preservation Act, NEPA, and other legislation.

Undertaking: "A project, activity, or program funded in whole or in part under the direct or indirect jurisdiction of a Federal agency, including those carried out by or on behalf of a Federal agency; those carried out with Federal financial assistance; and those requiring a Federal permit, license, or approval" (36 CFR 800.16(y)).

Value: The worth or importance of a site regardless of whether the site is **significant** from a National Register perspective.

9.6 SELECTED RESOURCES

Advisory Council on Historic Preservation. (1991). *Treatment of Archeological Properties: A Handbook*. Washington, DC: Advisory Council on Historic Preservation. http://libraryarchives.metro.net/DPGTL/archaeology/1980_treatment_archaeological_properties.pdf.

Anfinson, Scott F. (2018). *Practical Heritage Management: Preserving a Tangible Past*. Lanham, MD: Rowman & Littlefield Publishers.

Braithwaite, Rosemary, D. Hopkins and P. Grover. (2001). Archaeological and Other Material and Cultural Assets. In P. Morris and R. Therivel (Eds.), *Methods of Environmental Impact Analysis*, 2nd ed. (pp. 122–144). London: Spon Press.

Hardesty, Donald L. and Barbara J. Little. (2009). *Assessing Site Significance: A Guide for Archaeologists and Historians*, 2nd ed. Lanham, MD: AltaMira Press.

Hester, Thomas R., Harry J. Shafer and Kenneth L. Feder. (2008). *Field Methods in Archaeology*. Walnut Creek, CA: Left Coast Press.

Historic Map Works. www.historicmapworks.com/. Contains a large amount of historic maps.

King, Thomas F. (2000). *Federal Planning and Historical Places: The Section 106 Process*. Walnut Creek, CA: AltaMira Press.

King, Thomas F. (2005). *Doing Archaeology: A Cultural Resource Management Perspective*. Walnut Creek, CA: Left Coast Press.

King, Thomas F. (2007). *Saving Places That Matter: A Citizen's Guide to the National Historic Preservation Act*. Walnut Creek, CA: Left Coast Press.

King, Thomas F. (2013). *Cultural Resources Laws and Practice: An Introductory Guide*, 4th ed. Lanham, MD: AltaMira Press.

King, Thomas F. (Ed.). (2020). *Cultural Resources Management: A Collaborative Primer for Archaeologists*. New York: Berghahn Books.

McManamon, Francis P. (Ed.). (2017). *New Perspectives in Cultural Resources Management*. New York: Routledge.

National Association of State Archaeologists (NASA). https://sites.google.com/view/state-Archaeologists. The Association Provides Contact Information for Each State, Commonwealth, Protectorate, and District.

National Park Service. (1995). *How to Apply the National Register Criteria*. National Register Bulletin 15. Washington, DC: National Park Service.

National Trust for Historic Preservation. This Nation-Wide Non-Profit Organization Has a Leadership Role in Advising Communities on Historical, Architectural and Cultural Resources. www.nthp.org/.

Neumann, Thomas W., Robert M. Sanford and Karen G. Harry. (2010). *Cultural Resources Archaeology: An Introduction*, 2nd ed. Lanham, MD: AltaMira Press.

Neumann, Thomas W., Robert M. Sanford and Mary S. Neumann. (2022). *Practicing Archaeology*, 3rd ed. Lanham, MD: Rowman & Littlefield.

Parker, Patricia L. and Thomas F. King. (1992). *Guidelines for Evaluating and Documenting Traditional Cultural Properties. National Register Bulletin*. Washington, DC: National Park Service.

Society for American Archaeology. www.saa.org/. The Premier Professional Society. Provides Publications, Education and Outreach.

State and Tribal Historic Preservation Offices (SHPO/THPO). These People and Their Offices Are a Point of Early Contact for EIA Research and Project Review.

White, Gregory G. and Thomas F. King. (2007). *The Archaeological Survey Manual*. Walnut Creek, CA: Left Coast Pres.

9.7 TOPIC REFERENCES

Advisory Council on Historic Preservation (ACHP). (2022). Section 106 Archaeology Guidance. ACHP. Adwww.achp.gov/.

Council on Environmental Quality. (2013). *NEPA and NHPA: A Handbook for Integrating NEPA and Section 106*. Council on Environmental Quality (CEQ) and Advisory Council on Historic Preservation. Washington DC: CEQ.

King, Thomas F. (2000). *Federal Planning and Historical Places: The Section 106 Process*. Walnut Creek, CA: AltaMira Press.

Little, Barbara, Erika M. Seibert, Jan Townsend, John Sprinkle, Jr. and John Knoerl. (2000). Guidelines for Evaluating and Registering Archaeological Properties. National Register Bulletin 36. US Department of the Interior, National Park Service.

Neumann, Thomas W. and Robert M. Sanford. (2010). *Practicing Archaeology*, 2nd ed. Lanham, MD: Rowman & Littlefield.

Neumann, Thomas W., Robert M. Sanford and Karen G. Harry. (2010). *Cultural Resources Archaeology: An Introduction*, 2nd ed. Lanham, MD: AltaMira Press.

Parker, Patricia and Thomas E. King. (1992). *Guidelines for Evaluating and Documenting Traditional Cultural Properties*. National Register Bulletin 38. US Department of the Interior, National Park Service.

Sanford, Robert M. (2018). *Environmental Site Plans and Development Review*. New York: Routledge.

US Army Corps of Engineers. (2013). Archaeological Curation in the U.S. Army Corps of Engineers. Mandatory Center of Expertise for the Curation and Management of Archaeological Collections.

TOPIC 10
Energy

10.1 Environmental setting	139
10.2 Project description	140
10.3 Environmental impacts	142
10.4 Energy management and mitigation	144
10.5 Concepts and terms	147
10.6 Selected resources	148
10.7 Topic references	149

Energy issues are a big part of environmental impact assessment. NEPA supports the development of energy-producing projects and energy efficiency in general. Its premises are:

1. Responsibility for the future. Each generation is a custodian of the environment for future generations.
2. Promoting the general commerce (this is the constitutional basis of NEPA). Energy efficiency contributes to economic value.
3. Sustainable practices. These enhance the quality of renewable resources and seek to maximize recycling of nonrenewable resources.
4. Project impact assessments. These contribute to improved energy knowledge, use, and efficiency.

NEPA CEQ Regulations Section 1502.16(e) requires agencies to address "energy requirements and conservation potential of federal actions, including various alternatives and mitigation measures." Three aspects of energy for EIA project analysis and review are the exterior site conditions, internal use of energy for buildings, and the production of energy (Figure 10.1). The site itself refers to exterior conditions such as land use, exterior configuration of plantings, berms and other surface characteristics, wind speed and direction, climate, and other factors that potentially affect energy usage. Internal conditions refer to building performance, including energy consumption and efficiency. Energy production can refer to power plants, wind turbines, solar farms, and other sources. This includes primary generation, cogeneration, and transmission/distribution.

DOI: 10.4324/9781003030713-10

FIGURE 10.1 Three energy aspects for EIA review: exterior, interior, and production.

10.1 ENVIRONMENTAL SETTING

For energy production facilities, the environmental setting provides the major context for assessing impacts. All projects involve the use of energy to some degree or another. The environmental setting may be a source of passive energy gain for a project, particularly if it includes a habitable structure or work environment. Enhanced heating and cooling may be achieved through the placement on the landscape and through managing aspects of the landscape such as berms and tree planting.

The environmental setting may also form part of the project description as in the extraction of fossil fuels, tree growth for firewood, methane recovery, wind turbines, power generation dams, solar panel fields, or other integral project. The EIA is expected to address the suitability of the setting to meet the project needs as part of the description and in the impacts section.

The environmental setting should be conducive for the type of project proposed. Wind turbine projects need sufficient space for optimal placement of the turbines in sufficient numbers to be viable as "wind farms." Often, the conditions of the setting that make it good for wind and solar farms also make these facilities visible from a distance, raising concerns about aesthetic and other impacts.

10.2 PROJECT DESCRIPTION

Projects that include construction, conversion, or use of built structures will describe and assess internal energy aspects such as the efficiency of energy supply and use in buildings. Another energy aspect is projects that produce or transfer energy through primary generation, cogeneration, and transmission. The description helps form the basis for understanding environmental impacts and therefore needs to be comprehensive and thorough. Many categories of environmental impact suggested by a project description have energy implications. Common aspects include transportation facilities, air emissions, lighting, heating, air conditioning, facility design, landscaping, water use, waste generation and disposal, recycling, power transmission, resource extraction, and processing. How projects contribute to climate change are an aspect of review.

Executive orders from the president help guide the creation of energy projects and the evaluation of impacts from them. A few of the major executive orders are listed below.

EO 14008, Tackling the Climate Crisis at Home and Abroad: establishes climate considerations as an essential element of policy, including those for mineral leasing, fuel subsidies, and land conservation (2021).

EO 13990, Protecting Public Health and the Environment and Restoring Science to Tackle to Climate Crisis: directs a new analysis of the environmental impacts of the Coastal Plain Oil and Gas Leasing Program in the Arctic National Wildlife Refuge. Revokes presidential approval of TransCanada Keystone XL Pipeline (2021).

EO 13412, Strengthening Federal Environmental, Energy and Transportation Management (2007).

EO 13221, Energy Efficient Standby Power Devices (2001).

EO 13211, Actions Concerning Regulations That Significantly Affect Energy Supply, Distribution, or Use (2001).

EO 12902, Energy Efficiency and Water Conservation: directs all federal agencies and facilities to increase efforts to conserve energy and increase energy efficiency (1994).

Many federal laws involving energy relate to projects that involve EIA. Accordingly, components of the EIA report may also address these laws. A few of the major Acts are summarized below.

National Energy Conservation Policy Act (NECPA) of 1978: Applies to federal buildings, agencies, and related federal authorities. Regulates interstate commerce, reduces growth in energy demand, and conserves nonrenewable energy resources (without inhibiting economic growth). The Act addresses metering (and reducing) energy use, life cycle cost methods and procedures, incentives for energy efficiency improvements, energy audits, training, and procurement for federal agencies.

Energy Policy Act: Goal is to "ensure jobs for our future with secure, affordable, and reliable energy." It applies to federal programs, energy assistance and state programs, energy efficient products, and public housing. The Energy Policy Act covers monitoring of energy use and works towards 2% reductions by 2015, ENERGY STAR, or Federal Energy Management Program (FEMP) designated product. Energy strategies for this Act include:

- Managing energy intensity (energy per unit of physical output in an industrial process);
- Cement or concrete with "recovered mineral component";
- Energy performance standards from ASHRAE (American Society of Heating, Refrigeration, and Air Conditioning Engineers) standards or International Energy Conservation Code standards;
- Sustainable design principles;
- Water conservation technologies;
- Energy efficient motor vehicles (including zero emission vehicles [ZEVs]);
- Photovoltaic energy commercialization on federal buildings;
- New minimum energy efficiency standards for specific residential and commercial products;
- Manufacturer and consumer tax incentives for advanced energy savings technologies and practices.

Energy Independence and Security Act of 2007: Creates a comprehensive energy strategy. The Department of Energy (DOE) is to study integrating alternative energy technologies into regional electric transmission grids.

US Green Building Council (USGBC): Based in Washington, DC. A 501(c)(3) nonprofit organization committed to a prosperous and sustainable future for our nation through cost-efficient and energy-saving green buildings.

The USGBC is the driving force of a trillion-dollar industry. Its constituency includes builders, environmentalists, corporations and nonprofit organizations, elected officials and concerned citizens, and teachers and students. It promulgates the LEED green building certification program. LEED buildings represent significant energy savings and reduce emissions that contribute to climate change (approximately 50% compared to conventional buildings, according to Monzingo and Arens, 2014). Table 10.1 contains an abbreviated LEED checklist for structures.

TABLE 10.1 EIA energy component checklist based on LEED green building certification.

√	Energy Component
	Construction activity pollution prevention measures
	Energy efficient site selection
	Development density and community connectivity
	Brownfield redevelopment
	Alternative transportation components
	Site development for passive heat gain and cooling
	Maximize open space
	Energy efficient stormwater design
	Heat island effect mitigation
	Visible light pollution reduction
	Water use reduction during construction
	Water efficient landscaping
	Innovative wastewater technology
	Water use reduction for buildings and processes
	On-site renewable energy
	Green power
	Building reuse
	Construction waste management
	Materials reuse/recycling
	Locally and sustainably sourced materials
	Efficient ventilation and air exchanges
	Air Quality Management Plan pre- and post-construction
	Energy management and control systems
	Daylight and views

10.3 ENVIRONMENTAL IMPACTS

The analysis of environmental impacts requires significant attention to trade-offs. Wind turbines are a good example of this because although they are an alternative (non-fossil) energy source, there are potential impacts associated with (and not exclusive to) bat and bird mortality, visual aesthetics, electromagnetic interference, as well as the environmental effects of constructing and installing and maintaining turbines and associated infrastructure.

The EIA should address energy efficiency for projects that involve constructing and operating any facility. Broader energy policy and program impact considerations include a variety of items to review, as suggested by Gilpin (1995: 29–36). The EPA *Energy Efficiency Reference for Environmental Reviewers* (2010) provides guidance. Other agencies will also have specific guidance documents. Tables 10.2 and 10.3 are examples of the type of checklists of potential impact categories for EIA review.

TABLE 10.2 Energy policy and program impact assessment checklist.

√	Item
	Aesthetics
	Air pollutants: synergistic, health, cumulative, and climate change
	Capital for energy improvements
	Economic factors affecting location of electricity generating units
	Energy efficiency goals and targets
	Energy exploration
	Energy impact mitigation
	Energy needs for rural and outlying districts
	Effects on future planning and development
	Effects on human health
	Effects on wildlife
	Environmental and social factors of energy choices
	Environmental implications of transmission networks
	Fuel source
	Incentives for efficiency and conservation
	Integration of energy with development and environmental management policies
	Location and environmental implications of energy sources
	Noise pollution
	Physical and cyber security of facilities
	Probable validity of energy demand forecasts
	Research and development
	Size of energy source reserves
	Sustainable development
	Taxes and subsidies for energy consumption
	Transportation and transmission of energy
	Trends in vehicle fuel consumption
	Viable conservation measures
	Water pollution
	Waste disposal

Source: Modified from Gilpin (1995: 29–30).

Assignment

Wind turbines and solar farms often generate controversy about the degree and nature of environment impacts.
1. What are the known facts about the potential effects of wind turbines on human health?
2. What percent of energy in your state is supplied by wind and solar? Is this amount changing?

TABLE 10.3 Sample landscape checklist for energy efficient sites.

√	Item
	Berms and/or conifers to the north of structures to help block wind
	Buildings cluster towards north lot lines to allow yard space to the south
	Carbon capture and sequestration
	Colors selected for ability to absorb or reflect sunlight
	Distance between structures for effective solar access
	Energy codes to be met for structures
	Little or no development of north-facing slopes
	Lot orientation favors north-south direction
	Mature height and canopy size of trees considered for project build-out
	Natural ventilation
	Orientation of streets predominantly east-west rather than north-south
	Placement of deciduous trees for summer cooling
	Site protection and security
	South-facing slopes have the highest structure density
	Tall buildings are north of shorter ones

Source: Based on Sanford (2018).

3. Research the most recent wind or solar project in your state. What were the major impact issues?
4. The National Academies of Sciences, Engineering, and Medicine (2022) has reported on potential wind turbine generator effects on marine navigation. What are the major concerns?

10.4 ENERGY MANAGEMENT AND MITIGATION

The generation of electricity accounts for about 32% of all US emissions of carbon dioxide, the most prevalent greenhouse gas (US Energy Information Administration, 2021). According to the EPA (2010), electricity generation also accounts for 65% of US sulfur dioxide emissions. These pollutants, along with nitrogen oxides, have been decreasing as a result of increased efficiency and accountability. These techniques, which range from specific measures such as using compact, long-lasting fluorescent lighting as an alternative to incandescent bulbs and using fuel-efficient vehicles and alternative transportation methods to large-scale carbon sequestration measures, are having a significant impact on improving environmental quality.

A life cycle assessment approach provides a way to compare impacts for products, materials, and processes. Traditionally, it is a tool for examining the impacts of a specific product from raw material extraction (cradle), through production and use, to disposal (grave). It requires quantification of impacts so that "apples and oranges" can be compared. The three major components of a life cycle assessment are inventory analysis, impact analysis, and improvement (mitigation) analysis.

Cost benefit analysis is addressed in the Council on Environmental Quality Regulations, § 1502.23, and may be used as part of energy source comparisons and energy efficiency measures. Many of these factors, such as emission reduction, are addressed in LEED certification (Table 10.4).

TABLE 10.4 Checklist of energy-related emission reduction categories based on LEED green building certification.

√	*Lighting*
	Best available technology (BAT)
	Energy awareness for employees/site users
	Lighting assessment report
	Minimal external lighting
	Motion sensors or timers
	Use of natural sunlight
√	*Electrical Products and Equipment*
	Adjustable-speed motors
	Energy efficient appliances
	Timers
√	*Heating and Cooling*
	Building materials minimize energy use from heating and air conditioning
	Ceiling fans or other systems to move warm air down from the underside of the roof
	Chilled water line insulation
	Computerized energy management system (EMS)
	Insulated boiler steam/hot water lines
	Load sharing to reduce energy consumption
	Low emissivity window coatings
	Natural shading from trees and shrubbery
	Smallest, most-efficient HVAC system
	Solar panels
	Use of outside air for cooling
	Variable settings for off-hours
√	*Insulation*
	Doors that minimize energy losses
	Heat exchangers for heat recovery
	Insulated windows
	Insulation materials fabricated from recycled materials
	Insulation of equipment surfaces and building walls and roof
√	*Hot Water*
	Faucet aerators
	Heat capture from hot water pipes
	Insulated, energy efficient water heaters
	Water-efficient showerheads

(continued)

TABLE 10.4 (Continued)

√	*Fuel and Gasoline*
	Alternatively fueled vehicles
	Employee vehicle reduction plan
	Fuel-efficient, cleaner burning fleet vehicles
√	*Energy Production Facilities*
	Alternative energy comparisons
	Cogeneration activities
	Environmental impact trade-offs
	Physical and cyber security
	Rate structures that reduce peak loads

The Council on Environmental Quality Regulations, § 1502.16(e) call for "Energy requirements and conservation potential of various alternatives and mitigation measures." Subsection 1502.12(f) requires discussion of "Natural or depletable resource requirements and conservation potential of various alternatives and mitigation measures."

Mitigation will be considered as part of trade-off analysis, life cycle assessment, and other tools in addressing environmental impacts of energy creation and usage. Projects that have energy management plans will incorporate efficiency management and energy use monitoring; these should all be addressed in the EIA. Periodic review to accommodate changes in technology represent one strategy that could be built into project management, and as an enhancement of the stability of a project.

A safe assumption is that all projects will continue to be examined in light of contribution to climate change, and the need to incorporate ongoing best available technology (BAT) in energy efficiency and management. Relatedly, projects and attendant EIAs will be looked at from the perspective of cumulative impacts.

Assignment

1. One aspect or category of energy for EIA is the exterior conditions of the site, including landscaping. What are the other two main categories or aspects for EIA review?
2. How does the 1978 Energy Conservation Policy Act pertain to EIA?
3. How does the Energy Policy Act of 2005 pertain to EIA?
4. What are the primary means by which the Energy Independence and Security Act of 2007 applies under NEPA?
5. Energy efficiency under EPA review for the Office of Federal Activities includes a number of methodologies. Briefly, what are these and how are they different?
   a. Life cycle cost analysis
   b. Total cost assessment
   c. Life cycle assessment
   d. Demand-side management
   e. Integrated resource planning
6. How does your state relate climate change to energy management in its environmental assessment and review policies and procedures?

10.5 CONCEPTS AND TERMS

American Society of Heating, Refrigeration and Air Conditioning Engineers (ASHRAE): A source of design standards to promote the efficient use of energy resources in construction.

Annual Fuel Efficiency (AFUE): A measure of the efficiency of heating systems in a building. The AFUE is determined by a standard US Department of Energy test method for both gas and oil equipment.

Building envelope: This includes all components of a structure with enclosed spaces to which heating and cooling are applied. Building envelope components distinguish conditioned spaces from unconditioned spaces or from outside air. For example, walls and doors between an unheated garage and a living area are part of the building envelope; walls separating an unheated garage from the outside are not.

Demand-side management: An approach favoring the consumer side of energy usage, particularly among energy utilities. This approach can consider consumer load control, storage, conservation, interruptible loads, and rate design.

Emissivity: The ratio of energy from surface to energy radiated from a perfect emitter under the same conditions.

Energy rating: This is a uniform method of ranking houses based on energy efficiency. Energy ratings range from 1 to 100 points and from 1 to 5 stars. Eighty points, the beginning of the 4-star range, is "energy efficient."

ENERGY STAR: Joint EPA and Department of Energy (DOE) program to identify and promote energy efficient products and buildings.

Green building: Designated as having efficient resource use and reduced impacts through better design and use.

Green power: "Electricity produced from a subset of renewable resources, such as solar, wind, geothermal, biomass, and low-impact hydropower" (EPA, 2010: xviii).

Heating degree days (HDD): The number of degrees that a day's average temperature is below 65 degrees Fahrenheit.

HVAC equipment: Equipment and distribution network that provide the processes of heating, ventilating, or air conditioning to a building.

Integrated energy systems: "A system design that brings together gas-fired and electrically driven equipment to provide heating, cooling, dehumidification, and electrical service to commercial and public buildings" (EPA, 2010: xix).

Leadership in Energy and Environmental Design (LEED): A certification program for green building design available from the US Green Building Council.

Life cycle costing: An evaluation of trade-offs to determine the point of least cost. The analysis includes "cradle-to-grave" aspects of products and systems, including the environmental and other costs of final disposal. Taking the energy costs expected over time and applying them to the initial construction costs will result in a total cost for each option. Mitigative measures are considered in the analysis. The option that costs the least over a period of years is the most cost-effective and energy efficient choice.

R-value: A measure of thermal resistance, or how well a material resists the flow of heat. Insulation materials are commonly rated and labeled using an R-value. Higher numbers indicate superior performance. R-values are additive, meaning adding an R-19 layer to an R-11 layer creates an R-30 value.

Solar gain: Energy benefit from the sun. A structure oriented towards the south benefits from increased solar gain. A super-insulated house balances out heat loss with solar gain through efficient design and mechanisms of thermal storage.

Thermal value: The insulation level in the roof, foundation, and exterior walls.

U-value: A measure of how well a material conducts heat. Windows and doors are typically rated using the U-value rather than the R-value. Lower numbers mean better performance and less heat loss.

Whole building design: "A concept that forms the entire building stakeholder community into a team at the beginning of a project. The team examines project objectives, building materials, systems, and assemblies from various perspectives, and improves the final building product by integrating the components into a more energy efficient system" (EPA, 2010: xxi).

Xeriscape: Landscape design and maintenance that uses indigenous plans and site configuration to reduce the need for irrigation.

10.6 SELECTED RESOURCES

American Council for an Energy-Efficient Economy. http://aceee.org/. In Addition to Other Resources on This Site. http://aceee.org/sector/state-policy Allows a Comparison Among State and City Policies and Resources.

American Society of Heating, Refrigerating and Air Conditioning Engineers, Inc. (ASHRAE). www.ashrae.org. Standard 90.1 *Energy Standard for Buildings Except Low-Rise Residential Buildings*. This Can Be Found Online at Various Sites. It May Be Incorporated into Your State or Community's Code.

American Wind Energy Association. This Is a Trade Group for Wind Energy. www.awea.org/.

Bass, Ronald E., Albert I. Herson and Kenneth M. Bogdan. (201). *The NEPA Book: A Step-by-Step Guide on How to Comply with the National Environmental Policy Act*, 2nd ed. Point Arena, CA: Solano Press.

Building Codes Assistance Project (BCAP). This Non-Profit Advocacy Group Promotes "the Adoption, Enforcement, and Compliance of Building Energy Codes and Standards." In Addition to Many Other Resources, It Maintains a Table of the Current Status of Each State's Building Code. http://bcap-energy.org/code-status/state/.

Center for Energy Efficiency and Renewable Technologies. This Group of Environmental Groups and Clean Energy Companies Focuses on "Global Warming Solutions for California and the West." http://ceert.org/.

Center for Renewable Energy and Sustainable Technology (CREST), Renewable Energy Policy Project Washington, DC. www.crest.org/index.

Consumer Federation of America Foundation, Washington, DC. www.buyenergyefficient.org.

Eccleston, Charles H. (200). *Environmental Impact Statements: A Comprehensive Guide to Project and Strategic Planning*. New York: John Wiley & Sons.

Energy and Environmental Building Association. www.eeba.org.

Energy Information Administration, EI 30, 1000 Independence Avenue SW, Washington, DC, 20585. www.eia.doe.gov/.

Environmental Protection Agency. (2010). *Energy Efficiency Reference for Environmental Reviewers: Guidance for EPA Staff*. EPA Office of Federal Activities. EPA Pub No. 315-R-09-001. Washington, DC: US Environmental Protection Agency.

Federal Transit Administration, 400 7th St SW, Washington, DC, 20590. www.fta.dot.gov

International Code Council, International Energy Code. http://publiccodes.cyberregs.com/icod/iecc/2012/. The 2012 Code Increased Efficiency Over the 2006 Edition by 30%.

LEED. Leadership in Energy and Environmental Design, US Green Building Council. www.usgbc.org/leed.
Martinez, Daniel M., Ben W. Ebenhack and Travis P. Wagner. (2019). *Energy Efficiency: Concepts and Calculations*. Amsterdam: Elsevier.
National Research Council. (2007). *Environmental Impacts of Wind-Energy Projects*. Committee on Environmental Impacts of Wind Energy, National Research Council. Washington, DC: National Academies Press.
Northeast Sustainable Energy Association. 23 Ames Street, Greenfield, MA 01301.
Rocky Mountain Institute. 1739 Snowmass Creek Road, Snowmass, CO 81654–9199.
Science Applications International Corporation. (1994). *Energy Efficiency Reference for Environmental Reviewers. Office of Federal Activities, US Environmental Protection Agency*. Washington, DC: US EPA.
U.S. Climate Change Science Program/U.S. Global Change Research Program. Suite 250, 17117 Pennsylvania Avenue NW, Washington, DC, 2006. www.usgcrp.gov
U.S. Department of Energy, Federal Energy Management Program. CE-44, 1000 Independence Avenue, SW, Washington, DC 20585.
U.S. EPA, Energy Star Buildings Program. U.S. EPA, Energy Star Computers Program. U.S. EPA, Green Lights Program.
US Green Building Council (USGBC). This Non-Profit Organization Supports Cost-Efficient and Energy-Saving Green Buildings. It Is Responsible for the LEED Green Building Certification Program.

10.7 TOPIC REFERENCES

Energy Information Agency. (2021). Frequently Asked Questions. EIA.gov.
Environmental Protection Agency. (2010). *Energy Efficiency Reference for Environmental Reviewers*. Washington, DC: Environmental Protection Agency.
Gilpin, Aland. (1995). *Environmental Impact Assessment (EIA): Cutting Edge for the Twenty-First Century*. Cambridge: Cambridge University Press.
Monzingo, Louise and Edward Arens. (2014). Quantifying the Comprehensive Greenhouse Gas Co-Benefits of Green Buildings. Final Report for California Air Resources Board and California Environmental Protection Agency. Center for the Built Environment. UC Berkeley. https://escholarship.org/uc/item/935461rm.
National Academies of Sciences, Engineering, and Medicine. (2022). *Wind Turbine Generator Impacts to Marine Radar*. Washington, DC: National Academies Press.
Sanford, Robert. (2018). *Environmental Site Plans and Development Review*. New York: Routledge.
USGBC. (2021). US Green Building Council. www.usgbc.org/about/brand. Contains LEED Rating System.

TOPIC 11

Noise impact analysis

11.1	Environmental setting	150
11.2	Project description	151
11.3	Potential impacts	152
11.4	Mitigation and monitoring	155
11.5	Concepts and terms	158
11.6	Selected resources	159
11.7	Topic references	159

11.1 ENVIRONMENTAL SETTING

Environmental setting of a noise assessment includes most areas in and near a project that may receive or create sound. Sounds that are considered objectionable or harmful are usually identified as noise. Sounds that are of a continuous and desired volume may be called music, amplified speech, or another acceptable sound source. The *source* is the thing or action that generates the noise. The *receptor* category includes primarily humans but also could be wildlife populations, hospitals, schools, or other sensitive places. The environmental setting to be considered or evaluated includes land topography and the media through which noise transmits, which is usually air. Some sound settings convey noise more readily than others; canyons channels are clear examples. The field of "soundscape ecology" is proposed by researchers as a way to address the relationship between landscape and sound (Pijanowski et al., 2011).

Land development, various modes of transportation, and special events have led to general increases in noise among human settlements and the infrastructure that serves them. Background noise levels are a mix of several noise sources and are increasing in many communities or other inhabited places. This trend is affecting animal and human populations. Even marine mammals in ocean environments are affected by noise pollution (Weilgart, 2018). The US Environmental Protection Agency (1978) generally defines noise as unwanted sound. At its worst, noise moves beyond nuisance to become a form of physical pollution that raises stress levels, causes headaches, harm to hearing, and disrupts pets and wildlife. The Noise Control

DOI: 10.4324/9781003030713-11

Act of 1972 and the Quiet Communities Act of 1978 (PL 95–609) set federal policies to measure and control noises. Other federal and state legislation applies, but most outdoor noise issues occur locally and must be handled by local governments with local regulations, primarily through noise ordinances.

The setting portion of an EA contains information about varying degrees of sensitivities for measurement of noise impacts. Sometimes, Traditional Cultural Practices (TCPs) in designated areas or sites may engender a more significant potential for noise or even greater impacts such as prayer or meditation. Noise can interrupt or intersect with sacred activities and disturb the setting of sacred places. Such places may exist implicitly in scenic vistas or other special areas, even if not formally designated as such in agency resource management plans.

11.2 PROJECT DESCRIPTION

The EA project description should include all the potential sources of noise pollution that may be predicted. All phases of a construction project should be evaluated for noise impacts including planning, testing, building construction, operation, and closure or post-operations. Construction and operation are most likely to receive primary consideration. Projects that involve construction or physical operations (especially highways, rail routes, aircraft, mining, and industrial use) can expect noise as a potentially significant impact for sound coming from the project. Another category is noise received; sound coming to the project. A nature preserve proposed for an urban area comes to mind. A noise assessment includes the effects of noise on sensitive environmental resources (e.g., nesting birds) and other noise receptors (e.g., sensitive human populations, neighborhoods, health centers) at or adjacent to the project setting as part of predicting and evaluating noise impacts.

Strategic environmental impact assessments (SEAs) may be prepared for particular types of projects. The modeling for these SEAs will engender assumptions that might deviate from the estimations for specific project EIAs. This has been found to be true for airport noise modeling to develop noise contours, and indicates the need for on-site studies to verify individual predictions (Torija et al., 2018). Noise impact software is tailored for other specific project types, such as military training, national parks, highways, and underseas (e.g., see https://highways.dot.gov/public-roads/spring-2021/modern-tool-noise-analysis and www.nps.gov/subjects/sound/rm47-part-5-impact.htm.

Discussion of impacts and mitigation measures here is based on information in Sanford (2018). Impact assessment strategies include establishing noise monitoring stations and designating sensitive receptor locations (human and wildlife). Potential noise sources are identified from the project description and the type of project. Background noise levels are determined. Through the lens of noise regulations and standards, the project noise is assessed and potential impacts are determined. After applying mitigation via some combination of source reduction, environmental setting modification, and receptor management, the remaining impacts are the "residual impacts," which form the basis for determining the significance or acceptability of these final impacts.

11.3 POTENTIAL IMPACTS

A noise impact analysis seeks to answer the following questions:

1. What are the sources of noise from the proposed project, and are they from transportation, industrial processes, commercial activity, construction, cooling systems, households, people, or animals?
2. What kinds of noise are involved, and how are they characterized by conditions and occurrence?
3. Who is exposed to the noise? Are they sensitive receptors such as elderly persons, children, operators, passengers, bystanders, hospitals, schools, neighboring residences, or workplaces? Do they affect threatened or endangered species? What are the biodiversity implications? What is the length of exposure and the proximity to the source?
4. How are noise levels and impacts on receptors compared with accepted noise standards?
5. What mitigation measures are proposed by the project managers?
6. Are other mitigation measures recommended?
7. Are the residual noise impacts acceptable after mitigation?

Noise impacts are evaluated in three ways, often combined. These approaches are experimentation, modeling, and expert opinion. The first method is to conduct an on-site experiment with collection of data and analysis from "real world" noise sources. The experiment involves demonstrating a noise source similar to what is likely to come from the proposed development. In some cases, a similar noise test may not be practical for comparison and modeling or expert opinion is used instead. Modeling compares the proposed project with a similar one already built or by setting up a noise source substitute. The substitute must be a reasonable comparison. It may include sound maps with decibel contours, and computer simulations.

Experts, typically sound engineers, can ascertain the technical, quantitative, and qualitative aspects of potential noise impacts. Only a few states have formal licensing of professional sound engineers. Often, civil and other environmental engineers are the primary authors of noise studies as part of their other site assessment work. Other experts include planners, equipment operators, and project or construction managers. People who have lived near developments similar to the proposed project may have a certain degree of expertise. The perception of neighbors and "sensitive receptors," regardless of expertise, is valuable.

Traffic noise impact is a common concern in impact assessment of projects with large numbers of users. One technique is to develop a traffic noise index based on measurements of peak noise (the level exceeded 10% of the time) and background noise levels (the level exceeded 90% of the time). Noise indices can also be used for industrial activities on a property and are useful for OSHA compliance.

An agency preparing an EIA report or EIA may have some outcome objectives in the agency regulations or guidelines. These objectives should be addressed in the evaluation of impacts and associated mitigation. Agency noise outcome questions might include items such as:

1. Would there be exposure of persons to or generation of noise levels in excess of standards established in the local general plan or noise ordinance, or applicable standards of other agencies?

2. Would there be exposure of persons to or generation of excessive ground-borne vibration or ground-borne noise levels?
3. Would a substantial permanent increase in ambient noise levels in the project vicinity above existing levels occur without the project?
4. Would a substantial temporary or periodic increase in ambient noise levels in the project vicinity above existing levels occur without the project?
5. For a project located within an airport land-use plan or where such a plan has not been adopted, within 2 miles of a public airport or public use airport, would the project expose people residing or working in the project area to excessive noise levels?
6. For a project within the vicinity of a private airstrip, would the project expose people residing or working in the project area to excessive noise levels?

Assignment

Present an outline for a noise section of an EIS or EIA Report for expansion of a small city airport. What categories of impact should be in this outline?

Noise originates as sound in the form of pressure waves and is measured in decibels on one of three scales: A (low), B (intermediate), and C (high intensity). Most environmental assessments use the "A" scale, which simulates the response of the human ear but does not accurately measure impulse noise, such as a car crash, or high-intensity or low-frequency noises such as might emanate from explosives, large fans, engines, earthquakes, and avalanches. The "C" scale is equally responsive to low-pitched sounds and mid- and high-pitched sounds. Consequently, some communities across the country are converting to the C scale for ensuring compliance with local noise ordinances.

Decibel levels (measured as dBs) increase on a logarithmic scale rather than an arithmetic scale. For example, suburban street noise (70 dB) is not 10% louder than ordinary speech (60 dB); the difference will be perceived by a listener as perhaps a doubling of loudness (Table 11.1). Combining these sounds by adding suburban street noise to ordinary speech (70 dB to 60 dB) might make a new reading of 70.5 dB.

The Occupational Safety and Health Act established the Occupational Safety and Health Administration (OSHA), which regulates the construction of a project that involves construction or land alteration, and the operation of a project that involves employees. Under OSHA, the Environmental Protection Agency (EPA) sets permissible noise exposure levels (Table 11.1). These are based on an assumption that, after exposure, people have a quiet place to go to while they recover from the noise. Many European countries have more stringent standards, as may some US states and local governments.

Maximum exposure levels for humans are tolerable, but noise level preferences are the most common goals for builders. Both should be addressed in the EIA if workers or visitors will be present in any project phase. The US Department of Housing and Urban Development (HUD) sets noise preference levels for its projects (Table 11.2). These levels provide a guide to local governments, particularly in residential and urban areas. Noise preference levels depend on the time of day that sounds are heard. Most people expect and prefer that sound levels at night are less than daytime levels, so it is useful to have a single reading that considers this preference. The US Environmental Protection Agency (EPA) developed a day–night level (L_d) by using a formula that accounts for the increased annoyance effect of nighttime noises. The HUD online Day/Night Noise Level (DNL) Calculator determines DNL from roadway and railway traffic.

TABLE 11.1 EPA/OSHA standards for permissible noise exposure limits on construction sites.

Hours per Day	Decibels (A Scale)	Equivalent
8	90	Jackhammer
6	92	Heavy city traffic, large truck
4	95	Subway train at 200 feet
3	97	Boeing 737 at 1 nautical mile
2	100	Bulldozer, factory
1.5	102	Motorcycle
1	105	High school dance
0.5	110	Average pain threshold
0.25	115	Rock concert, leaf blower

Source: Occupational Noise Exposure, 29 CFR § 1926.52(d)(1)).

TABLE 11.2 HUD noise preference levels.

Clearly unacceptable	Exceeds 80 dB for 1 hour per day
	Exceeds 75 dB for 8 hours per day
Normally unacceptable	Exceeds 65 dB for 8 hours per day
	Produces loud, repetitive sounds
Normally acceptable	Does not exceed 65 dB for more than 8 hours per day
Clearly acceptable	Does not exceed 45 dB for more than 30 minutes per day

Source: US Department of Housing and Urban Development (www.hud.gov/).

It is available to the public, so even a small project EIA can take advantage of this resource (www.hudexchange.info/environmental-review/dnl-calculator/).

An EIA noise study should also address noise exposure for other species, such as pets. Animal populations will have different sensitivities, of course, and these will be affected by factors such as reproductive sites (notably nurseries and nesting areas). Kunc and Schmidt (2019) examined the effects of noise on 109 species, finding a range of significant responses to animals from across all taxonomic groups (amphibians, arthropods, birds, fish, mammals, mollusks, and reptiles). The type of project directly relates to the effects on species. For example, heavy shipping may affect marine mammals and fish, particularly species that use sound to locate spawning grounds or to aid in communicating during spawning (de Jong et al., 2020).

Environmental noise studies have a standard format, which evolved after years of concern about how transportation and industrial noise sources are managed. "Isochrone" graphs can show noise levels at specific distances from a source, thereby creating contour maps (Figure 11.1). For help in understanding a noise study, one should retrace the steps typically undertaken in preparing a noise impact report, starting with a definition of the project, then the gathering of background data and predicting findings. All this is to consider the necessity of formulating noise mitigation practices.

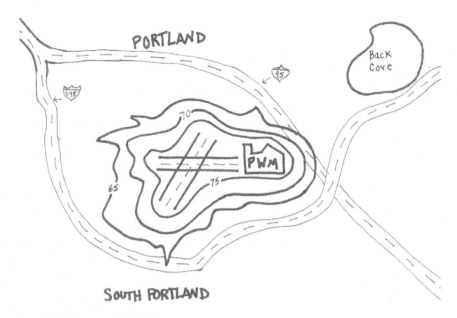

FIGURE 11.1 Airport noise contours.

A noise contour diagram shows noise level boundaries and works much like a lighting level diagram. The noise assessment report accompanying the diagram should have clear standards, with an explanation of how the predicted noise levels were derived and what the implications there are for the community. Sensitive receptors and sound monitoring stations may be indicated on the contour maps to help reviewers interpret potential impacts. Noise contours are useful ways to indicate the differences between unmitigated and mitigated impacts and for comparison among project alternatives. Since a 3 dB increase corresponds to a doubling of sound energy, the contours may be in intervals of three. A 10 dB increase represents a tenfold increase and may be used as well. An increase of 3 dB cuts the average person's "listening area" in half, and a 10 dB increase cuts it by 90%.

Assignment

Locate the nearest city airport.
1. What is the city?
2. How busy is the airport?
3. Does it have a noise contour diagram associated with it? If so, obtain it and comment on it.
4. If not, make a potential noise contour diagram that could be used for managing noise impacts; explain your suggested contours.

11.4 MITIGATION AND MONITORING

Sound pressure waves do not become noise until they are heard. A person's perception of noise involves both psychological as well as acoustical factors. Psychologically, noise measured below the levels at which adverse health impacts result is "annoyance" noise. There is no single way to create a scale that measures annoyance noise, such as loudness or decibel levels. Noise has both acoustic and non-acoustic components. Aspects of annoyance noise include: (1) the source of the noise; (2) differing individual perceptions of noise and emotional responses to noise; (3) the time of day the noise occurs; and (4) the degree of interference with personal activities.

The measurable features of noise include loudness, intermittency, duration (impact versus steady state) and frequency. Thus, the degree of annoyance noise does not depend entirely on the loudness of the sound. The duration and intermittency also matter. People often judge impulse noises, such as gunshots, as "noisier," or more unwanted, than non-impulse noise having the same total energy. Duration may be represented in community noise standards in the form of "sound level equivalents" (Leq), an averaging of the sound level for either 1 minute or 1 hour. Sound frequency is measured in hertz and is closely related to pitch.

A project proponent should propose measures to reduce or mitigate noise impacts as part of the impact assessment. Reviewers can make suggestions and regulators can impose conditions as part of the approval process.

The source of noise can be modified, the environmental setting of noise transmission can be modified, and, at least theoretically, the receptors can be modified. Modifications to site design and the use of berms (mounds of dirt), plantings of trees and shrubs, and sound walls can reduce the impacts of a development on neighboring properties. Table 11.3 shows an example of possible mitigative measures for a modest-sized gravel pit operation.

Assignment

1. Suggest additional mitigation options for Table 11.3.
2. What are some likely seasonal differences that could affect mitigation?

The assessment analysis may include recommendations for monitoring noise once construction has begun and after the project is completed or in full operation. This may include noise level readings at the site property line at appropriate times and comparing them to the conditions of approval in the project review.

Sound meters can be used by almost anyone, and they provide immediate information. People with smartphones can download a free application and measure decibels with the phone. This can be a quick assessment to determine whether or not more selective testing should be done. Even so, sound meters on smartphones are accurate to ±1 decibel (Celestina et al., 2018).

A qualitative measure of noise easily deployed is the "walk-away" noise test, which may be reported in an EIA from site visits that establish a noise level baseline after construction of a new project or at an existing facility contemplating an addition. The test requires only two people with average hearing and average voice levels, a 100-foot tape measure, and some unfamiliar reading material. One person reads while the other backs slowly away until he or she can only hear a word or two over a period of 10 seconds or more. The distance is then measured. The people switch roles and the distances are averaged after at least a few trials.

TABLE 11.3 Sample options for noise mitigation at a gravel pit with a crusher and screener.

Changes to Project	Changes to Site	Changes to Operation
Insulate crusher jaws	Recess crusher/screener base to reduce elevation	Pick a shorter time for more intensive operation
Use rubber-lined belt for screener	Periodically move crusher to reduce trip distance to and from the excavation site	Limit hours, days, seasons
Off-site processing of materials	Use berm as sound barrier	Reduce number of vehicles/equipment operating at any one time
Change size or type of crusher, screener	Plant conifers	Replace standard backup beepers with motion-sensing "smart alarms"; both are acceptable to OSHA
Reduce truck size	Relocate noisy operation further away from neighbors	Use "scoop-bed" trucks that do not have tailgates
Reduce number of truck trips	Sound barrier fence	Re-route or reschedule truck trips
Reduce on-site processing rate	Construct stockpile barrier	Change non-noise related aspects of the project to improve image
Reduce extraction rate		Use "white noise" generation devices
		Monitor noise levels before and after changes
		Neighbor notification policy that can include mediation

Mitigation of nighttime noise is a common practice for impact reduction. People tend to like their evening hours at least 10 decibels below daytime noises. Sleep hours are conservatively defined as 10:00 pm to 7:00 am.

Assignment

Explore the mapping of sound. Go online to www.nps.gov/subjects/sound/soundmap.htm. These sound maps are based on the L50 sound pressure level (decibels on the A scale, 20 µPa).
1. What is L50, and how is this term used?
2. What is the quietest region of the United States?
3. To answer question 1, should you use the "map of existing conditions" or the "map of natural conditions"? Explain your answer.
4. Assume you are preparing to do a noise assessment for a proposed 20,000-acre national park in the quietest area of the United States. Identify potential noise sources, receptors, and mitigative measures using Table 11.4.

TABLE 11.4 Potential noise for a national park.

Possible Noise Source	Potential Sensitive Receptors	Potential Mitigation

The US Department of Housing and Urban Development (HUD) has an online labeled calculator for Day/Night Noise Level (DNL). It can be used to determine DNL from roadway and railway traffic. It is designed for the environmental review of HUD-assisted projects and is available to the public (www.hudexchange.info/environmental-review/dnl-calculator/). The noise analysis should lead to a determination of whether or not changes to the type of project, the site design, or operation of the project are appropriate to mitigate the potential noise impacts on receptors. Monitoring and enforcement of the recommended conditions should be addressed necessary to ensure that the noise levels fall at or under the anticipated conditions of impact. Best available techniques (BAT) should be incorporated into the noise management scheme. The noise impact assessment modeling system should generate measures for mitigation and monitoring.

11.5 CONCEPTS AND TERMS

Ambient noise: Background level, used for comparison of sounds.
Associational noise: This is noise associated with unpleasant events or that conveys displeasing information. A good example is the sounds of breaking glass and crumpling metal.
Aviation Environmental Design Tool (AEDT): Replaced the Integrated Noise Model (INM) in 2015 as a software system used by the Federal Aviation Administration to estimate noise, fuel consumption, emissions, and other aspects of aircraft performance for airports.
Best available technology (or **techniques**) (BAT): The best practicable means to control pollution. A cost-benefit analysis is usually included as a way of determining what is reasonable for BAT.
Listening area: The spatial area in which a person can hear natural sounds—the sounds of nature.
Noise level isobar: A line on a map connecting points that mark the boundary of a particular decibel level.
Pure tone frequencies: Constant frequency tones may call attention to themselves, even at 10 to 15 decibels below a noise preference level, because such sounds resonate. Electrical utilities and factory whining sounds may produce pure tone frequencies, and mitigation should be considered.
Sudden or startling noises: Sudden noise actions like railroad air horns are efficient attention-getters. Such sounds as whistles and fire alarms are often a necessity but some sudden noise sources may allow or even require mitigation.
Visibility of noise source: The sight of a noise producer influences perception. There may be other factors associated with seeing the source. For example, vegetation screening is usually

perceived as a noise reduction technique, but in fact, such a measure has no effect on decibel levels. A large exhaust stack may be perceived as an unsightly symbol of neighborhood blight, thereby making the sounds of exhaust seem more irritating or louder. The perception need not always be negative; a noise source could be a reminder that much-needed jobs are being brought to the community.

11.6 SELECTED RESOURCES

Berger, Elliot H., Larry H. Royster, Julia D. Royster, Dennis P. Driscoll and Marty Layne (Eds.). (2000). *The Noise Manual*, 5th ed. Falls Church, VA: AIHA Press.

Canadian Centre for Occupational Health & Safety. (n.d.). OSH Answers Fact Sheets. www.ccohs.ca/oshanswers/phys_agents/noise_measurement.html.

Federal Aviation Administration (FAA). www.faa.gov/airports/environmental/airport_noise/. This Site Provides Checklists for the Effects of Airports on Communities and for Airport-Related Projects. It Includes Related Information for Noise Planning.

Federal Highway Administration (FHA). www.fhwa.dot.gov/environment/noise/index.cfm. This Site Pertains to Highway Traffic Noise and Mitigation. It Covers Everything from Effects on Wildlife to Construction Barriers and General Planning for Noise Management.

Institute of Noise Control Engineering (INCE). www.inceusa.org/. This Non-Profit Professional Organization Publishes the *Noise Control Engineering Journal*, and Promotes Noise Solutions.

Journal of the Acoustical Society of America. https://asa.scitation.org/journal/jas.

Noise Pollution Clearinghouse. www.nonoise.org//. This Is a National Non-Profit with a Comprehensive Listing of Resources and Materials for Dealing with Noise.

Schultz, T. J. and N. M. McMahon. (1971). *HUD Noise Assessment Guidelines*. Washington, DC: Government Printing Office.

Singapore National Environment Agency. (2016). Technical Guide for Land Traffic Noise Impact. www.nea.gov.sg/docs/default-source/our-services/technical-guidelines-for-noise-impact-assessment-.pdf.

Sordello, R., F. Flamerie De Lachapelle, B. Livoreil and S. Vanpeene. (2019). Evidence of the Environmental Impact of Noise Pollution on Biodiversity: A Systematic Map Protocol. *Environmental Evidence*, 8, 8. https://doi.org/10.1186/s13750-019-0146-6.

Therivel, R. and M. Breslin. (2001). Noise. In Peter Morris and Riki Therivel (Eds.), *Methods of Environmental Impact Assessment*, 2nd ed. (pp. 65–82). London: Spon Press.

US Department of Housing and Urban Development (HUD). (n.d.). HUD Exchange DNL Calculator. www.hudexchange.info/programs/environmental-review/dnl-calculator/. Contains an Online Noise Calculator Along with General Environmental Assessment-Related Information for HUD-Assisted Projects.

US Department of Labor. (n.d.). Occupational Noise Exposure [29 CFR 1926.52(d)(1)] Occupational Safety and Health Administration. www.osha.gov/pls/oshaweb/owadisp.show_document?p_table=STANDARDS&p_id=10625.

Zwerling, E. M. (2000). *Regulation of Amplified Sound Sources. Proceedings of Noise-Con 2000*. Newport Beach, CA: Acoustical Society of America/Institute of Noise Control Engineering.

11.7 TOPIC REFERENCES

Celestina, M., J. Hrovat and C. A. Kardous. (2018). Smartphone-Based Sound Level Measurement Apps: Evaluation of Compliance with International Sound Level Meter Standards. *Applied Acoustics*, 139, 119–128. https://doi.org/10.1016/j.apacoust.2018.04.011.

de Jong, Karen, Tonje Nesse Forland, Maria Clara P. Amorim, Guillaume Rieucau, Hans Slabbekoorn and Lise Doksaeter Sivle. (2020). Predicting the Effects of Anthropogenic Noise on Fish Reproduction. *Reviews in Fish Biology and Fisheries*, 30, 245–268. https://link.springer.com/article/10.1007/s11160-020-09598-9.

Environmental Protection Agency. (1978). *Protective Noise Levels*. Washington, DC: EPA Office of Noise Abatement & Control.

Kunc, Hansjoerg P. and Rouven Schmidt. (2019). The Effects of Anthropogenic Noise on Animals: A Meta-Analysis. Biology Letters. *The Royal Society*. November 20, 2019. https://royalsocietypublishing.org/doi/10.1098/rsbl.2019.0649,

Occupational Noise Exposure, 29 CFR 1926.52(d)(1)).

Pijanowski, B. C., L. J. Villanueva-Rivera, S. L. Dumyahn, A. Farina, B. L. Krause, B. M. Napoletano, S. H. Gage and N. Pieretti. (2011). Soundscape Ecology: The Science of Sound in the Landscape. *BioScience*, 61(3). March 2011, 203–216. https://doi.org/10.1525/bio.2011.61.3.6.

Sanford, Robert. (2018). *Environmental Site Plans and Development Review*. New York: Routledge.

Torija, Antonio J., Rod H. Self and Ian H. Flindell. (2018). Airport Noise Modelling for Strategic Environmental Impact Assessment of Aviation. *Applied Acoustics*, 132, 49–57.

US Department of Housing and Urban Development. Noise Preference Levels. www.hud.gov/.

Weilgart, Lindy. (2018). The Impact of Ocean Boise Pollution on Fish and Invertebrates. Ocean Care & Dalhousie University. OceanCare.Org. www.oceancare.org/wp-content/uploads/2017/10/OceanNoise_FishInvertebrates_May2018.pdf.

TOPIC 12
Aesthetics and visual impact analysis

12.1	Environmental setting	161
12.2	Project description	162
12.3	Potential or predicted impacts	162
12.4	Mitigation and monitoring	168
12.5	Concepts and terms	168
12.6	Selected resources	169
12.7	Topic references	170

12.1 ENVIRONMENTAL SETTING

The setting for a project has physical aspects (land and water, vegetation, structures) and cultural aspects (uses, perceptions, associations, values, expectations, regulations; Sanford, 2018: 155). An EIA assesses a development project in the context of its setting. We all make conscious or unconscious assessments of the physical conditions in a setting. The setting may have been previously evaluated or designated in terms of aesthetic or scenic quality. An agency may have guidelines or procedures for describing and assessing settings. The Bureau of Land Management (BLM) was an early adopter of visual assessment techniques. It uses Visual Resource Management Classes, with specific objectives for land in each classification (BLM, 1984). A viewshed analysis may be part of such classifications to identify scenic resources on a map. Such viewsheds may provide a basis for comparison with particular projects. However, the setting may be evaluated for the first time for a project-specific EIA report and result in new viewsheds.

An environmental setting may contain a Traditional Cultural Practice (TCP) that has sensitivity to visual and noise intrusions. A helicopter or low-flying aircraft over an outdoor Native American ceremony is an obvious example of a project affecting a TCP. The area of potential effect (APE) may vary quite a bit depending on the range of aesthetic concerns and sensitive receptor sources.

DOI: 10.4324/9781003030713-12

The setting may include foreground, middle ground, and background, which may be collectively assumed as the entire area of visibility within a 5-mile radius, since that is a distance zone historically referenced by the US Forest Service (1973). Large or high-impact projects may require analysis of impacts for a much greater background area; all this should be explained in the individual EIA.

12.2 PROJECT DESCRIPTION

The description of a project, in conjunction with the setting, forms the basis for developing a range of potential aesthetic impacts. Visual assessment studies should be prepared for physically large projects, highly visible projects, visually sensitive or controversial projects, and projects in sensitive areas. Certain types of projects are associated with specific types of impacts that require special treatment in an EIA. Wind energy projects come quickly to mind (Palmer, 2015). The Bureau of Land Management (BLM, 1986) suggests classifying projects on the basis of landform and water features, vegetative features, and structural features. Landform would include roads, mining, gravel pits, and landfills. Water features emphasize surface waters and projects such as shoreline stabilization, water impoundments, and channelization. Projects that emphasize vegetative features include timber harvest, grazing management, plantings, and other vegetative "manipulations." Structural features include wind turbines, buildings, water tanks, transmission lines, oil and gas developments, recreational facilities, and microwave stations.

The Visual Impact Assessment (VIA) should address the four potential phases of a project: planning/testing, construction (if there are built environment or land alteration components), operation, and closure/reclamation. For most projects, emphasis will be on the middle phases of construction, and operation.

12.3 POTENTIAL OR PREDICTED IMPACTS

Aesthetic impacts can include most of the human senses, particularly smell, hearing, and vision. If they are severe enough, impacts of smell, hearing, and vision can transcend from matters of aesthetics to matters of human health. We address sounds and hearing under a separate unit: noise.

The setting may be described in the EIA in terms of viewsheds of and from a particular project or parcel of land containing a project. A municipality or agency plan may contain identified scenic areas on a map or in a description. Such maps and descriptions form the basis for comparison with impacts and in developing mitigations. Viewsheds are also used in visual impact studies by developers and consultants. Viewshed areas can be identified for specific types of management and thereby provide a background or context for the review of projects. A viewshed map (Figure 12.1) may include viewing angles, distance of views, recipients of views, and duration of views (when adjacent to a transportation route). Computer simulations, GIS mapping, photomontages, drawings, narratives, assessment forms, demonstrations in the field, and physical models can all be used to show the environmental setting with or without the project.

BLM terms reflect VIA practices that evolved from experience with large, sensitive landscapes and projects (BLM, 1986). The BLM suggests a time frame for views, with "short-term" referring to the first 5 years and "long-term" for the duration (life) of a project. Highway

AESTHETICS AND VISUAL IMPACT ANALYSIS 163

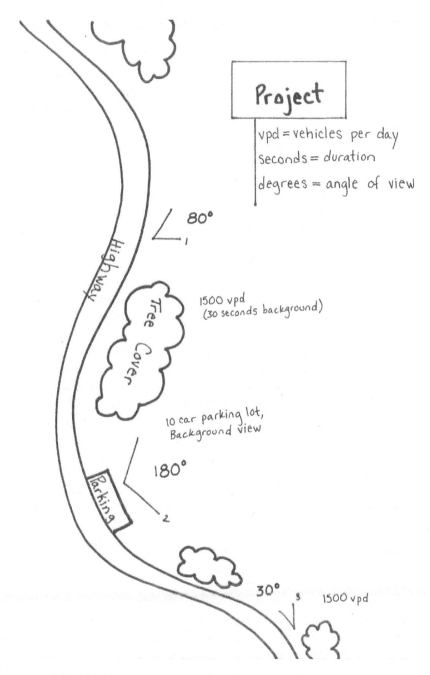

FIGURE 12.1 Viewshed area.

projects might have very short-term views with narrow angles of view, as in seeing a project when driving past it, and long-term views, as in living within view of a project. The BLM also evaluates projects based on the degree of contrast it provides with the setting. Color, line, form, and texture are attributes of contrast, as are distance, angle of observation, scale, viewing time, seasonality, spatial context, atmospheric conditions, and lighting conditions. Concepts for VIA are well established (Smarden et al., 1986). Most visual assessment processes use some kind of rating scheme involving the environmental setting, the project itself if it involves construction or other land alteration, and the social context of the project. Evaluations may be by professionals, but it is also appropriate to involve members of the public, as they may be primary consumers of project views and their values may inform any mitigation that may be proposed.

The VIA will address visual elements individually and collectively as a whole for an overall image or landscape setting. These elements may be scattered or they may have a pattern, such as a row of lines that has some degree of complexity and diversity. Complexity refers to the arrangement of parts in a visual element pattern. Diversity refers to the number and variety of parts that make up a pattern of elements. There should be a balance of these elements to keep the project visually interesting but still have it blend in to the setting.

The visual elements will be examined and described in terms of dominance and compatibility. The visual elements of an object, in comparison with a setting, describe the object's dominance in the landscape. Visual dominance will contribute to the importance associated with specific features of a landscape. The severity of a potential visual impact will be a combination of three factors: visual compatibility of elements, especially color, form, line, and texture; visual dominance of elements, particularly scale and space; and the relative importance of the individual elements.

Landscaping is as much an aesthetic concern as it is one of erosion control, site utility, and other physical concerns. Assessment and approval of landscape plans is, to a certain degree, an act of faith, but it can be verified after project build-out. Two-dimensional drawings of landscapes and buildings may require a significant level of sophistication (or imagination) to interpret as real-world scenarios. A logical response is to ensure that the intentions of the design are articulated as well as the actual details of how things will appear, and the mitigation checks that will occur. Table 12.1 provides some common items addressed in landscape plans.

TABLE 12.1 Items to address in assessing a landscape plan.

√	Site Plan Features	Comments
	Plan authors and qualifications provided on the plan.	
	Does the landscape plan have objectives?	
	Is the plan clear and easy to understand?	
	Are there appropriate details on the plan?	
	Dates for installing and inspecting plantings and features.	
	Accommodations to use and maintain durable indigenous species.	
	Is the plan realistic in terms of the applicant's ability to carry it out?	
	Is the landscape plan appropriate for the setting?	
	How will the site appear in different seasons?	
	What mechanisms are in place to ensure the plan will be carried out?	
	Are the plantings of sufficient size to address aesthetic concerns?	
	Is the street architecture detailed and appropriate?	
	Do the landscape plans and lighting promote safety and security?	
	Is there escrow or bonding to ensure adequate landscaping?	

Source: Based on Sanford (2018).

TABLE 12.2 Lighting impact and mitigation.

Lighting Feature	Measurement	Mitigation	Residual Impact
Details of lighting fixtures			
Intensity or brightness of illumination			
Addressed light "trespass" (intrusion of light onto neighboring properties)			
Description of light emitted into the night sky			
Quality of light and appropriateness in setting			
Documentation of lighting impacts			
Adequacy of lighting for public safety			
Energy efficiency			
Monitoring of lighting compliance and maintenance			
Emergency and security lighting			
Community and neighborhood aesthetics values			
Special interest group or stakeholder concerns			
Interior lighting spill-out onto exterior areas			
Effects on wildlife			

Exterior lighting can be a big concern. Lighting is usually necessary for security and normal work operations. All potential sources such as light poles; walkways, building and other security lamps; signs; reflective surfaces; and interior lights that shine through windows should be considered. Large projects will often have lighting studies, but many smaller projects rely on a brief narrative or incorporate lighting into the landscape plan. Yet even a single bulb glaring into a neighboring property can be a significant source of impact if you are the neighbor. The EIA report should address interior and exterior lighting concepts and issues (Table 12.2). The authoring agency's requirements or standards should be provided in context with any local community standards as part of describing visual impacts of illumination.

12.3.1 Quechee Analysis

The following two-stage process (based on the "Quechee Analysis") used in Vermont (Vermont Natural Resources Board, 2021; summarized in Sanford, 2018: 170–173) has received attention across the country and provides a useful framework for assessing aesthetic impacts. Part I of this assessment process focuses on the setting of the project. First, determine what specific scenic or visual resources exist on the site, including within its landscape context. In so doing, note whether these resources are of statewide significance. At this stage, details of the project itself are not an issue. Instead, the scenic values of the existing site and area are considered. Determining whether the site and/or context contains scenic resources of statewide significance in Part I involves the following four steps:

1. Preliminary review: Is the site or its context located in a sensitive area?
2. Identify and describe the landscape context in which the project will occur.
3. Is the landscape context of statewide significance?
4. Is the site of statewide significance?

Visually sensitive areas typically include ridge lines or hilltops, especially those silhouetted against the sky, panoramic views, shorelines, views along natural or constructed corridors, agricultural districts, areas of scenic quality, "special" areas, and areas in or adjacent to historic sites, districts, designated natural areas, and significant recreational areas.

If a proposed development is in a sensitive area, then identification of specific visual elements becomes critical. The area considered in determining context varies widely. It might be only a few city blocks, or it might be a scenic valley traversed by an interstate. The context depends on the scale of the project, the way in which it will be seen or experienced by the public, and the characteristics that define its setting. Some agencies and municipalities will have designated scenic areas. Other sources of scenic areas arise from individual assessments or case by case reviews.

Part II of the Vermont Agency of Natural Resources' Quechee Analysis process seeks to determine how a proposed project would adversely affect the scenic resource. The agency also considers whether or not there are any special circumstances that should be considered when reviewing the project. These circumstances might include taking notice of decisions made at the local level to allow projects in designated areas due to concerns that outweigh aesthetics. Among the things the state considers in determining adverse impacts are the planning efforts at the local level, the contrast of the project with its surroundings (this can be positive or negative), building styles, landscaping, signs, scale, and duration of impacts. After determining that there are adverse impacts, the agency decides if the adverse impacts are undue. The final determination the three-pronged test of the Quechee Analysis:

1. *Does the project violate a clearly written community standard (local, regional or state) regarding aesthetics?* Usually, if a project violates a standard, this becomes apparent through reviewing relevant planning documents and contacting local and regional officials.
2. *Does the project offend the sensibilities of the average person?* Decision makers can consider themselves "average" for this purpose and need not conduct surveys or interviews. If the analysis in Part I is appropriately documented then it will give clear support to the nature and degree of aesthetic offense.
3. *Has the applicant failed to take reasonable mitigation measures to reduce the visual impacts of the project?* Mitigation can include off-site accomplishments. It can be difficult to determine what mitigation measures are reasonable. Creative site and building design remedies rather than standard measures such as more landscaping or earth tone colors should be encouraged.

Assignment

Use this Visual Impact Assessment worksheet (Table 12.3) to assess a view (incorporates the rating scheme used by the Federal Bureau of Land Management, Manual 8431, Visual Resource Contrast Rating, 1986). Record the date. Describe the site location sufficiently for someone else to locate the visual perspective you assessed.

Odor impact assessment methods vary by use of sensorial, instrumental, and mathematical models (Bax et al., 2020). Where odor impacts relate to hazardous air pollutants, they are addresses in EIA sections on air pollution rather than aesthetics.

TABLE 12.3 Visual Impact Assessment worksheet.

A. Landscape compatibility	different **color:** hue, chroma, value	Severe = 3 Moderate = 2 Minimal = 1 None = 0	Color score _____
	Form: Shape compatibility may vary in comparison with other shapes	Severe = 3 Moderate = 2 Minimal = 1 None = 0	Form score _____
	Line: Edges, bands, outlines	Severe = 3 Moderate = 2 Minimal = 1 None = 0	Line score _____
	Texture: grain, density, pattern	Severe = 3 Moderate = 2 Minimal = 1 None = 0	Texture score
B. Scale contrast	Major intrusion/scale	Severe = 12	Score for B
	One of several big/major	Moderate = 8	
	Significant	Minimal = 4	
	Small	None = 0	
C. Spatial dominance	Dominates	Severe = 12	Score for C
	Co-dominates	Moderate = 8	
		Subordinate = 4	
		None = 0	
A + B + C = Total severity of visual impact	Severe	Total of 27 to 36	TOTAL:
	Strong	Total of 18 to 26	
	Moderate	Total of 9 to 17	
	Minor	Total of 0 to 8	

Source: Based on BLM (1986).

TABLE 12.4 Henning odor classification.

Odor Name	Example	Project
Spicy	Cinnamon	Bakery
Flowery	Jasmine	Greenhouse
Fruity	Peaches	Orchard
Resinous	Turpentine	Paint store
Foul	Rot, hydrogen sulfide	Paper mill
Burnt	Tar, scorched material	Chemical plant

Source: Henning (1916).

The Henning Odor Classification dates back to 1916 and only covers six categories, but it is a useful approach to the dimensions of odor (Table 12.4). There are other, more advanced classifications that might be used in an EIA, but these will be similar and context for them will normally be provided in the report.

12.4 MITIGATION AND MONITORING

If the visual assessment shows that an adverse aesthetic impact will occur, especially if it is likely to be severe or undue, then the next step examines mitigation. If the assessment impacts are described in terms of the visual elements, then these terms can be a guide in developing and evaluating mitigation. Aesthetic impacts can be mitigated through a variety of approaches that may largely be sorted into four categories: changes to the project design, changes in operation of the project, changes to the setting, and off-site compensation/mitigation.

A common mitigative change to a project setting, especially in the middle and foreground, is to provide screening. Three ways to soften or reduce visual impacts through screening of projects are changes to viewing distance, changes to the site or setting by grading, and plantings. An extra deep setback can provide a buffering distance from a visual intrusion. Grading can include berms, dikes, walls, and other artificial and natural barriers. In neighborhoods, there may be fence and wall ordinances, because perimeter barriers do not tend to reinforce a feeling of neighborhood identity. Mitigative plantings should be shown and described on a landscape plan. Their use should contemplate seasonality, particularly if there is a contrast between winter and summer conditions. Thus, conifers may be needed for year-round screening. "Staggered" plantings may help provide a more natural appearance. These screening methods may also reduce light glare, noise, and other intrusions from the project.

The EIA report should describe the mitigation and the monitoring. It should identify the "residual" impacts resulting after the mitigative measures. The public involvement processes will likely play a significant role in assessing mitigation and addressing monitoring.

12.5 CONCEPTS AND TERMS

Chroma: An attribute of color that refers to the relative purity, brilliance, or saturation. It ranges from pure (high chroma) to dull (low chroma). It is used along with hue and chroma in describing color in visual impact assessment.

Form: The two-dimensional shape or a three-dimensional mass of an object or feature. The form can also be described in terms of its geometry, its complexity, and its orientation (e.g., vertical, horizontal, compass directions). Form is affected by the viewing angle, by lighting, and by movement (e.g., smoke, steam, falling water). The perception of form is affected by how the form contrasts with the setting. A geometric form in a non-geometric setting may be more noticeable. Form may also be affected by cultural values associated with shape. The most obvious example is in the distinct concave shape of a nuclear cooling tower; its visual quality will be affected by the knowledge of what it represents.

Hue: An attribute of color based on the wavelength of visible light. It refers to the familiar primary red, blue, or yellow color or some combination. It is used along with value and chroma in describing color in visual impact assessment.

Illuminance: The amount of visible light in an area; "foot-lamberts" or "foot-candles" and measured in lumens per square foot. It may also be portrayed as "lux" and measured in lumens per square meter in the SI system. Illumination can be calculated by knowing the lumens of a light source, and its relationship to the area being brightened. If a pole light has a lamp of 2,000 lumens radiating in a hemisphere 20 feet above the pavement, its lumens on the ground is 2,000 divided by the area of the hemisphere (1,257 square feet).

Lamp efficiency: The efficiency of a lighting source is its lumens per watt.

Line: There are three types of lines: edge, band, and silhouette. An edge joins two contrasting areas. A band is two edges that together divide an area, as in a road. A silhouette is the outlined edge of a mass, as in a skyline. Lines have properties of boldness, complexity, and orientation. Lines are affected by distance, atmospheric conditions, and lighting.

Lumen: A measurement of the quantity of light flowing from a source, also called "luminous flux." A typical 100-watt incandescent bulb emits about 1750 lumens, or about 17.5 lumens per watt.

Scale: The proportionate relationship between an object and its setting determines the scale of the object. Scale is an absolute term when the object is described by dimensions. Scale is a relative concept when it is based on the apparent size relationship between an object and its surroundings. The field of view and the contrast amongst objects influence the perception of scale. Scale is often described in terms of dominance. Large, massive objects in a small area are more dominant than small, delicate objects in broader settings. Viewer angle, atmospheric conditions, distance, and position in the setting all influence the perception of scale.

Space: The three-dimensional arrangements of objects and empty spaces (voids) in an environmental setting determine the spatial characteristics of the landscape. Landscape spatial characteristics can be described as five composition types: panoramic, enclosed, featured, focal, and canopy. The backdrop of sky, water, and features or contours affects the spatial character of a landscape. The observer positions of inferior (below), normal, or superior (above) affects the perceived degree of enclosure or panorama. Longer distances tend to reduce the sense of enclosure and dominance.

Texture: Composed of small forms or color combinations in a continuous surface area. We perceive texture through variation of color, light, and shade. Texture can be described in terms of grain, density, regularity, and contrast. A hayfield would appear to be fine-grained with dense surface variations. Trees scattered on a hillside might have a coarse-grained texture, with sparse density; a cornfield or stone building would have an ordered regularity. Variation within the colors or values making up the texture provide contrast. For example, multi-colored leaves in the fall have much greater textural contrast than trees in summer. Texture, like other elements, is affected by environmental factors such as distance, atmospheric conditions, and illumination.

Value (in reference to color): The relative lightness or intensity of color. Value ranges from black to white. It, along with hue and chroma, describes color in visual impact assessment.

Viewshed: The area that is visible from a particular site or location. It often can be described as an angle with sides of a certain distance. GIS may produce a terrain analysis that includes a viewshed delineation and viewshed analysis.

12.6 SELECTED RESOURCES

American Society of Landscape Architects (ASLA). www.asla.org/.
Churchward, Craig, James F. Palmer, Joan I. Nassauer and Carys A. Swanwick. (2013). *Evaluation of Methodologies for Visual Impact Assessments*. Washington, DC: The National Academies Press.
Illuminating Engineering Society of North America. www.iesna.org/.

International Dark-Sky Association. Http://www.darksky.org/index.html.
New England Light Pollution Advisory Group (NELPAG) Http://cfa-www.harvard.edu/cfa/ps/nelpag.html.
New York State Department of Environmental Conservation. (n.d.). Impact on Aesthetic Resources, Full Environmental Assessment Form. www.dec.ny.gov/permits/91750.html.
Smarden, Richard C., James F. Palmer and John Felleman (Eds.). (1986). *Foundations for Visual Project Analysis*. New York: John Wiley & Sons.
US Bureau of Land Management. (1986). *Visual Resources Contrast Rating*. Manual 8431. Washington, DC: US GPO. www.blm.gov/style/medialib/blm/wo/Information_Resources_Management/policy/blm_handbook.Par.79462.File.dat/8431.pdf.
US Department of Transportation, Federal Highway Administration. (n.d.). Environmental Review Toolkit, Guidelines for the Visual Impact Assessment of Highway Projects. www.environment.fhwa.dot.gov/guidebook/documents/VIA_Guidelines_for_Highway_Projects.asp.
Vermont Natural Resources Board. (2021). District Environmental Commission Training Manual. Criterion 8 Aesthetics. https://nrb.vermont.gov/regulations/commission-manual.

12.7 TOPIC REFERENCES

Bax, Carmen, Selena Sironi and Laura Capelli. (2020). How Can Odors Be Measured? An Overview of Methods and Their Applications. *Atmosphere*, 11, 92. doi: 10.3390/atmos11010092.
Federal Highway Administration. (n.d.). Environmental Review Toolkit. www.environment.fhwa.dot.gov/guidebook/documents/VIA_Guidelines_for_Highway_Projects.asp.
Henning, H. (1916). *Der Geruch*. Leipzig (Germany). Verlag von Johann Ambrosius Barth.
New York Department of Environmental Conservation. (n.d.). Impact on Aesthetic Resources, Full Environmental Assessment Form. www.dec.ny.gov/permits/91750.html.
Palmer, James F. (2015). Effect Size as a Basis for Evaluating the Acceptability of Scenic Impacts: Ten Wind Energy Projects from Maine, USA. *Landscape and Urban Planning*, 140, 55–66.
Sanford, R. M. (2018). *Environmental Site Plans and Development Review*. New York: Routledge.
Smarden, Richard C., James F. Palmer and John Felleman (Eds.). (1986). *Foundations for Visual Project Analysis*. New York: John Wiley & Sons.
US Bureau of Land Management (BLM). (1984). *Manual 8400, Visual Resource Management*. Department of the Interior. Washington, DC: US GPO.
US Bureau of Land Management (BLM). (1986). *Visual Resources Contrast Rating. Manual 8431*. Washington, DC: US GPO. www.blm.gov/style/medialib/blm/wo/Information_Resources_Management/policy/blm_handbook.Par.79462.File.dat/8431.pdf.
U.S. Forest Service. (1973). Agricultural Handbook Number 434, February 1973.
Vermont Natural Resources Board. (2021). District Environmental Commission Training Manual. Criterion 8 Aesthetics. https://nrb.vermont.gov/regulations/commission-manual.

TOPIC **13**

Social impacts and environmental justice

13.1	Environmental setting	171
13.2	Project description	172
13.3	Environmental justice	174
13.4	Identifying potential impacts	176
13.5	Mitigations and monitoring	178
13.6	Concepts and terms	179
13.7	Selected resources	180
13.8	Topic references	180

13.1 ENVIRONMENTAL SETTING

Most development projects have a geographic area in which they occur. This setting usually includes the demographics of human settlement. The project setting description constitutes the baseline conditions of the human environment in the EIA. A variety of approaches may be taken in addressing the human aspect of EIA. It may treat social impacts from an economic perspective, from a cultural perspective, or separately as a community services or fiscal impact perspective. We have somewhat arbitrarily chosen to describe socioeconomic impacts here, and we have reserved the infrastructure and community services aspects of socioeconomic impacts for Topic 14. The important thing to consider in any approach is that, thanks to the comprehensive nature of EIA, the range of potential impacts addressed will include social impacts, cultural impacts, and economic impacts. Environmental justice is evaluated elsewhere and is about fairness in how projects affect people. Thus, it became a major subset of social impact analysis and is part of the analysis which was specifically incorporated into the Council on Environmental Quality (CEQ) regulations of 1979.

If the project occurs on or is likely to affect tribal lands, the social and cultural aspects of the project could be a large part of the EIA. Clearly, environmental justice may be a significant factor due to social differences in behavior, economic status, access to goods and services, and other factors. Projects not on tribal land or in another cultural group's setting may still have social impacts, particularly in cases of economic development and infrastructural services projects. All projects should be reviewed for potential social and cultural impacts.

DOI: 10.4324/9781003030713-13

As we deal with climate change and consequential resettlement or infrastructure relocations, we will see more social impact assessment aspects in new projects. The EIA report is expected to promote understanding of the setting in all its aspects. There is an obligation to include stakeholders in the process of understanding and assessing the setting. The information-sharing protocols should reflect the social groups and the demographics of the setting.

13.2 PROJECT DESCRIPTION

Project descriptions usually include employment as a social aspect. The description should also list the nature of jobs to be created, including permanent, temporary, full-time, part-time, skilled, and unskilled. The project description may also include social programs that are part of the project housing provided by or generated from the project.

The Interorganizational Committee on Guidelines and Principles for Social Impact Assessment (1994), sponsored by three US federal agencies, set the stage internationally for SIA as a stand-alone report and for incorporation of social impact assessment (SIA) into EIAs. SIA can be done for a project that either does not require an EIA/EIS or is so large in terms of potential social impacts that it merits *both* an EIS and an EIA report. Social impacts are the consequences of human actions and developments that may alter the way in which people live, work, plan, and interact. The EIA might treat social impact assessment as its own category, or it might include it into a "socioeconomic impact" assessment. A SIA should include the norms, beliefs, and values of social groups within a culture group and of different cultures within a project area.

A project could affect a natural or cultural heritage site. Historical, archaeological, and local cultures are understood to have social impacts that must be addressed. They will also have a cultural resource or an archaeological/historical resource section in the report. The connection to past and present peoples will need to be addressed, as some impacts may fall into multiple categories. Cultural heritage and historical aspects of it also raise social justice issues. These issues can be included for multiple phases of a project (see section 13.4). The Interorganizational Committee on Guidelines and Principles for Social Impact Assessment suggested variables for comparison with the impacts that could arise in any of the four phases of a project (1994: 12–13). This list (Table 13.1) was updated and sorted into five overarching categories: population change, community and institutional structures, political and social resources, community and family changes, and community resources (2003: 243). Various federal agencies use their own categories similar to these.

TABLE 13.1 Variables for social impacts.

Variable	Planning and Construction	Operation	Closure
1 Population change			
Ethnic and racial composition and distribution			

Variable	Planning and Construction	Operation	Closure
Relocating people			
Influx and outflows of temporaries			
Presence of seasonal residents			
2 Community and institutional structures			
Voluntary associations			
Interest group activity			
Size and structure of local government			
Historical experience with change			
Employment/income characteristics			
Employment equity of disadvantaged groups			
Local/regional/national linkages			
Industrial/commercial diversity			
Presence of planning and zoning activity			
3 Political and social resources			
Distribution of power and authority			
Conflict newcomers and old-timers			
Identification of stakeholders			
Interested and affected publics			
Leadership capability and characteristics			
Interorganizational cooperation			
4 Community and family changes			
Perceptions of risk, health, and safety			
Displacement/relocation concerns			
Trust in political and social institutions			
Residential stability			
Density of acquaintanceship			
Attitudes toward project			
Family and friendship networks			
Concerns about social well-being			
5 Community resources			
Change in community infrastructure			
Indigenous populations			
Land use patterns			
Effects on cultural, historical, sacred, and archaeological resources			

Source: Interorganizational Committee on Guidelines and Principles for Social Impact Assessment (1994: 12–13; 2003: 243).

Assignment

1. Do you agree with the utility of the items in Table 13.1? What might you want to add, remove, or change?
2. Examine the social impact categories used by the US Forest Service, US Bureau of Land Reclamation, or another federal agency and compare them to Table 13.1.
3. Do these items capture the full range of potential impacts that might assess social justice issues?

13.3 ENVIRONMENTAL JUSTICE

If they can help it, most people do not choose to live downwind of toxic fumes from a chemical plant, nor would they eat plants and fish contaminated with heavy metals. Proximity to contaminated waters from old landfills and other such hazards would (and should) be avoided. Similarly, some citizen groups may be better able than others to resist undue hazardous burdens in local neighborhoods.

A variety of federal executive orders (EOs) pertain to environmental justice. Executive Order 12898, Federal Actions to Address Environmental Justice in Minority Populations and Low-Income Populations (1994), deserves particular mention. This EO gave each federal agency an environmental justice mission. Several ways in which this rule affects NEPA and the EIA process include recognizing differential patterns of subsistence consumption of fish and wildlife. Consumers may have differential access to public information. Prior to this EO, environmental justice was implicit in social impacts, but now the policy became explicit. Agencies had to examine the potential impact of projects on low-income people and minorities. Environmental justice under NEPA is included in the development of projects ("actions" or "undertakings"), mitigation efforts, information sharing, and the role of the EPA in commenting. The EPA considers environmental justice in its roles as data and standards sources and as reviewers in the EIA process.

The Federal Interagency Working Group on Environmental Justice & NEPA Committee (2016) recommends a variety of approaches to identifying minority populations, including US census data. Environmental justice assessment methods of data collection and analysis can include all social groups. Socioeconomic differences can exist in access to resources and access to government services. Different socioeconomic and cultural groups have different proximity to hazards, exposure rates, exposure vectors, as well as treatments and responses to exposures. Differential effects, real or perceived, in addition to maters of justice, lead to controversy. Controversy itself is a consideration in the CEQ regulations as part of the formulation or evaluation of the intensity of an impact.

The steps in addressing environmental justice in NEPA and other environmental assessments are progressive but not strictly linear. They overlap and are an iterative process using feedback while advancing toward the final assessment report. Table 13.2 projects a checklist of these steps for use in evaluating the environmental justice component of an EA report. It would be based on the Federal Interagency Working Group on Environmental Justice & NEPA

SOCIAL IMPACTS AND ENVIRONMENTAL JUSTICE

TABLE 13.2 Checklist of steps for evaluating the environmental justice component of an EA report.

√	Step	Comments/Examples From the EA/EIS
	Meaningful engagement	
	Effective scoping	
	Description of affected environment	
	Selection of alternatives	
	Identification of minority populations	
	Identification of low-income populations	
	Impact analysis	
	Disproportionately high adverse impacts	
	Mitigation	
	Monitoring	

Committee (2016). The EPA has a National Environmental Justice Advisory Council, as does the White House (White House Environmental Justice Advisory Council). Other federal agencies also have offices and implementation plans dedicated to advancing environmental justice and are available for input into the creation and review of environmental assessment reports.

Assignment

1. How did NEPA address environmental justice (or not) before Executive Order 12898 was enacted? Support your answer with documentary references.
2. What are four ways Executive Order 12898 intersects with (or is applied under) NEPA?
3. Do environmental justice rules and regulations apply even if there is no EA or EIS? Explain your opinion.
4. Choose one of the following EIA development process steps—scoping, public participation, determining the affected environment, analysis, alternatives, and record of decision—and outline the potential roles of environmental justice in them.
5. The Environmental Protection Agency has a technical guidance document that is particularly applicable for EIA work (2016). Factors for environmental justice include minority and low-income populations, access to information, disproportionate effects (differential impacts: direct and indirect), cumulative effects, subsistence consumption of fish and wildlife, exposure and risk, and meaningful involvement (CEQ, 1997). Use those and other terms for column and row headings to design a matrix for an EIA to address these concerns for operation of the Williams Transco Pipeline, which runs from the Gulf Coast to the Northeast (natural gas, 10,200 miles; online sources show the general route). The matrix is a placeholder for the areas of inquiry suitable for effective environmental justice review of one zone of the project. The EIA team would go into the field with this guidance to inform their investigations as they create an EIS for review. The good news is this means we do not have to *solve* the problems; we just have to point out a framework for the team to use to *look* for problems.

6. Look at the Environmental Protection Agency's Toxics Release Inventory (TRI) Program (2021) for a community known to be low income. Report on your findings. Are there any apparent environmental justice issues that might seem to arise? (www.epa.gov/toxics-release-inventory-tri-program/tri-for-communities)

Environmental justice directly pertains to how government responds to disasters, whether natural or human-caused. Rapid Environmental Assessment of the placement and suitability of settlement camps requires quick decision-making based on a qualitative assessment of potential environmental impacts (see Topic 17). We can expect to see more of this in an era of environmental change and global conflicts.

13.4 IDENTIFYING POTENTIAL IMPACTS

Many of the phases of a project can be associated with specific impacts. The planning phase of a project can raise issues when there is exploratory testing. Mining, power and industrial plants, solid waste facilities, roads, utility corridors, dams and reservoirs, and similar projects are particularly known for addressing impacts to social groups. They may be from lower-income households. In the planning stage, project rumors can lead to land speculation and higher real estate costs. The construction phase of a development project can involve socially disruptive land clearing, traffic congestion, dust, noise, and an influx of temporary workers driving up the costs of housing, goods, and services. Operations of projects can produce another set of impacts, such as from traffic, waste disposal, noise, light pollution, and air emissions, that can affect social groups with increased exposures and reduced access to exposure remedies. Closure of projects such as landfills and industrial lands can lead to costly contamination issues that affect people with varying abilities to reduce the consequences of those impacts. For example, minorities and low-income households may have less access to potable water and may rely more on subsistence gardens in soils with heavy metals.

Cultural impacts can include effects on subsistence, particularly for indigenous peoples. Declines in traditional hunting, fishing, and food gathering may result. Other cultural impacts can include linguistic changes, loss of traditional ecological knowledge, changes in core values such as sharing of resources, effects of western media influence, medical services, emergency services, educational services, effects of housing, energy sources, recreational patterns, and transportation need and access. The Precautionary Principle is particularly apt for evaluating potential social impacts. Under this principle, responsibility should be placed on the proponent to show its project does *not* adversely affect other people. This conclusion should form the basis for proper scoping of impacts.

Social impact assessment includes quantitative and qualitative methods of analysis. It can use ethnographic methods, community participation, and reactions of review panels. The SIA should focus on key elements of the human environment such as social and cultural issues. It should produce quality information for use in decision-making. Data should be robust and secure.

Impacts can be viewed in the context of socioeconomic values. A common approach to values assessment is to divide them into three types: life support, material goods, and amenity services. Life support includes things essential for health and survival. Material goods are represented by extractive and processing sectors, such as manufacturing, agriculture, forestry, market goods, fisheries, and mining. Amenity services are for recreation and aesthetics—things that range from essential for social well-being and mental health to things that are "nice to have." They can be consumptive or non-consumptive. These types are not necessarily mutually exclusive, as they may overlap and may represent continuums of goods and services.

Ecosystem services are often addressed in EIA reports, particularly in the biological sections. Ecosystem services are ways to view the environment as a source of economic, social, and ecological value. These can be direct use, such as plants and animals for food and forests for wood. The direct use can be consumptive or non-consumptive, as in the case of visual amenities of landscapes. Ecosystem services can also be of indirect social and economic value. Examples of indirect use of ecosystem services include pollination, nutrient cycling, carbon storage, genetic diversity, water, and food for domestic livestock. Ecosystems services can be of simultaneous direct and indirect use, as in the case of ecotourism—an ever-increasing area of emphasis in many regions and communities.

Areas of social impact can be summarized as categories of project influence in the setting in terms of demographic change, economic change, environmental change, and institutional change. The impacts can be felt in lifestyle, social behavior, quality of life, amenity services, culture, community, and health. Agencies may have particular checklists of social impacts to employ for projects under their purview.

Assignment

1. A square mile of coastal ocean (plus peripheral areas on shore) aquaculture station is proposed for your latitude position on the eastern coast of the United States (or choose a large lake). The station will include research but is intended to commercially supply Atlantic salmon, 20 tons per day in full operation. The station is federally licensed. It includes 50 employees (5% managers/scientists, with the rest blue-collar). Complete Table 13.3 for potential social issues and impacts, including sensitive populations and other stakeholders for the aquaculture station.

TABLE 13.3 Potential social impacts of a near-shore aquaculture station.

Stakeholder/ Affected Group	Type of Social Impact	Magnitude of Impact	Potential Types of Mitigative Measures

2. For the above project, what are likely to be the biggest environmental impact issues?

Risk factors deserve special mention in dealing with social impacts because different social groups have different exposures to hazards and different responses to exposures. Risk as part of decision-making is also discussed in Topic 17. In both discussions, risk is approached in two formats: human health risk assessment and ecological risk assessment. The first step in either format is hazard identification. This means looking at known conditions, contaminants, concentrations of contaminants, potential routes of exposure, and potentially affected populations of humans and non-humans. In addition to direct impacts, the environment can cause synergistic interactions and reactions with these materials, making hazard identification complex. A variety of other interactions can occur, including the creation of new organic contaminants. The next step is to evaluate the hazards. Next, a risk is computed from the hazard evaluation.

Risk is evaluated based on exposure and dose response. The EIA should describe significant potential exposures and consequences. It is normal to expect different toxic levels and different rates of change in effects among different compounds and materials. For example, the linear non-threshold (LNT) dose-response model was originally intended for ionizing radiation. It spread to use in chemical carcinogens, but there are many forms of dose response models. Ultimately, we can characterize risk in a variety of terms: health, environmental/ecological factors, economic factors, and safety/security. Disadvantaged groups often are disproportionately exposed and may have disproportionate access to remedies. Different groups also have different perceptions, beliefs, and reactions to risk. The results will also affect how risk is communicated and mitigated. Large projects and those with potentially significant issues will have designated risk managers.

13.5 MITIGATIONS AND MONITORING

Mitigation should generally be associated with monitoring, as there is usually a longitudinal aspect that can be easily interrupted. The EIA should address mechanisms for evaluating and implementing mitigation measures. There should be a structure for overall planning of mitigation actions and for reporting the results of monitoring, both periodically and continuous. Mitigation measures in an EIA should address the specific social and cultural groups that potentially benefit or are otherwise affected by them. The residual impacts left over after mitigation should be evaluated in terms of who they affect and what the consequences are. This may be addressed as a component of environmental justice or as part of generalized socioeconomic impacts. Systematized planning, tracking, and responding to these factors may be documented in the form of a Mitigation Action Plan (MAP). A checklist or matrix can be used to examine the MAP. Such a list should reflect the agency values and mission (Table 13.4).

Assignment

1. Are the United Nations "good practices" in Table 13.4 sufficient to support environmental justice in an EIA? Explain your answer.
2. Do any other international organizations supply recommendations for social impact assessment?

TABLE 13.4 United Nations social impact assessment good practices in impact mitigation.

√	Practice	Project Phase
	Identify impact measures for each impact	
	Customize mitigation measures for each group affected	
	Give priority to avoiding social impacts	
	Minimize social impacts	
	Use compensation as a last resort	
	Ensure no disproportionate impacts accrue to any particular group	
	No one should be worse off than before	
	Treat relocation/resettlement as a last resort/special case	
	Livelihoods of those displaced should be improved	
	Enhance benefits for local people through job training and development	

Source: United Nations (2002, Topic 13, p. 18).

Monitoring of impacts and impact mitigation should fall under three categories according to Chadwick (2001: 58–59). The first such category is an administrative system that provide flows of information about key parameters. The second category is a periodic survey of the workforce, including types of jobs, number of workers, travel patterns, and expenditures. The third category is the monitoring of social and economic trends and patterns in and adjacent to the project setting. Chadwick suggests these can include house prices, rentals, occupancy rates, school enrollments, health and medical data, and other demographic variables. Usually, a regional planning commission of some sort will have periodic reports that can be used for this information gathering along with specific on-site data collection.

13.6 CONCEPTS AND TERMS

Cumulative impacts: These result from "the incremental impact of the action when added to past, present, and reasonably foreseeable future actions," regardless of who undertakes the action or past actions. "Individually minor but collectively significant actions taking place over a period of time" can lead to cumulative impacts (40 CFR § 1508.7).
Direct impact: "caused by the action and occur at the same time and place" as the action. 40 CFR § 1508.8.
Disbenefit: Adverse effect of a project.
Ecosystem services: The gains acquired by humankind from surrounding ecosystems. These may be described in terms of provisioning services, regulating services, cultural services, and supporting services (Millennium Ecosystem Assessment, 2005). Often used in evaluating trade-offs of impacts and mitigation.
Ethnicity: Socially identifiable group based on common national or cultural tradition. Ethnicity is not the same as "race."
Hazard: Source of potential harm.

Health Impact Assessment (HIA): A process or report on human health impacts of a project.

Impact monitoring: A process of evaluating the likely environmental impacts of a proposed project taking into account interrelated socioeconomic, cultural and human-health impacts, both beneficial and adverse.

Indirect impacts: These are "caused by the action and are later in time or farther removed in distance, but still reasonably foreseeable results of the action." 40 CFR § 1508.8.

Intensity: The severity of a project's impact on the environment. Agencies are required under 40 CFR § 1508.27(b) to address intensity in terms of a variety of factors, including public health, controversy, risk, and cultural resources.

Race: Generally considered to be some combination of physical cultural, and behavioral attributes. This non-scientific term can be quite divisive. However, it is referred to in EO 12898 and other EIA-related places probably because it is also referenced by the US Census Bureau and has social meaning, rather than biological or anthropological meaning.

Risk: Likelihood of harm occurring. Two major concepts in risk assessment are the probability of a harm and the magnitude of the harm. Additional issues are raised in terms of who gets exposed and what remedies are available.

Social Impact Assessment: The process of research, planning, and the management of social change, particularly in the context of a proposed project, regulation, or policy.

13.7 SELECTED RESOURCES

Canter, Larry. (1996). *Environmental Impact Assessment*, 2nd ed. New York: McGraw-Hill.

Chadwick, Andrew. (2001). Socio-Economic Impacts 2: Social Impacts. Water. In P. Morris and R. Therivel (Eds.), *Methods of Environmental Impact Analysis*, 2nd ed. (pp. 42–64). London: Spon Press.

Environmental Protection Agency (EPA). Environmental Justice. www.epa.gov/environmentaljustice. Site contains multiple links and the EPA annual Environmental Justice Report.

Environmental Protection Agency (EPA). Risk Assessment. www.epa.gov/risk.

Federal Interagency Working Group on Environmental Justice & NEPA Committee. (2019). *Community Guide to Environmental Justice and NEPA Methods*. Federal Interagency Working Group on Environmental Justice & NEPA Committee.

Glasson, John. (2001). Socio-Economic Impacts 1: Overview and Economic Impacts. In P. Morris and R. Therivel (Eds.), *Methods of Environmental Impact Analysis*, 2nd ed. (pp. 20–41). London: Spon Press.

McAllister, Donald M. (1995). *Evaluation in Environmental Planning: Assessing Environmental, Social, Economic, and Political Tradeoffs*. Cambridge, MA: MIT Press.

Online Interactive SIA Bibliography. www.iaia.org/sia-bibliography/Search.aspx.

Therivel, Riki and Graham Wood. (2009). *Methods of Environmental and Social Impact Assessment*, 4th ed. New York: Routledge.

13.8 TOPIC REFERENCES

Chadwick, Andrew. (2001). Socio-Economic Impacts 2: Social Impacts. In Peter Morris and Riki Therivel (Eds.), *Methods of Environmental Impact Assessment*, 2nd ed. (pp. 42–64). London: UCL Press.

Council on Environmental Quality (CEQ). (1997). *Environmental Justice: Guidance Under the National Environmental Policy Act*. Washington, DC: Executive Office of the President. https://ceq.doe.gov/docs/ceq-regulations-and-guidance/regs/ej/justice.pdf.

Environmental Protection Agency (EPA). (2016). *Technical Guidance for Assessing Environmental Justice in Regulatory Analysis.* Washington, DC: Environmental Protection Agency.

Environmental Protection Agency (EPA). (2021). Toxics Release Inventory (TRI) Program. www.epa.gov/toxics-release-inventory-tri-program.

Federal Interagency Working Group on Environmental Justice & NEPA Committee. (2016). *Promising Practices for EJ Methodologies in NEPA Reviews.* Federal Interagency Working Group on Environmental Justice & NEPA Committee.

Interorganizational Committee on Guidelines and Principles for Social Impact Assessment (SIA). (1994). *Guidelines and Principles for Social Impact Assessment.* US Department of Commerce, NOAA, and National Marine Fisheries Service.

Interorganizational Committee on Guidelines and Principles for Social Impact Assessment (SIA). (2003). Principles and Guidelines for Social Impact Assessment in the USA: The Interorganizational Committee on Principles and Guidelines for Social Impact Assessment. *Impact Assessment and Project Appraisal,* 21(3), 231–250. doi: 10.3152/147154603781766293.

Millennium Ecosystem Assessment. (2005). *Ecosystems and Human Well-Being: Synthesis.* A Report of the Millennium Ecosystem Assessment. Washington, DC: Island Press.

United Nations. (2002). *Environmental Impact Assessment Training Resource Manual,* 2nd ed. United Nations Environmental Programme. https://wedocs.unep.org/handle/20.500.11822/26503.

Infrastructure, fiscal impacts, and community services

14.1	Environmental setting	182
14.2	Project description	184
14.3	Potential impacts	184
14.4	Mitigation	189
14.5	Concepts and terms	191
14.6	Selected resources	191
14.7	Topic references	192

14.1 ENVIRONMENTAL SETTING

The setting of a project in terms of infrastructure, fiscal impacts, and community service is best examined from the perspective of population demographics and government structures. A project, even if on public lands, will be adjacent to one or more municipalities even if it is not within a particular municipality. The environmental setting of concern here is the landscape of settlement patterns and communities and what burdens, goods, and services this entails (Figure 14.1).

Infrastructure refers to the physical, economic, and social networks that serve a project or are essential to the setting in which a project is located and thus potentially affected by it. This includes roads, canals and transportation routes, utility services such as electrical power lines, natural gas, sewer and water, wastewater, bridges, airports, cell towers, emergency services, law enforcement, healthcare, recreational facilities, educational services, housing, labor force, social groups, customers clients, and visitors.

Within infrastructure, four lifeline functions are essential to the security of the public: transportation, water, energy, and communications. But many other aspects of infrastructure are important, if not critical, and should be addressed in an impact assessment. The Department of Homeland Security (2019) identifies critical US infrastructure sectors as chemical, commercial facilities, communications, critical manufacturing, dams, defense industrial base, emergency services, energy, financial services, food and agriculture, government facilities, healthcare and public health, information technology, nuclear reactors and related transportation systems, and

DOI: 10.4324/9781003030713-14

INFRASTRUCTURE, FISCAL IMPACTS, AND COMMUNITY SERVICES

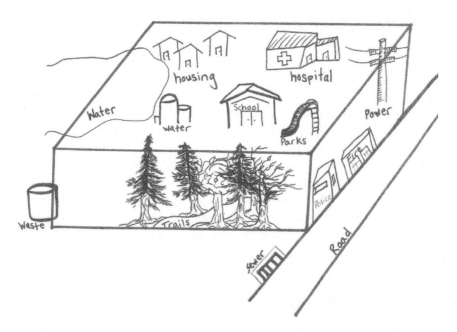

FIGURE 14.1 Projects can involve a variety of services.

water supply and wastewater systems. This is a broad arena from which to draw impacts. Much of infrastructure impact assessment is based on evaluating what municipalities that contain or are adjacent to the project need to operate, and how this operation is affected by the project.

Frameworks of resilience and sustainability apply to assessing infrastructural and fiscal impacts. A list of ten "smart growth" principles provides a useful context for evaluating and mitigating the effects of projects on infrastructure, especially for municipal and community services. These principles descend from generations of general guidelines for sound planning and are articulated by the Smart Growth Network (2022). The principles can be applied to the "fit" of a project either formally through applying the language if incorporated into municipal or regional planning documents or informally through expressed views and participation at public meetings and review processes.

- Mix land uses.
- Take advantage of compact building design.
- Create a range of housing opportunities and choices.
- Create walkable neighborhoods.
- Foster distinctive, attractive communities with a strong sense of place.
- Preserve open space, farmland, natural beauty, and critical environmental areas.
- Strengthen and direct development towards existing communities.
- Provide a variety of transportation choices.
- Make development decisions predictable, fair, and cost-effective.
- Encourage community and stakeholder collaboration in development decisions.

14.2 PROJECT DESCRIPTION

Projects to be described and analyzed will be in a variety of forms. If the projects are in the form of regulations and policies, their impacts might be indirect consequences rather than a direct altering of the environment. If they are physical projects, such as quarries, mines, dams, and other structures, they will have direct physical impacts and attendant direct infrastructure and service requirements. Occupied buildings and work sites will have service demands for employees and visitors. Housing projects will have their own set of impacts.

A general rule of thumb is that residential units cost more for municipal services then they provide in revenues ("tax negative"). Commercial and industrial projects are much more likely to bring in more revenues than they cost ("tax positive"). Secondary houses may not cost much in terms of services, but there is always the possibility that the second homes become primary housing, and if that likelihood exists, they might be evaluated accordingly in fiscal and environmental impact assessments.

14.3 POTENTIAL IMPACTS

Any physical project could have a demand for fire and emergency services. Some projects plan to supply their own fire and emergency services but may depend on a public agency for back-up services. Fire departments are concerned with the adequacy of the water supply in terms of capacity, location, and access to the project. Planners ensure that there is room to bring in emergency vehicles. Structures are required to have sprinklers, alarms, and related measures. In addition to these major categories, the municipal fire departments will want to know if the proposed occupancy presents any special hazards. The emergency service providers will also want to know if service to the project requires special resources over and above that of a typical commercial or residential use. Table 14.1 provides typical questions for assessment and review.

Police review is very similar to fire and emergency review, with more emphasis on security. Proper security should enhance the value of the property and its proposed use, protect people from harm, and mitigate burden on municipalities and entities that provide services. Federal police and security services will also have jurisdiction in national parks, on military bases, and on other federal properties, but collaboration with local services may be appropriate. Table 14.2 lists typical features that police departments review.

TABLE 14.1 Fire and emergency department questions for project review.

Check	Fire and Emergency Service Questions
	Are there sufficient hydrants?
	Do hydrants have sufficient flow?
	Are hydrants spaced appropriately?
	If there is no municipal water, is there accessible cistern or pond?
	What is the static water supply volume?
	Will there be sufficient water supply for any required sprinkler systems?

Check	Fire and Emergency Service Questions
	Are structures accessible by fire department apparatus?
	What is the expected fire flow for the structures? (conform to NFPA standard)
	Is a fire pond or other infrastructure needed for the project, and if so, what details are needed to ensure its success for the life of the project?
	Are fire lanes necessary?
	Do lanes comply with acceptable NFPA standards?
	Is the road constructed with sufficient base and foundation to carry heavy fire apparatus (water tankers and ladder trucks)?
	Are the road width, radii, and slope adequate for fire apparatus?
	Are there any dead ends or cul-de-sacs? (A grassed right of way may be used to provide a second access that discourages use by non-emergency service providers.)
	How will the normal traffic patterns affect fire department response to this development?
	Are there road signs? Does each street have a unique name?
	Are the houses and other buildings numbered clearly?
	Is acceptable mitigation proposed?

Source: Based on Sanford (2018).

TABLE 14.2 Police department checklist for project review.

Check	
	Commercial projects
	Does access extend all the way around the project?
	Foot-candle/lumen information: Is the lighting sufficient? Does it properly illuminate shadows?
	Is visibility of the site good, or do various features obscure it?
	Is the ingress/egress safe for vehicles and pedestrians?
	Risk profile: How does the proposed use match with the community? Does it raise security issues?
	Are alarm systems specified?
	Residential projects
	Is there an open view approaching the development, and is it easy to get in and out of the property?
	Are the stop signs adequate?
	Can the local highway system handle the projected traffic counts from the project?
	Are the streets wide enough?
	Is the lighting adequate for public safety?
	Does the landscaping create a public safety problem?
	Does the project create an attractive nuisance or other potential liability for community safety?
	Are the streets named?
	Will buildings have identifying numbers?

Source: Based on Sanford (2018).

Large federal projects and properties may have their own internal transportation services. Other projects may use state and local services or at least collaborate with them. Highway departments review development projects against the town's road standards or public works specifications (also see Topic 15). Typical features to consider are listed in Table 14.3.

Sewer and water supply review is conducted by the utilities section of a municipality or by specialized departments at the state or federal level. Table 14.4 shows the level of detail likely to occur in the review.

If the project involves creation of solid waste as a result of construction or operation, the impacts from this should be addressed. If there are staging areas for dumpsters, these should be shown on the site plan along with landscaping and fencing details. If the project involves

TABLE 14.3 Highway department review questions.

Check	Road, Access and Transportation Related
	Is there a traffic impact assessment study?
	Does the project include construction drawings of all proposed public improvements?
	Are typical cross sections of the proposed grading, roadways, and sidewalks shown?
	Describe the parking lot size, grade, surface, access, and maintenance.
	Are all proposed culverts shown and provisions made for collecting and discharging stormwater drainage?
	Are the locations, widths, and names of existing and proposed streets, easements, building lines, and alleys pertaining to the proposed project and to any adjacent properties shown?
	Do the site plans include the location and dimension of existing driveways, curb cuts, parking lots, loading areas, or any other vehicular use area?
	Are the locations of all existing access points onto city streets shown?
	Do the plans include the location and dimensions of existing and proposed sidewalks, pedestrian walkways, and other paved surfaces?
	Are the length of all straight lines, the deflection angles, the radii, and the length of curves and central angles of all road curves shown?
	Are minimum sight distances per AASHTO standards shown?
	Are roads within maximum allowed grades?
	Are the required road widths met? What additional duties and services will accrue to the highway department?
	Maintenance: What is the description of the management organization or homeowner's association? Is there a disclosure statement for prospective purchasers detailing responsibility for services, such as maintenance and plowing of roadways within the development?
	Have access permits been applied for?
	Are there alternative transportation plans/linkages?
	Is there a ZEV provision and other sustainable use plans?
	What are the acceptable mitigation measures?

Source: Based on Sanford (2018).

TABLE 14.4 Sewer and water checklist.

Check	Sewer and Water Service Questions
	Are profiles of proposed sanitary and storm sewers shown on the plans?
	Are the locations and size of existing sewer and water lines on the plans shown, including hydrants, gates, valves, and "blowoffs"?
	Are the location and width of all sewer and water line easements shown on the plan?
	Are details of proposed connections with existing water supply or alternative means of providing water supply to the project shown?
	Are required pipe widths and materials for water and sewer lines shown on the plans?
	Are service connections shown at the required minimum depth?
	Are manholes shown on the plans, and are they the appropriate size and of the proper materials?
	Are the horizontal and vertical separation for sewer and water lines to be designed and installed in accordance with required standards?
	Do pump stations adhere to appropriate standards regarding materials, capacity, and force main velocity?
	What is the maintenance plan for on-site facilities?
	What additional demands are needed for sewer and water accommodations? How will these demands affect the ability to provide municipal utility services?
	What are the impact fees or other mitigation?

Source: Based on Sanford (2018).

handling solid wastes, there might be designation of a handling area with an impervious surface, possibly with a berm to protect the surrounding ground from contamination. The site design should also be able to accommodate the providers of solid waste and recycling services. Generally, if the proposed roadway can accommodate a fire truck, it will be adequate for a waste hauler.

Federal and other government projects should set a standard of sustainable practices. Will the project handle its own recycling? Will its recycling needs and plans mesh with local recycling programs, including collection and storage?

Social impacts are addressed in Topic 13, but social programs certainly include fiscal impacts. Socioeconomic values are also a factor. These are commonly addressed in term of three categories: life support, materials, and amenity services. Life support refers to essential factors of health, ecological services, weather, and climate. Materials are the commodity items important to society. These are things produced through manufacturing, agricultural forestry, market goods, fisheries, mining, and so forth. Amenity services are broadly defined goods and services that are nice to have, and are often associated with recreation and aesthetics.

Amenity services may be consumptive or non-consumptive. Amenity values are important to the quality of life of residents. Amenity values are influenced by cultural factors, life stage, socioeconomic status, and personal preferences. Natural amenities can include things such as the number of sunny days in a year, temperature averages, percentages of open or green space, water features, topography, and mountain views. These values can be assessed and impacts to them measured in quantitative and qualitative ways. The three categories of socioeconomic values factor into fiscal impacts and in determining what is acceptable as proposed or when mitigation or compensation is required.

The options for development of any given parcel of land have varying impacts on a municipality's costs and revenues. Some projects may require expensive sewer and water line extensions, stress school capacity, and provide few tax rewards to the town. Government officials and citizens, especially those in areas with significant growth pressure (or perhaps quite the opposite: desperate for growth), may want to know the financial burdens or benefits of development projects and seek an estimate of what they will be. Such an estimate is called a fiscal impact analysis (FIA) and can play a large role in a review board's consideration of a project. Review boards often use the results of a fiscal impact analysis as part, not all, of their consideration. Developments that may be expensive to the town may also confer non-financial benefits such as preserved open space, bicycle and pedestrian trails, and affordable housing.

The economic context or fit of a project can be derived from fiscal impact assessment. Fiscal impacts can be determined using marginal costing methods (e.g., case study, comparable city, and employment anticipation) or, most commonly, the average costing methods (e.g., per capita multiplier, service standard, and proportional valuation). The average costing methods tend to be faster and cheaper, and are used when the situation and city/town represent "average" conditions (where services are relatively "in tune" with levels of demand). Proportional valuation and employment anticipation are only used for commercial projects. Economic benefits include number of jobs and type of job (blue-collar, white-collar, temporary, and permanent). Assessment also looks at labor sources: whether or not the area can supply the workers or if they come from somewhere else. Economic impacts need to consider primary and secondary effects. Cost-benefit analysis uses various combinations of fiscal impact assessment tools.

Fiscal impact analyses help reviewers determine what is an "unreasonable burden." There is no standard format for agencies or developers to follow when undertaking an analysis, but there are standard methods as described by Burchell et al. (1985). The "per capita multiplier method" and the "case study method" are two of the most versatile and frequently used methods. The former is often used in situations where the municipal infrastructure is closely related to the service demand in such a way that the average costs of providing services to current users is a good approximation of the costs to provide similar services to future users. This method is most typically used in contexts of mid-sized, established suburban areas or secondary cities experiencing slow or moderate growth. The case study method is well suited for the development impacts seen in large or secondary stable or declining cities or small, rapidly growing rural-fringe areas (Burchell et al., 1985).

The per capita method is the traditional average costing approach for projecting the impact of population change on local, municipal, and school district costs and revenues. The sequential steps for this type of analysis (Burchell et al., 1985) are:

1. Obtain information on budgets, current populations, and assessments.
2. Categorize local expenditures into five municipal service categories plus the school district function (general government, public safety, public works, health and welfare, and recreation and culture).
3. Obtain total annual municipal (by service category) and school district expenses (include operating and debt service costs).

4. Assign a share of annual municipal expenditures to existing nonresidential uses.
5. Calculate net annual per capita and per pupil expenditures.
6. Calculate anticipated total resident and school population by housing type.
7. Calculate residentially induced total annual municipal and school district expenses.
8. Calculate municipal costs for inclusive non-residential uses.
9. Determine total annual public costs and distributive costs both by municipal service and by municipal and school district operating versus debt service expenditures.
10. Project total annual public revenues.
11. Calculate the cost–revenue surplus or deficit.

The case study method is the traditional marginal cost approach to projecting the effect of population change on municipal and school district costs. Burchell et al. (1985) recommend the following steps while performing this analysis:

1. Contact local officials and elicit their support.
2. Categorize public service functions and delineate responsibilities.
3. Determine excess and deficient service capacity.
4. Project population increases and population-induced demand.
5. Determine anticipated local service response.
6. Project total annual public costs.
7. Project total annual public revenue.
8. Calculate the cost revenue impact.

Every fiscal impact analysis should reflect basic elements. The cost side includes operating costs and capital costs. The revenue side includes real property revenues and operating revenues. The net fiscal impact is the total revenues minus the total costs.

Federal and state agencies are expected to reflect principles of good socially and environmentally efficient growth. Smart growth principles help reduce the net fiscal impact. It is less expensive for an agency or municipality to provide services to compact development than to sprawl or low-density development. This is due mostly to reduced costs for infrastructure such as sewer and water lines, roads, and service deliveries.

Projects that provide housing may directly affect schools. Projects that cause people to move to the area may indirectly affect schools through this action. Usually, local or regional planning agencies will have data and reports that help determine the amount of children generated directly or indirectly from a project. State education departments have requirements for schools that can be interpreted on a per-child basis to get an idea of the amount of capital improvements necessary to accommodate an increase in children. If it can be shown that the local schools are at or near capacity, then there may be a request for mitigation.

14.4 MITIGATION

Mitigation may be in the form of impact fees or establishing a phasing schedule for covering costs or for building out the project so that the services are gradually increased in accordance with the municipality's ability to accommodate growth. Some municipalities, fearful of losing a tax-revenue development to an adjacent community, may waive mitigation and may even grant tax concessions. Such an approach may be legitimate in encouraging in-field development in a depleted downtown or in encouraging a struggling local business, but it is easy for developers to play one community against another in seeking tax increment financing or other creative venues.

The imposition of impact fees may partially offset the fiscal burdens of development projects. They are only allowed under certain circumstances. Exempt agencies like those of the federal government may have a policy of payment in lieu of taxes (PLT) instead. Towns may adopt an impact fee ordinance for a particular service, such as recreation, or they may adopt an ordinance that applies to a spectrum of services, including schools, transportation, and sewer services. Most impact fee ordinances contain provisions for refunds, waivers of the fee, and a credit of in-kind contributions.

Impact fees and phasing of projects can be effective forms of mitigation. They require monitoring to evaluate their effectiveness as the project unfolds. Federal projects that make PLTs are providing a form of mitigation. Cooperating service agreements also are compensatory. Government planning efforts of federal projects can have beneficial spillover effects on local community services.

Assignment

Assume you have a project to add a new 180,000-square-foot Veterans Affairs building with outpatient counseling and veteran services located adjacent to the Jetport in Portland, Maine. Assume a building cost of $350 per square foot. There will be 100 new employees. The project includes 200 new parking spaces. Each space typically costs $1,000. Blue collar jobs average $20 per hour. Assume white-collar jobs will average $45,000 per year. Assume ten blue collar jobs for every white-collar construction job and an 8:1 ratio for the VA employees. Fifty construction workers spend 6 months making the facility. Address the following by estimating the scope and direction of your investigation for an EIA, including what parameters you think would apply. This is a critical thinking exercise, so just outline a reasonable approach and support your conclusions with calculations and explanations.

1. Assume 75% of construction materials and labor benefit the local economy. What is the economic contribution of construction from materials and salary?
2. What are the direct long-term economic benefits of the facility in terms of employee salaries?
3. What are some of the indirect short-term economic benefits likely to occur from construction?
4. What are some of the indirect economic benefits likely to occur from operation?
5. Would it be reasonable for the VA to provide the municipality with a payment in lieu of taxes (PLT)? Why or why not?
6. What other information might be provided to more fully address fiscal impacts?

14.5 CONCEPTS AND TERMS

Capacity: Ability of a municipality to deliver services. For new projects, it can be inferred by comparison with other similar municipalities of comparable size and growth rate.

Capital costs: Long-term debts incurred to provide physical improvements (buildings, water and sewer lines, and equipment) necessary to provide local services.

Infrastructure: Government systems and assets, real or virtual; roads, streets, and sidewalk networks; drainage facilities, sewer and water services; fire, police, and emergency services, schools, internet services, power generation and transmission, telecommunications, mass transit, hazardous waste removal and storage, other utilities, and the capital works required to provide public services.

Marginal abatement costs (MAC): The economic value or cost to reduce an additional unit of emission.

Payment in lieu of taxes (PLT): Federal payments to local governments to help offset the lost property tax that might have been received from a private or taxable development or property (31 USC ch. 69).

Public works: The combination of physical assets, management practices, policies, and personnel necessary for government to provide and sustain structures and services essential to the welfare and acceptable quality of life for its citizens (American Public Works Association, 2022).

Real value: The assessed property value of a municipality containing a project city. It is used to determine revenue generated by a project.

Revenues: For a municipality, primarily tax on real property and other taxes. Also income from various fees and licenses and PLTs.

Social cost of carbon (SCC): Monetary value of the costs of climate change. Estimated by the present value of future economic damage of greenhouse gases (GHG).

Shadow price: The true marginal value of a good or opportunity cost of a resource, whether or not reflected in the market price.

14.6 SELECTED RESOURCES

Burchell, R. W. and D. Listokin. (1992). *Fiscal Impact Procedures: State of the Art: The Subset Questions of Nonresidential and Open Space Costs*. New Brunswick, NJ: The Center for Urban Policy Research.

Burchell, R. W., D. Listokin and W. R. Dolphin. (1985). *The New Practitioner's Guide to Fiscal Impact Analysis*. New Brunswick, NJ: The Center for Urban Policy Research.

Duncan Associates. (2015). Impact Fees.com: A Comprehensive Gathering of Impact Fee Information and Resources. www.impactfees.com/.

Glasson, John. (2001). Socio-Economic Impacts 1: Overview and Economic Impacts. In P. Morris and R. Therivel (Eds.), *Methods of Environmental Impact Analysis*, 2nd ed. (pp. 20–41). London: Spon Press.

Hill, Elizabeth, John Bergstrom, H. Ken Cordell and J. M. Bowker. (2009). *Natural Resource Amenity Service Values and Impacts in the US*. Internet Research Information Series (IRIS). Athens, GA: USDA Forest Service.

Kotval, Zenia and John Mullin. (2006). Fiscal Impact Analysis: Methods, Cases, and Intellectual Debate. Lincoln Land Institute Working Paper. www.lincolninst.edu/subcenters/teaching-fiscal-dimensions-of-planning/materials/kotval-mullin-fiscal-impact.pdf.

14.7 TOPIC REFERENCES

American Public Works Association (APWA). (2022). What Is Public Works? Kansas City, MO: APWA. www.apwa.net.

Burchell, Robert W., David Listokin and William R. Dolphin. (1985). *The New Practitioner's Guide to Fiscal Impact Analysis*. New Brunswick, NJ: The Center for Urban Policy Research.

Sanford, Robert. (2018). *Environmental Site Plans and Development Review*. New York: Routledge.

Smart Growth Network. (2022). Smart Growth Principles. Smart Growth Online. https://smartgrowth.org/smart-growth-principles/.

US Department of Homeland Security. (2019). *A Guide to Critical Infrastructure Security and Resilience*. November 2019. Cybersecurity and Infrastructure Security Agency. Washington DC: Department of Homeland Security.

TOPIC 15

Traffic and transport systems

15.1 Environmental setting	193
15.2 Project description	194
15.3 Potential or predicted impacts	195
15.4 Mitigation and monitoring	201
15.5 Concepts and terms	202
15.6 Selected resources	204
15.7 Topic references	204

15.1 ENVIRONMENTAL SETTING

The setting of a project includes the transportation network in which the project occurs and the internal flow of traffic within a project itself. Travel and traffic are about much more than just cars. Traffic and its associated impacts include mass transit, airplanes, pedestrians, bicycles, horses, goats, and many other components. Traffic impact analysis addresses many issues, including capacity, safety, sustainability, and resilience, in addition to effects on environmental, social, and infrastructural resources.

The setting provides the context for determining if a traffic study is needed, and if so, what should be in it. A traffic study is required if federal, state, or local regulations or guidelines call for one—often a study is triggered if a threshold is reached, such as 150 residential units, 15,000 square feet of commercial space, or a large parking lot. A traffic study looks at accident rates and at levels of service (LOS) for intersections and other critical parts of a road system. They observe internal and external traffic flows and use a Trip Generation Manual to estimate trips. Traffic engineers also look at similar sites, similar traffic studies, and statistical models used by local and regional planning agencies. An outline of what a traffic study should contain is provided further on, but the essential topics a traffic study should address are background conditions, contributions from the project, a capacity analysis, a safety analysis, and recommendations.

DOI: 10.4324/9781003030713-15

Changes in numbers and rates of travel might be rapid, but associated changes to infrastructure tend to be quite slow and require considerable advanced planning. An EIA should consider effects of traffic and growth on other impact categories, notably infrastructural services, local economy, travel safety, water runoff, and aesthetics.

15.2 PROJECT DESCRIPTION

Project that deal with transport tend to be policy projects, such as federal procedures for waste materials; a physical development such as federal highway improvements; or constructing a federal facility that accommodates employees and visitors or that affects transportation sectors. The project itself may require federal or other approvals. States issue entrance permits and access permits for driveways and connections to state roads. States and agencies require traffic permits for large projects. For example, Maine issues a "traffic movement permit" for any project that generates 100 or more passenger car equivalents at peak hour (23 USC ch. 13, § 704).

Traffic impacts should be examined from three levels or perspectives: internal traffic (within the project), local external (entrances and exits, area capacities), and regional (networks, regional patterns). People tend to feel strongly about traffic issues, particularly in their neighborhoods, and the issue may become contentious. To best prepare for this, the collection of data for a traffic study should focus on a thorough understanding of the impacts quantitatively and qualitatively. The focus of traffic studies tends to be on vehicles and pedestrians, but all aspects of traffic and modes of transportation that could be affected by the project should be addressed.

In the earliest stages of project planning and long before designers start designing, the lead agency should examine and understand the capacity of a site and its surrounding region to accommodate the transportation sector. This understanding should be shared with all participants. For that task, checklists like the ones below may be used to guide project design by an approving agency or the EA authoring agency (Tables 15.1 and 15.2).

TABLE 15.1 Sample checklist for applications involving transportation review or permits.

√	Contained in Project
	Sufficient sets of legible, stamped plans of all development, including landscaping, drainage
	Erosion control, mitigation, and details
	Location map of the site relative to local roadways and state rights of way
	Written description of the current and proposed use of the property
	Amount of traffic the development is expected to generate and basis for this prediction
	Discussion of access management, including efforts to use secondary rather than primary road access, maintain effective flow of traffic and the safety of state roadways; and accommodate the access needs of the project
	Discussion of how project complies with local and regional plans
	Evidence that the intensity of the proposed land use will not overburden the regional transportation system (e.g., will not result in congestion; will maintain the current level of service [LOS] on highways)
	Background data and assumptions used in traffic study

Source: Based on Sanford (2018).

TABLE 15.2 Sample checklist for transportation construction plans.

√	Contained on Project Plans and Blueprints
	Date, title, scale, north arrow, and date/stamp for revisions
	Legal descriptions of land, monuments, and bearings
	Labels and descriptions for state and other rights of way
	Dimensions from the center line of proposed roads to connection with existing roadways and public rights of way
	Contours (existing and proposed), drainage boundaries, and flow patterns
	Existing and proposed drainage facilities
	Drainage calculations, including stormwater runoff rates, paths, and treatment
	Construction, building area, staging area footprints
	Show how project complies with local comprehensive drainage plans
	Copies of permits and approvals already obtained, including local municipalities, state agencies, watershed organizations, and US Army Corps of Engineers
	Street construction layout with dimensions and radii
	Street cross sections or contours and profile grades
	Time schedule for work
	Peak hour vehicle trips, traffic study information
	Parking areas, including overflow parking and snow storage, as needed
	Driveways and street access
	Turn lanes/movements, sight distances, and geometrics
	Number of residential units and/or square feet of commercial or industrial building space

Source: Based on Sanford (2018).

15.3 POTENTIAL OR PREDICTED IMPACTS

A project with the potential for significant impacts will include a traffic study. The traffic study, sometimes called Traffic Impact Analysis (TIA), examines the ability of transportation systems to accommodate a project. The primary focus addresses effects on drivers of vehicles and pedestrians, but traffic includes any and all modes of transportation involved as part of the project or potentially affected by the project: private roads, airports, boat docks, canals, railroads, trucks, highways, parking lots, bicycle lanes, pedestrian paths and sidewalks, bus routes, trolleys, emergency service vehicles, handicap accessibility, and helicopter pads. In transportation planning, changes in demand can occur rapidly, while changes in physical infrastructure tend to occur slowly—hence the need for planning. Fortunately, traffic analysis is a well-established field, and most large projects will contain traffic studies that can be reviewed by expert officials at the federal, state, and municipal levels. Nevertheless, regulatory review may not focus on all of the transportation-related aspects of a project. A comprehensive framework to understanding transportation studies and plans will help ensure that all relevant elements are addressed. Anything related to the movement of goods, services, and people is a potential transportation issue.

A transportation system has a number of components that should be considered in the environmental impact assessment. These include infrastructure (vehicles, roads, pathways, canals); resources (financial, energy, environmental); flow (levels of service, volumes); demand (desire or need for services); and land use (adjacent, population levels, employment, neighborhoods).

Outline of a Typical Traffic Study

1. Introduction
   Purpose of the study
   Regulatory requirements
   Objectives of the study
   Authors of the study
2. Description of proposed project or development
   Transportation uses and requirements
   Densities
   Location of project and access points
   Project phasing schedule for build-out or operation
3. Background conditions
   Limits of the study area (including rationale)
   Land uses, current and anticipated (based on approvals and on growth projections)
   Site access for roadway system, mass or public transit, pedestrian, and other
   Existing traffic volumes and conditions
   Transit and other traffic services
   Any other transportation modes not previously addressed
4. Projected traffic volumes
   Site traffic for each analysis year or phase
   Trip generation: sources of and types of trips
   Trip distribution
   Modal splits (e.g., passenger, truck, pedestrian, mass transit)
   Trip assignment to the roadway network
   Roadway traffic (for each analysis year or phase)
   Method of projection
   Area transportation plan
   Trends and growth rates
   Assignment of the projected trips to the roadway network
   Total traffic volumes (for each analysis year or phase)
5. Traffic analysis
   Site access capacity and level of service
   Intersection of access road or driveway
   Roadway
   Regional
   Site circulation and parking
   Turning radii
   Access widths, grades, and surfaces
   Emergency vehicles access
   Loading and storage facilities (including snow loads)

Traffic safety
On-site: design standards, sight distances, widths, grades, and geometry
Off-site: accident data, sight distances, speed limits, configurations, and standards
6. Improvement analysis
Improvements to accommodate background (base) traffic conditions
Improvements to accommodate base and site traffic
Alternative improvements
Funding and other mechanisms for improvements
Evaluation
7. Findings
Site accessibility
Traffic impacts: noise, aesthetics, air pollution, runoff, and safety
May also address historical resources and other environmental criteria
Necessary improvements
Compliance with local codes or accepted standards
Traffic safety
8. Recommendations
Site access plans
Internal circulation
Roadway improvements: on-site, off-site, and phasing
Signal warrants and phasing
May include alternative transportation (ZEV and sustainability)

Some agencies and reviewers use a checklist of questions for reviewing a project. Others go a step further because they have to make formal findings of fact and conclusions of law in which specific questions must be answered. Still others leave it up to individual discretion. Some transit study review questions to consider are in Table 15.3.

Government projects are expected to set a good example for growth. Several patterns associated with urban sprawl include the proliferation of curb cuts in areas characterized as "strip development" and large expanses of pavement in the form of new, wide roads and parking lots. The center of town may have a coherent network of streets, but newer development in suburban areas has looping roads with dead ends and no connectivity to adjacent parcels or roadways. During review, regulatory boards may have some opportunities to guide the project in a direction that minimizes curb cuts, places parking to the rear of buildings, reduces the width of internal roads, and encourages transportation options in the form of public transit. However, it is far better to start out with good site design that does not need remediation.

The traffic study will supply the context and references for threshold impacts and other basis for comparison. Usually, regional data is available in addition to municipal statistics. Table 15.6 shows some sample thresholds for small projects at which 100 peak hour trips or 750 daily trips might be expected; often these are the levels at which a formal traffic study is required.

TABLE 15.3 Questions to ask about a traffic study.

√	Contained on Project Plans and Blueprints
	Can you determine who prepared the study and what qualifications they have?
	Is the traffic study easy to understand? If not, this indicates something is wrong with the study or the way in which it is reported, and it should be a red flag about the project.
	Is the data used in describing the background or the impacts available?
	Does the accident history from official records mesh with local perceptions of how traffic in the area works?
	Are the data from appropriate sources?
	Was an accepted standard computer model used? What assumptions were made in the model?
	Have yearly trends and seasonal, weekly, and daily variations been accounted for?
	Are design hours, peak use, and other terms clearly defined and applied in the study?
	Do clear diagrams and tables accompany the narrative?
	Are mass transit and other modes addressed?
	Does the study address ZEV accommodation and other sustainability-related matters?
	Is the treatment of access management adequate for a project of this size?
	Are the recommended improvements and mitigations drawn on site plans?
	How do the recommendations mesh with local planning, zoning, and subdivision ordinances?

Source: Based on Sanford (2018).

TABLE 15.4 What Smart Growth transportation site planning principles does the project meet?

√	Smart Growth Principle
	Provides options for bicycle users, pedestrians, and the transportation disadvantaged and provides connections to adjoining development and regional on- and off-road networks.
	Provides opportunities for linking internal roads with an existing road network to improve connectivity and accessibility (instead of designing dead end streets and cul-de-sacs).
	Include sidewalks along at least one side of the street in urban and suburban areas. Rural subdivisions can incorporate a network of on- and off-road systems.
	Include recreational paths to provide access to regional trail networks and open land.
	Use traffic calming devices in high-density areas. Devices include road geometry (roundabouts, T-intersections), abrupt changes in road alignment, short blocks, on-street parking, raised crosswalks, street trees, and "bump-outs."
	In commercial development, design lanes or service alleys that provide rear or mid-block access to lots and parking areas.

Source: Based on Vermont Forum on Sprawl (2001).

Assignment

Assume you have a project to add a new 180,000-square-foot Veterans Affairs building with outpatient counseling and veteran services at the Portland Jetport. The project includes 200 new parking spaces. Fifty construction workers spend 6 months making the facility. Assume ten blue-collar jobs for every white-collar construction job, and an 8:1 ratio of these for the VA employees after the project is completed. The project is accessed from Route 9 and via a feeder route from the interstate.

TABLE 15.5 General transit-related questions to ask about a project.

√	Transportation-Related Questions
	Has the project minimized dangerous traffic movements?
	Does the project meet the standards in the current edition of the *Institute of Traffic Engineers' Transportation and Traffic Engineering Handbook* and other local sources of authority as adopted by resolution?
	Does the project road configuration encourage or discourage through traffic?
	Is there a mechanism in place to ensure that future traffic devices and road improvements caused in part by the project can be constructed or installed when they are warranted?
	Does the project design have an efficient site layout for access, parking, and open space?
	Will the project promote sprawl or poor growth management?
	How does it affect overall quality of life for the area?
	Are there rights of way on hand to serve the needs of the project?
	Does the project adversely affect adjacent public investments? If so, to what degree?
	Are the layout and intended use compatible with adjacent and nearby properties?
	Does the project conserve and protect valuable natural features and amenities?
	Does the project promote the efficient provision of public services?
	Does the project address social equity and environmental justice?
	Does the project address cultural, historical and archaeological resources?
	Does the project promote sustainability and reflect smart growth principles?
	Does the project address alternative and multimodal transportation?
	Does the design preserve existing healthy and long-lived trees whenever possible?
	Do proposed drainage facilities promote the use and conservation of natural watercourses and patterns of drainage?
	Has the project been designed to reduce alterations to existing topography in environmentally sensitive areas?
	Are the proposed plantings compatible with the climate of the region and micro-climate conditions on the site?
	Does the location, size, design, and orientation of signage fit with local requirements and standards, and do they avoid causing a distraction to motorists?
	Will the signage minimize obstructions and hazards to pedestrians?
	Does the project provide adequate and unrestricted access for fire and emergency vehicles?

Source: Based on Sanford (2018).

1. Estimate the number of daily trips (one-way). Supply your reasoning.
2. Does this project require a traffic study? Why or why not?
3. What are the likely major traffic concerns associated with this project?

The level of service (LOS) is a common tool for evaluating traffic conditions preparatory to assessing impacts. The LOS provides a qualitative measure based on ranges in three critical variables: average travel *speed*, *density*, and maximum service *flow* rate. Their relation depends on the prevailing roadway segment and traffic conditions. LOS reads like a grade report, with A being the highest and F representing "failing." It is not possible to have a high level of service for all intersections and flow-through conditions; however, lower levels may still be safe. LOS is

TABLE 15.6 Sample threshold trip levels based on type and size of project.

Land Use Development Type	Amount Assumed to Generate 100 Peak Hour Trips	Amount Assumed to Generate 750 Daily Trips
Single family houses	150 units	70 units
Apartments	245 ft^2	120 units
Condominium/townhouses	295 ft^2	120 units
Mobile home park	305 ft^2	150 units
Shopping center	15,500 ft^2	2,700 ft^2
Fast food restaurant	5,200 ft^2	1,200 ft^2
Convenience store w/gas	1,300 ft^2 and/or 5 pumps	-
Bank w/drive-through	4,400 ft^2	2,800 ft^2
Hotel/motel	250 rooms	90 rooms
General office	55,000 ft^2	45,000 ft^2
Medical/dental office	37,000 ft^2	26,000 ft^2
Research and development	85,000 ft^2 or 4.5 acres	70,000 ft^2 or 4 acres
Light industrial	115,000 ft^2 or 8 acres	115,000 ft^2 or 11.5 acres
Manufacturing	250,000 ft^2	195,000 ft^2

Source: ITE (2021).

not quite the same thing as a safety evaluation or rating, but there is some interplay between the two, which is largely due to motorist behavior. The Transportation Research Board, in addition to providing standards and guidelines for design in its *Highway Capacity Manual* (2015), defines LOS:

LOS A: Free flowing traffic. Drivers, passengers, and pedestrians have a high level of comfort and convenience. Each individual is unaffected by other users and is able to maneuver about safely.

LOS B: Stable flow. Other drivers are noticeable because there is some effect on behavior. Less freedom to maneuver than in level A.

LOS C: Still in the range of stable flow, but actions of others may significantly affect the individual. Safe speeds are determined by factoring in the behavior of others. Maneuvering about is more difficult, and the levels of comfort and convenience decline.

LOS D: High-density traffic but still has a stable flow. Maneuverability and speed are restricted by other traffic. Users experience low levels of comfort and convenience. Pedestrians have difficulty at times with crossings.

LOS E: Operations are at or near capacity level and are often unstable. All speeds are greatly reduced. Maneuverability is significantly hampered. Levels of comfort, convenience, and safety are dangerously low, and frustration runs high.

LOS F: Forced or breakdown flow occur because the system is over capacity. Lines queue up and move in stop-and-go waves that are highly unstable. This is a failing grade for the intersection or road segment. Service breaks down, and gridlock can occur.

A single intersection can experience various measurements of LOS for different turns and maneuvers and at different times of the day or year. Road segments can experience a wide range of LOS as well. For example, LOS drops due to congestion when motorists enter the Green and White Mountains of New England during leaf-peeping season. The same thing occurs when motorists leave a ski resort, cruise through a tourist town, or attend a large concert event.

Field observation, studies, and computer simulation programs such as HCM-94, NCAP, and ICU can inform LOS determination. The reviewing agency will list approved methods for assessing traffic LOS and safety evaluations.

15.4 MITIGATION AND MONITORING

The complexity of transportation issues makes it important to plan when traffic improvements such as signals and signal upgrades should be made. A new project can contribute to the need for a change in signals by accelerating the need for a particular change. The goal is to have a plan at the outset so that a solution to an anticipated problem is already at hand and approved.

Determining the number of vehicles per hour is a big part of figuring out what kind of traffic accommodations are needed to mitigate the impacts of a project. Anyone can conduct a simple count to get an idea of basic conditions or to check on the data reported in a permit application or assessment report. Begin the count at least a half hour before the peak hour, the hour that likely has the maximum volume of traffic. In most areas, maximum traffic volume occurs during the morning and evening weekday rush hours, typically between 6:30 and 9:30 am and between 4:00 and 7:00 pm.

Mitigation of traffic impacts may include on-site considerations such as parking. Although the public might tend to assume it to be for private automobiles, it includes service vehicles, public and mass transit, bicycles, and other modalities. Space dimensions and parking lot configuration depends on the local or state jurisdiction. Trip generation programs and manuals help in determining the number of spaces recommended. Usually, there will be a ratio of number of spaces per square foot of space, per number of restaurant seats, per hospital bed, or per dwelling unit. If local regulations require nine spaces per 1,000 square feet, this would equate to a 3:1 parking ratio.

Each project should be carefully examined for adequacy of the parking. It may help to keep in mind that the parking should be capable of supporting complete use of the site. The parking should also be adequate during winter conditions when there are piles of snow blocking some of the spaces.

Each parking configuration has its advantages and disadvantages. For example, angled parking is convenient to pull into and out of but takes up more room than right-angle parking. A parking lot is not just a place to park cars. A bare-bones parking lot may have little to commend itself except a low initial cost. Landscape and pavement islands to control turns and guide flow may be seen as an expense, but in the long run they will pay off in safety and aesthetics. The review of parking lot plans can involve many issues such as storm-water runoff, pedestrian flow, handicap accessibility, snow storage, municipal services, vehicle circulation, lighting levels, public safety, aesthetics, landscaping, physical infrastructure, and neighborhood character.

15.5 CONCEPTS AND TERMS

Traffic engineering and traffic impact assessment are huge areas of study. There are many, many terms involved—far more than we could provide here. A familiarization with the following list of technical terms will be helpful in communicating with the project traffic engineers, designers, and promoters.

American Association of State Highway and Transportation Officials (AASHTO): The nonprofit reference group for traffic and highway design standards, safety, and all transportation modes. Dedicated to fostering an integrated, safe, and effective national transportation system.

Arterials and freeways: Large public highways that convey traffic from one community to another and to major interstate highways.

Average daily traffic (ADT): The total number of vehicles passing a point or segment of a highway facility in both directions for 1 year divided by 365. While there are seasonal, weekly, daily, or hourly variations, ADT is adequate for low and moderate volume facilities where it is certain that only one or two lanes are needed.

Collector: A major residential or commercial street with two or more travelways and 36 feet in paved width, usually serving at least several thousand vehicle trips per day.

Density: The number of vehicles occupying a given length of lane or roadway averaged over time. Density is reported in terms of vehicles per mile.

Design hourly volume (DHV): Usually the 30th highest hourly volume for the design year, commonly 20 years from the time of construction completion. DHV is the total traffic in both directions of travel for two-lane rural highways. If there are more than two lanes or if there are major intersections on two-lane roads, the directional distribution of traffic during the design hour (DDHV) is also used in designing highway improvements. The percent of ADT occurring in the design hour (h) may be used to convert ADT to DHV. To do this, use DHV = (ADT)(h).

Flow (q): Defined as the number of vehicles traversing a point of roadway per unit time. Flow is reported as vehicles per hour. As such, flow doesn't tell one whether those vehicles are large trucks or passenger cars. Flow is determined by multiplying density (vehicles/mile) times speed (miles/hour).

Handicap accessibility and pedestrian flow: Federal regulations such as the Americans with Disabilities Act (ADA) govern much of handicap accessibility issues like parking spaces, sidewalk and crossing design, and wheelchair maneuverability. A consequence of sprawl and traffic increase is more pressure to improve the flow of vehicles. Sometimes the cost is borne by non-vehicle traffic, such as reduced time for pedestrians to get across intersections. Good pedestrian flow means accommodating the *least* able of our population of users, just as good highway design means being able to accommodate older vehicles.

Lane: A short, local road that may connect several places and lead to a longer street. This road may be paved 20 feet wide and carry up to 400 trips per day.

Level of service (LOS): A measure of the quality of service provided to motor vehicle traffic by a road or intersection. Measured like a letter grade: A, B, C, D, E, and F.

Peak hour traffic: This is the traffic at the busiest time of day and busiest season. Congestion has become a major problem in many parts of the country, including small towns and rural areas. A traditional response has been to widen streets and take other steps to increase flow capacity, which in turn increases peak hour traffic volume. An alternate response, "traffic calming," seeks control though other means.

Place: A short, local road, usually a cul-de-sac, with one travel lane and a paved or gravel width of 18 feet. Average daily traffic for this road is likely to be less than 100 vehicles.

Road geometry: Engineers design according to regulatory and professional standards. They often refer to the AASHTO "Green Book," *A Policy on Geometric Design of Highways and Streets* (2018, periodically updated), for commonly used standards. Occasionally there may be several factors at work, especially in the interior of a site, that operate as constraints to good road design. These factors may be the result of client needs, site conditions, or economics. Delving into these factors may be necessary in evaluating mitigation.

Road profiles: Part of the road geometry includes road profiles. The road profile exaggerates the vertical dimension, usually by a factor of ten, in order to make the diagram more readable.

Sight distance (SD): The minimum distance needed for a motorist to safely make maneuvers such as entering a road or turning. A rule of thumb is that each mile per hour of posted speed limit requires at least 11 feet of sight distance. For example, if the posted speed limit is 45 miles per hour (mph), then the minimum sight distance is 495 feet. The major SD factors are posted road speed, the speed at which drivers actually tend to drive, road grade, road alignment, road condition, and weather. Traffic engineers use the 85th percentile speed in addition to the other factors to determine sight-distance requirements. Sight distance for an existing intersection is measured from the point where the driver of a passenger car would be sitting while waiting to pull out onto the road. For a proposed intersection, the sight distance is measured from the centerline of the intended access at a height of 3.5 feet (1.06 m) above the proposed road surface. Sight distance can become an issue when background conditions change, such as adjacent road speeds and with poor maintenance of vegetative clearings.

Signal warrant: Anticipated condition or circumstances for traffic signal installation or upgrade. (See *Manual on Uniform Traffic Control Devices*, Federal Highway Administration, 2009, revised 2012.)

Stopping distance: A function of many things, including road conditions, grade, weather, type of vehicle, tire condition, highway alignment, driver reaction time, and speed of travel. The AASHTO calculations for stopping sight distance (SSD) assume a driver eye height of 3.5 feet (1.06 m). Populations of elderly drivers tend to have lower driver height, but 3.5 feet remains the standard for passenger vehicles. AASHTO (2018) uses a brake reaction time of 2.5 seconds in its calculations. Some experts have suggested that a longer time of 3.5 seconds should be used to accommodate the elderly with diminished visual, cognitive, and psychomotor capabilities (Gordon et al., 1984).

Subcollector: A local road that serves more than one parcel and leads to a larger street. It may have a paved width of 26 feet with parking on both sides of the travelway. A subcollector might carry up to 1,000 trips per day.

Transportation Impact Assessment (or **Traffic Impact Analysis**) (TIA): A study that assesses the impact of a proposed development project.

Vehicle trip: A one-way trip; if you drive to the store and back, that is two trips. A common rule of thumb is that the average household generates ten trips per day. This ensures that residential subdivisions and condominiums have adequate road design.

15.6 SELECTED RESOURCES

AASHTO. (2018). *A Policy on Geometric Design of Highways and Streets*, 7th ed. American Association of State Highway and Transportation Officials (AASHTO). [Often Referred to as the "Green Book"].
Alexiadis, Vassili, Krista Jeannotte and Andre Chandra. (2004). *Traffic Analysis Toolbox Volume 1: Traffic Analysis Tools Primer.* FHWA-HRT-04-038. June 2004. Washington DC: Federal Highway Administration. www.library.northwestern.edu/libraries-collections/evanston-campus/transportation-library Contains many references and links for transportation-related projects.
Homburger, K. and D. Perkins. (1992). *Fundamentals of Traffic Engineering*, 13th ed. Berkeley: Institute of Transportation Studies, University of California.
Wolshon, Brian and Anurag Pande. (2016). *Traffic Engineering Handbook*, 7th ed. Hoboken: ITE (Institute of Transportation Engineers).

15.7 TOPIC REFERENCES

AASHTO. (2018). *A Policy on Geometric Design of Highways and Streets*, 7th ed. Arlington VA: American Association of State Highway and Transportation Officials (AASHTO).
Federal Highway Administration. (2009, revised 2012). Manual on Uniform Traffic Control Devices. https://mutcd.fhwa.dot.gov/.
Gordon, D. A., Hugh W. McGee and Kevin G. Hooper. (1984). Driver Characteristics Impacting Highway Design and Operation. *Public Roads*, 48(1), 12–16.
Institute of Transportation Engineers (ITE). (2021). *Trip Generation Manual*, 11th ed. Washington DC: ITE.
Sanford, Robert. (2018). *Environmental Site Plans and Development Review*. New York: Routledge.
Transportation Research Board. (2015). *Highway Capacity Manual*, 6th ed. A Guide for Multimodal Mobility Analysis. Washington, DC: Transportation Research Board.
Vermont Forum on Sprawl. (2001). *Growing Smarter – Best Site Planning Practices for Residential, Commercial, and Industrial Development*. Burlington, VT.

TOPIC **16**

Writing the report

16.1	Advice on writing	205
16.2	Writing responses to comment letters	211
16.3	Concepts and terms	213
16.4	Selected resources	215
16.5	Topic references	215

16.1 ADVICE ON WRITING

EIA reports filed in response to NEPA are "detailed statements" in the form of an Environmental Assessment (EA) or an Environmental Impact Statement (EIS). The EA is a brief analysis to see if any impacts are potentially significant. It may be used as a step in preparing an EIS, or it may be the final document if there are no potentially significant impacts. The EIS is an in-depth analysis for actions that may cause significant impacts of some sort. Both documents need to be efficient. The CEQ Regulations state that the main body of the FEIS shall "normally be less than 150 pages and for proposals of unusual scope or complexity, shall normally be less than 300 pages" (CEQ § 1502.7). Even though detailed, the EIS, like the EA, should be written in "plain language" and "clear prose" per CEQ § 1502.8.

In preparing the report, it helps to picture three sets of readers: government officials and other professionals, supporters of the project, and opponents of the project. These sets are not mutually exclusive. Government officials and other professionals may have a pro- or anti-project stance or bias, and proponents or opponents may have ambiguities or be torn by various aspects. Supporters may not want the EA or EIS to be contentious or controversial. Opponents will want the EIA report to be thorough. Opponents may include skeptics, neighbors, non-government groups, and critics who are looking for any grounds procedural or content-wise to challenge the project. The EIA report has to stand up to public review and scrutiny. Its defense and refuge is in its professionalism, its scientific accuracy, and its objectivity. It must follow the agency format for NEPA compliance. Essentially, an EIA report will contain the components identified in Table 16.1.

DOI: 10.4324/9781003030713-16

TABLE 16.1 Checklist for EIA report completion.

√	Component
	Cover sheet (title page)
	Summary/abstract
	Table of contents
	Purpose/reason for the project (statement of need)
	Jurisdiction (legal reason) for the EIA (relevant laws and rules)
	Description of project
	Description of environment and field conditions
	Description of alternatives, range of reasonable alternatives including the proposed action, and no-action
	Description of consequences of the action: the environmental impacts
	Description of mitigation
	Description of residual impacts after mitigation
	Conclusion/recommendation
	List of preparers
	List of agencies and other involved groups (may include a distribution list)
	Appendices, references and index

Detwiler (2005: 2) calls for an "explicit statement of the purpose and need" that addressed a true planning purpose rather than merely to comply with NEPA. The Department of Energy "Green Book" (2004a) provides guidance on identifying the range of alternatives, including the no-action alternative, and other components of an EIS. The Green Book guidance on writing includes the environmental effects on human health, biological resources, transportation, accident analysis, environmental justice, cumulative impacts, and compliance with other acts and regulations, notably the Endangered Species Act, Clean Air Act, Clean Water Act and related floodplain and wetlands regulation, and the National Historic Preservation Act.

Detwiler (2005: 6–11) provides excellent writing advice to Department of Energy employees:
Authors and editors should follow these guidelines, which are listed in order of importance, as they write and revise EISs and EAs.

1. *Use the active voice.* The most effective way to increase clarity and shorten sentences is to use the active voice. Compare the following: A groundwater extraction system was installed in 1983 by DOE. DOE installed a groundwater extraction system in 1983. The second sentence is clearer and shorter. The focus is on the actor and its act, not the object that was acted upon. Every reviewer of EISs cites use of the passive voice as the most serious problem in the writing. . . .
2. *Eliminate freight trains.* "Freight trains" are strings of three or more nouns, adjectives and gerunds unbroken by a verb or conjunction or preposition. They are a hallmark of technical jargon. The problem with freight trains is identifying which word is the subject and which are modifiers. The reader quickly tires of keeping all these words in the air until

something comes along to indicate which word is the important one. Studies show that the reader's eyes must flash back and forth over these word strings to determine which word is the subject. This is the main reason that technical writing is described as "tiring," "wordy" or "verbose." The following are some examples of freight trains taken from EISs:

"Resource Conservation and Recovery Act permit regulations"
"South Carolina Department of Health and Environmental Control
"('SCDHEC') hazardous waste regulations"
"high dissolved iron concentrations"
"expected case waste generation forecast"
"well sampling purge water"
"nonradioactively contaminated lead shielding"
"waste management activities"

Technical writers use freight trains for the sake of precision. There are, however, several ways to eliminate freight trains without sacrificing precision. Authors should use these guidelines whenever possible to modify any freight train of three or four words. Freight trains of five or more words are unacceptable and must be revised.

    a. *Eliminate unnecessary words.* Authors can shorten "waste management activities" to "waste management" in almost every instance without the loss of any information.

    b. *Use prepositions.* Prepositions indicate which words are the modifiers: "purge water from well sampling"; "high concentrations of dissolved iron"; "the expected forecast of waste generation."

    c. *Use the possessive form.* The possessive also indicates modifiers: "the Savannah River Site's environmental impact statement for waste management."

    d. *Use infinitive and gerund phrases.* "activities to manage wastes"; "activities for managing wastes"; "trenches engineered for the greater confinement of intermediate-level wastes."

    e. *Use acronyms and abbreviations.* The use of common, easily recognized acronyms and abbreviations can shorten freight trains: "RCRA permit regulations." But see guideline 5 below.

    f. *Use pronouns and shortened forms.* In the event that one must use a mouthful such as "the Savannah River Site's environmental impact statement for waste management," there is no reason to repeat it verbatim several times in the same paragraph. One can use shortened phrases such as "this statement," "the EIS for waste management," or the pronoun "it."

    g. *Think.* What is "nonradioactively contaminated lead shielding?" Is it shielding that is contaminated with something other than radioactivity, or shielding that is not contaminated with radioactivity or anything else? The former is "lead shielding contaminated with nonradioactive materials," the latter is "uncontaminated lead shielding."

3. *Use consistent and appropriate nomenclature.* Authors and editors should use consistent names for items that appear frequently in their documents. . . . In addition, authors should select appropriate names for recurring items: an EIS should not use "scenario" or "case" to refer to an alternative. . . . [B]e consistent in their use of shortened references. For example, an

EIS can refer to a "greater confinement trench" simply as a "trench" where the meaning is clear because of prior references. However, the EIS should not refer to it as a "trench" in one section of the document and a "vault" in another.

4. *Avoid multiple subjects, verbs and objects in a single sentence.* There are limits to the amount of information a single sentence can convey. Sentences with strings of nouns, verbs and objects often exceed these limits: "For reactive, corrosive, toxic and/or ignitable wastes, RCRA requires treatment, storage and disposal in compliance with applicable laws, regulations and/or EPA guidelines." Such sentences should be split into pieces. "RCRA defines four types of characteristic wastes: reactive, corrosive, toxic and ignitable. The act requires that persons treat, store and dispose of these waste in accordance with RCRA's requirements, which are set forth in regulations, EPA guidelines and the statute itself."

5. *Use acronyms and abbreviations sparingly.* Authors should spell out every acronym (and most abbreviations) the first time (and only the first time) it is used. Acronyms and abbreviations that are unique and widely known can shorten sentences without sacrificing precision: "CERCLA," "RCRA," "WIPP," "DOE,""EPA." Acronyms for mundane items—such as TWG for "technical working group" or PMP for "project management plan"—should be avoided, especially in combination. "The EIS team is responsible for the PMP, and consists of the PM, the CPMs and the TWG," is a sentence guaranteed to send the reader to the glossary. A few more sentences like this, however, and even the most diligent reader won't bother.

6. Words and phrases that are misused and overused in EISs.
    a. *"comprises"*—Invariably misused; it means "embraces" or "includes." "An EIS comprises many chapters"; not, "An EIS is comprised of many chapter"; and not, "Several chapters comprise the EIS."
    b. *"and/or"*—Awkward; "or" alone is usually sufficient. The slash "/" is not a punctuation mark and should not be used in text; use it only and sparingly in tables, forms, figures and acronyms.
    c. *"as amended"*—Laws are amended frequently. There is no need to note this common occurrence every time a statue is mentioned in an EIS or EA. The only time it is necessary is when the document discusses in detail the effects of a recent amendment on DOE. For example "RCRA, as amended by the Federal Facility Compliance Act, imposes penalties on DOE for. . . ." Even in these cases, the document should not note the amendment every time it mentions RCRA.
    d. *"associated with"*—"In," "at," "near," and "of" are shorter.
    e. *"provide"*—"Create," "generate," "give," and "make" are good substitutes.
    f. *"considers"*—Substitute "examines," "evaluates," or "analyzes."
    g. *"presents"*—Can often be omitted. "Presents a description" can be shortened to "describes."
    h. *"implements"*—It seems that DOE never does something; instead, the Department "implements a decision to" do something. This word can often (but not always) be omitted. Similarly, the following phrases are useless filler that weaken verbs:
        1. *"resulted in the . . ."* For example, "resulted in the generation of waste" should be shortened to "generated waste."
        2. *"with the objective of . . ."* For example, shorten "with the objective of reducing costs" to "in order to reduce costs" or "to reduce costs."

3. *"consisted of . . ."* For example, "consisted of an assessment" can be replaced with "assessed."
4. *"was initiated . . ."* Replace with "started" or "began."
5. *"serves to . . ."* For example, "serves to provide" is easily shortened to "provides."

i. *"utilizes"*—"Uses" is better and shorter.
j. *"interfaces"*—Computers interface; people talk, cooperate, work together, discuss, communicate or inform.
k. *"interacts"*—Like "interface," this is a weak, vague verb that covers everything from cooperating to litigating. Use a more precise verb to describe the relationship between the parties or objects.
l. *"deinventoried"*—"Emptied" or "removed" is better.
m. *"available information"*—The adjective is usually unnecessary; it goes without saying that one cannot use, analyze or rely on information that is unavailable.
n. *"implementability"*—"Feasibility" or "workability" are better words.
o. *"generated from"*—Wastes are generated by or in a process, and at or in a facility. They may come from a process or facility, but they are not generated from a process or facility.
p. *"disposed"*—Wastes are "disposed of"—the preposition is mandatory, even at the end of a sentence.
q. *"proactive"*—This is not a word, it is management jargon. "The Department is being proactive on this issue" means little. Try instead: "The Department wants to resolve this issue" or "The Department is seeking a solution to this issue and would appreciate suggestions from stakeholders."
r. *"release"* (a document)—One "issues" or "publishes" a document.
s. *"prior to"*—Use "before."
t. *"following"*—Use "after."

7. *Nouns that do not make good verbs.*

There is a growing and noisome trend in the turning of good nouns into bad verbs. Stop it. Some examples of good nouns gone bad when used as verbs: "partner," "team," "pilot" (unless one is talking about directing a boat or plane), "dialogue" and "task." There are verbs that already describe what you are trying to say. For example, "we can discuss the issue," not "we can dialogue about the issue."

In writing the EIA, picture a specific reader, intelligent but perhaps uninformed about the specifics of the project. Know your purpose. Evaluate the contribution of each word, each sentence, each paragraph, and each chapter. Use simple, concrete, and familiar language. Each section should have a lead-in sentence or paragraph. Follow the rule of three—briefly identify what you are going to say, say it in the main text, then summarize what you said. Follow convention in units, citations, format, and all other aspects. The EIA report should be visually attractive and readable. The CEQ Regulations call for plain language and readily understandable graphics. The RFP that leads to conducting the EIA and writing the report will have specifications for how this is to be done and presented.

The EIA is a planning document and process that provides the background for use in making decisions. EIA report authors must "earn" the conclusion/results by building up to them through data collection and analysis. Many readers will first read the summary then go into the report for details.

Expect to provide a summary or a somewhat longer "executive summary." The executive summary should start with the project description, include the environmental setting, the alternatives, the major impacts, mitigations, public involvement, and a conclusion or recommendation. Most agencies will have formats or procedures for creating the summary or executive summary, along with the report itself.

The report will be distributed according to the requirements specified in agency procedures or in the RFP. The recipients will have already been on the distribution list for the notices, meetings, and documents in the EIA process. Some entities will receive the report to comment on it, others will be receiving it for archival purposes. Document repositories will receive copies. These keepers of EIA reports usually accomplish three objectives in this regard: record keeping, public access, and curation.

Assignment

Complete Table 16.2 with your critique of an EIA writing sample in accordance with what is expected for good writing in an EA or EIS. Use the EPA former evaluation system and Mr. Detwiler's advice to inform your critique. Decide on a rating system to use. It might be numerical, or a presence/absence checklist.

TABLE 16.2 EIA writing sample critique.

Desirable Writing Trait	Your Rating of Sample (Include the Scale/Range Used)	Your Comments, Supporting Information/ Examples
Assumptions are stated and appropriate		
Effective scoping of issues brings appropriate emphasis on them		
Varied sentence structure that effectively conveys points		
Balanced use of acronyms and abbreviations		
Active voice		
"Freight trains": inefficient word strings that are hard to interpret		
Meets CEQ § 1502.8 call for "plain language" and "clear prose"		
Conventional and consistent units for weights and measurements		
Appropriate significant figures for numbers. Example: 5 feet 10 in. rather than 5.83333 feet		

Desirable Writing Trait	Your Rating of Sample (Include the Scale/Range Used)	Your Comments, Supporting Information/ Examples
Regulatory terms are consistent with their regulatory definitions		
Definitions supplied as needed		
Appropriate "tone" of communication among equals (not pedantic; lateral information sharing)		
Continuity and consistency of writing among all sections of the report		
Tables and figures suitable to topic and appropriately used		

16.2 WRITING RESPONSES TO COMMENT LETTERS

Agencies with jurisdiction or "special expertise" have a "duty to comment" (40 CFR § 1503.2).

In preparing a final EIS, agencies must "assess and consider comments both individually and collectively" and respond to these comments on the draft EIA by modifying alternatives; developing and evaluating alternatives not previously given serious consideration; supplementing, improving, or modifying analyses; making factual corrections; or explaining why comments do not warrant further agency response (§ 1503.4(a)).

Tact is required, particularly in dealing with the public. Your role in NEPA if you are an agency employee or consultant might be to serve as an objective fact supplier or as a neutral facilitator of the EIA process. Perhaps the public can be biased and unfair, but you cannot. Your objectivity should be blended with empathy. You might be responding to one person or to a great many. Some EISs generate many thousands of letters and the final EIS (FEIS) can only incorporate categorical responses. Other EISs have only a few letters—these may be reproduced along with their individual response in the FEIS. The Department of Energy provides guidance for comment responses (2004b). This guidance addresses the overall process, including how to organize, track, and respond to comments. Other agencies have NEPA guidance and procedural documents that address this as well.

Assignment

1. The letter below was received in response to the Lake Champlain Lampricide DEIS, issued in 1987 by a state and federal partnership of lead agencies (New York State Department of Environmental Conservation, US Fish and Wildlife Service, and Vermont Department of Fish and Wildlife). It was reproduced in the FEIS (1990:N-307). If you were charged with writing a response to Mr. Zarzynski for inclusion in the FEIS, what might you say? Use the principles of good writing and address the substantive issues he raised.

Lake Champlain Phenomena Investigation
PO Box 2134, Wilton, New York 12866 USA

Mr. Robert P. O'Connor
New York Department of Environmental Conservation

50 Wolf Road
Albany, New York 12233

September 18, 1988

Dear Mr. O'Connor:
I recently read in the BURLINGTON (VT) FREE PRESS newspaper an article entitled "Use of lampricides in lake to be weighed" (Sept. 15, 1988).

Though I will not be able to attend any of the public hearings I did want to write to give you my personal thoughts on this important matter.

I am certainly not an expert in chemical Lampricides so it is difficult for me to comment directly on their use to control the sea lamprey problem. However, I urge your dep't consider all avenues to this problem. Will the use of such chemicals do more harm than good? I am not sure! What affect will such chemicals have upon the ecology of the lake? And should Lake Champlain be the habitat of a colony of large unidentified animals (commonly referred to as "the Lake Champlain monsters"), how might the chemical Lampricides possibly affect their well being?

I would appreciate be on your mailing list for information related to the sea lamprey problems that affect Lake Champlain. I also realize you have a very difficult job ahead of you and I wish you the very best of luck, too.

Sincerely.
/s/

Joseph W. Zarzynski
"Fellow"—The Explorers Club

Joseph W. Zarzynski, B.A., M.A.T.
Founder—Lake Champlain Phenomena Investigation
Author of Books CHAMP: BEYOND THE LEGEND
MONSTER WRECKS OF LOCH NESS AND LAKE CHAMPLAIN

2. Critique your response to Mr. Zarzynski using Table 16.3. If you are part of an EIA course, compare and critique the responses of others in your class.

TABLE 16.3 Response letter critique.

Evaluation Category	Yes	no
Exhibits desirable writing traits of grammar per Table 16.2		
Does not condescend to the writer but does make him feel "heard"		
Does not convey an agency or government endorsement or belief in Pleistocene megafauna		
Addresses potential responses to reader's questions		
Short and efficient		
Does not contradict writer		
Use of tact and discretion		

16.3 CONCEPTS AND TERMS

Categorical exclusion (CE): An action or category of actions that do not individually or cumulatively have a significant effect on the environment pursuant to NEPA (§ 1508.4).

Cooperating agency: A federal, state, or local agency or tribal government other than the agency preparing the NEPA review (lead agency), that has jurisdiction by law or special expertise with respect to environmental impacts related to a proposal and that has been deemed a cooperating agency by lead agency (§ 1508.5).

Cumulative impact: The incremental environmental impact of an action, when added to the impacts of other past, present, and reasonably foreseeable future actions (§ 1508.7).

Effect (synonymous with impact): A direct result of an action which occurs at the same time and place; or an indirect result of an action which occurs later in time or in a different place and is reasonably foreseeable (§ 1508.8).

Environmental Assessment (ES): A concise public document, prepared in compliance with NEPA that briefly provides sufficient evidence and analysis of impacts to determine whether to prepare an EIS or FONSI (§ 1508.9).

Environmental Impact Statement (EIS): A detailed written statement required by section 102(2)(C) of NEPA (§ 1508.11).

Environmentally Preferable Alternative: The alternative required by 40 CFR § 1505.2(b) to be identified in ROD, that causes the least damage to the biological and physical environment and best protects, preserves, and enhances historical, cultural, and natural resources (§ 46.30).

Extraordinary circumstances: Conditions under which a categorical exclusion may not be used and an EA or an EIS must be prepared (§ 46.205).

Finding of No Significant Impact (FONSI): A decision document prepared in compliance with NEPA, supported by an EA that presents the reasons why an action will not have significant impacts on the human environment (§ 1508.13).

Freedom of Information Act (FOIA): People have a right to government information under most circumstances. The government must justify why records should not be disclosed. Under this act, people can file Freedom of Information Requests (often called FOI or FOIR).

Human environment: The natural and physical environment and the relationship of people with that environment (§ 1508.14).

Impact topics: Headings used in a NEPA document that represent specific resources that would be affected by a proposed action or alternatives under consideration.

Jurisdiction by law: Agency authority to approve, veto, or finance all or part of a proposal (§ 1508.15).

Lead agency: The agency or agencies responsible for preparing an EA or EIS (§ 1508.16). Major federal actions with adverse effects that may be significant and which are potentially subject to federal control and responsibility (§ 1508.18).

Memorandum to File: Documentation of a determination that an existing NEPA review provides complete and accurate NEPA documentation for a specific proposal.

Mitigated FONSI: A FONSI that relies on mitigation to avoid or lessen potentially significant environmental effects of proposed actions that would otherwise need to be analyzed in an EIS.

Mitigation: Planning actions taken to avoid an impact altogether to minimize the degree or magnitude of the impact, reduce the impact over time, rectify the impact, or compensate for the impact (§ 1508.20).

NEPA Document: Generally refers to an EA or EIS and can also refer to documentation prepared for a CE.

NEPA Pathway: Level of analysis and documentation for a NEPA review. CEs, EAs, and EISs are all specific NEPA pathways.

NEPA Process: All measures necessary to comply with the procedural requirements of NEPA for a specific action (§ 1508.21).

NEPA Review: Applies broadly to all levels of NEPA documentation, whether it is a CE, EA, or EIS.

No-Action Alternative: Either "no change" from a current management direction or level of management intensity or "no project" in cases where a new project is proposed for implementation (§ 46.30).

Notice of Intent: A notice that an EIS will be prepared (§ 1508.22).

Notice of Availability: A notice submitted to the Federal Register announcing that a draft EIS, final EIS, ROD or other document is available to the public.

Preferred Alternative: The alternative identified in draft and final EISs, and most EAs, that the authoring agency believes would best accomplish the purpose and need of the proposed action while fulfilling its statutory mission and responsibilities, giving consideration to economic, environmental, technical, and other factors (§ 46.420). If the agency has a preferred alternative, it should identify it.

Proposed Action (synonymous with proposal): A federal activity or undertaking consideration (§ 46.30).

Reasonably Foreseeable Future Action: Federal and non-federal activities not yet undertaken, but sufficiently likely to occur, that a Responsible Official of ordinary prudence would take such activities into account in reaching a decision. Reasonably foreseeable future actions do not include those actions that are highly speculative or indefinite (46.30).

Record of Decision (ROD): The document that is prepared to substantiate a decision based on an EIS (§ 1505.2).

Scope: The range of actions, alternatives, and impacts to be considered in an EIS (§ 1508.25). This term can also apply to EAs.

Scoping: An early and open process for determining the scope of issues to be addressed and for identifying the significant issues related to a proposed action (§ 1501.7).

Significant: A subjective interpretation of the level of impact that will result to the human environment if an action is implemented, taking into account the context and intensity of an impact (§ 1508.27).

Tiering: The coverage of general matters in broader EISs (or EAs) with subsequent narrower statements of environmental analysis, incorporating by reference, the general discussions and concentrating on specific issues (§ 1508.28).

16.4 SELECTED RESOURCES

National Park Service. (2015). *NEPA Handbook*. Washington, DC: Department of the Interior.

United Nations. (2002). *Environmental Impact Assessment Training Resource Manual*, 2nd ed. United Nations Environmental Programme. https://wedocs.unep.org/handle/20.500.11822/26503.

US Department of Transportation and Oregon Department of Transportation. (2010). Environmental Impact Statement Template. USDOT, FHWA, ODOT. www.oregon.gov/odot/GeoEnvironmental/Docs_NEPA/EIS_Annotated_Template.pdf.

US Government Publishing Office. (2016). *Style Manual: An Official Guide to the Form and Style of Federal Government Publishing*. Washington, DC: US Government Printing Office. www.govinfo.gov/collection/gpo-style-manual?path=/gpo/U.S.%20Government%20Publishing%20Office%20Style%20Manual/2016/GPO-STYLEMANUAL-2016.

16.5 TOPIC REFERENCES

Detwiler, R. P. (2005). *The Environmental Style: Writing Environmental Assessments and Impact Statements*. Department of Energy. Washington, DC: Office of NEPA Policy and Compliance.

NYSDEC, USFWS and VTDFW. (1987). *Use of Lampricides in a Temporary Program of Sea Lamprey Control in Lake Champlain with an Assessment of Effects on Certain Fish Populations and Sport Fisheries Draft Environmental Impact Statement*. Ray Brook, NY: NYSDEC.

NYSDEC USFWS and VTDFW. (1990). *Use of Lampricides in a Temporary Program of Sea Lamprey Control in Lake Champlain with an Assessment of Effects on Certain Fish Populations and Sport Fisheries Final Environmental Impact Statement*. Ray Brook, NY: NYSDEC.

US Department of Energy (DOE). (2004a). *Recommendations for the Preparation of Environmental Assessments and Environmental Impact Assessments*, 2nd ed. DOE. Washington, DC: Office of NEPA Policy and Compliance.

US Department of Energy (DOE). (2004b). *The EIS Comment-Response Process*. DOE. Washington, DC: Office of NEPA Policy and Compliance.

TOPIC 17

Making and implementing the decision

17.1	A comprehensive approach	216
17.2	Impacts, hazards, and risk	219
17.3	Decision-making tools	222
17.4	NEPA and fast decision-making	227
17.5	Using EIA to improve the EIA process	232
17.6	Concepts and terms	232
17.7	Selected resources	233
17.8	Topic references	233

17.1 A COMPREHENSIVE APPROACH

Section 102 in Title I of NEPA requires federal agencies to use a systematic interdisciplinary approach. Decision-making in EIA occurs all along the way from the very inception of the project through all the aspects of discussing, scoping, and reviewing. Early decisions occur in the selection of the project and alternatives (Steinemann, 2001). The next big decisions are whether or not to accept the EA/EIS as complete (is it sufficient to satisfy NEPA?). Next, the agency moves forward with the final EIS (FEIS). The results of the EA or FEIS will suggest whether or not to go ahead with implementation of the project. The final decision on the project typically falls into one of the following four categories: unconditional yes, conditional yes, conditional no (potential to resubmit a modification), and unconditional no as a final action. The agency must wait 30 days after the FEIS has been filed with the EPA. The entire process of EIA is documented in the Record of Decision (ROD). Only after the ROD is issued can an agency take any actions that have been found to have negative consequences. The ROD explains why the lead agency is doing whatever it proposes. Also, the ROD must tell what mitigation measures are to be used. The ROD has to be made available to the public, although it might not necessarily be put into the Federal Register.

Some agencies such as the Federal Highway Administration treat NEPA as an "umbrella" process under which they treat all other environmental laws, executive orders, and regulations. The ultimate decision under NEPA thus incorporates these other considerations and compliances. Umbrella processes usually require comparisons of different impacts as part of a trade-off analysis of some sort. Successful comparisons require assignation of significance or importance to different impacts. This can be done in terms of economics, ecological, or social values, as well as other factors. Impact significance assessment (ISA) examines data characteristics with terms of quantification, magnitude, frequency, duration, consequences, permanence, reversibility, probability, synergistic (catalytic), geographic area, regulatory thresholds, accumulation, mitigative ability, and other factors used in subsequent decision-making processes.

NEPA decisions must also adequately consider project alternatives. In choosing among alternatives, a decision is made largely based on two factors: how well an alternative fits the project's purposes, and which option poses the least negative environmental consequences. In examining mitigation and the development of alternatives, one approach is to describe and compare the ways in which projects can vary. Table 17.1 lists some common ways projects and project alternatives can differ. These and other categories can be made into a checklist or matrix to develop and evaluate ways to modify a project.

There is a difference between ranking and rating as tools for decision-making. Evaluation can involve use of ranking, which is an internal sorting, or weight-scaling of options. If alternatives are compared to an external scale or set of criteria, this constitutes a form of rating. A matrix is an often used for an impact trade-off analysis. They may include rankings, ratings, or a combination. The literature abounds with descriptions of these tools and how they are used (Canter, 1996). The EIA should explain the method selected and how the various analyses occurred.

Ideally, most impacts are quantifiable. However, quantifiable impacts have qualitative aspects that appear in decision-making processes. Regulatory decisions concerning thresholds of acceptable impact often reflect socio-political determinations. Resource associated impacts can have a wide range of values in addition to general ecological consequences. Some common values that might arise for decision-making analysis are cultural, symbolic, aesthetic, religious, economic, medicinal, intrinsic, educational, personal, scientific. We can also measure impacts based on market value or other scaled value determents. Goods and services can be considered in terms of consumptive or non-consumptive, renewable or non-renewable, individual, or general benefit, and many other measures for comparison. Socioeconomic values of impacts and of the potential consequences of impacts are commonly sorted into three aspects:

TABLE 17.1 Some ways in which a project can be modified.

Site (selection, location, placement)	Construction
Operation	Size and scale
No-build/build options	Configuration and layout
Costs	Timing, scheduling, phasing, master planning
Demand for goods and services	Effects on stakeholders
Governing regulations	Effects on environment

1. Life support: health, ecological, climate
2. Amenity services: recreation and aesthetics
3. Materials: manufacturing, agricultural forestry, fisheries, mining, etc.

Assignment

Tagliani and Walter (2018) developed an impact significance decision tree for use in Brazil, but they suggest it for others as a general tool. Use their method (Figure 17.1) to evaluate the impact importance assessments of the PVT Land Company Integrated Solid Waste Management Facility Relocation Project FEIS (2020), Oahu, Hawai'i.

Agencies and parties may have different levels of responsibility in moving a project forward with making and accepting decisions. Legal processes highlight the nature of this as burden. The **burden of proof** (*onus probandi*) refers to the job of producing the evidence that will shift the conclusion away from the default position to one's own position. The **burden of persuasion** is often used interchangeably with burden of proof (especially in the US Supreme Court), but it can be separated out in other courts and systems. Burden of persuasion is somewhat broader

YES →							
NO ↓ Permanent? →	Irreversible? →	Diffuse? →	High Synergy? →	High Magnitude? →	Highly Cumulative? →	**Very High Impact**	
↓	↓	↓	↓	↓	↓		
Irreversible? →	Diffuse? →	High Synergy? →	High Magnitude? →	Highly Cumulative? →	**High Impact**		
↓	↓	↓	↓	↓			
Diffuse? →	High Synergy? →	High Magnitude →	Highly Cumulative? →	**High or Medium Impact**			
↓	↓	↓	↓				
High Synergy? →	High Magnitude? →	Highly Cumulative? →	**Medium Impact**				
↓	↓	↓	↓				
High Magnitude? →	Highly Cumulative? →	**Low or Medium Impact**					
↓	↓						
Highly Cumulative? →	**Low Impact**						
↓							
Very Low Impact							

FIGURE 17.1 Decision tree for impact significance in Brazil (Tagliani and Walter, 2018).

than burden of proof and may be less legalistic. In the case of an environmental complaint or objection to a process, it applies to the person who raises the issue it is their burden of persuasion to make their case. The **burden of production** may be considered as an initial burden of proof. It presents the need to show that there is enough of a case to proceed. This is perhaps the most frequent application in an EIA process.

17.2 IMPACTS, HAZARDS, AND RISK

Decision-making for EIA includes consideration of whether a project constitutes an acceptable level of risk for the environment and for the public. In the public domain, the language of risk is often less precise than as used in other assessment methods. Risk for the public may be thought of in at least six ways:

1. Activities that can be a source of risk, such as oil exploration;
2. Specific hazards that pose a threat, such as an oil spill;
3. Security (bioterrorism and related);
4. Exposure to hazards, such as oil adhering to wildlife after a spill;
5. Harm that might result from exposure, such as bleeding in the stomach if oil is ingested;
6. Loss of value placed on these consequences by society, such as temporary bird population decline from exposure.

The public tends not to think of risk and its consequences in a full range of variations and with the variety of does reactions that can occur. A linear non-threshold (LNT) dose-response model was originally intended for ionizing radiation. It spread to use in chemical carcinogens, and now often dominates the way the public thinks in terms of risk. An EIA has to account for this practice and must account for environmental justice aspects in decision-making (see Topic 13).

Emerging risks often generate public concern because they may be assumed to be uncontrollable. They may not be well understood or not competently managed. Therefore, it is important to strive for an environment of "no surprises" and to consider social issues during all stages of risk assessment and management.

Risk assessments normally follow a four-step process:

1. Identifying a risk hazard is done through already known components of a project or through specific on-site data collection. Typical questions in looking for risk may be:
   What are the features of the site?
   Are there vulnerable areas such as surface waters or delicate ecosystems?
   What are the external processes acting on the site such as weather, tide, and flooding?
   What processes currently occur on the site, such as manufacturing, resource extraction, and waste disposal?
2. Degree of reaction (how severe), and the probability or certainty of reaction (how likely).
3. Exposure. What are the most sensitive groups of participants? Are they residents? Pregnant/nursing mothers? Workers who are on site for long shifts?
4. Risk characterizations. Assigning risks to categories: (a) health; (b) economic; (c) environmental.

Once the risk is characterized, the assessment moves to considering mitigation. The mitigation is examined in terms of environmental impacts, socioeconomic factors, and environmental justice.

The EPA reviews may comment in EASs and EISs in terms of risk issues and other factors. The EPA provides a peer review guide (U.S. Environmental Protection Agency, 2015) for use by reviewers. The questions and comments can apply to NEPA documents prepared by the EPA or to the underlying work products prepared by other agencies. Other agencies also may have their own guides to peer review. The US General Accounting Office (GAO) reported on agency peer review variations, largely concerning scientific studies (1999). The Office of Science and Technology Policy provides general recommendations for conducting peer review procedures.

Assignment

Assume an EA exists to evaluate feasibility of a military readiness activity of the US Navy for training. The activity would use active sonar and explosives in the Atlantic (www.AFTTEIS.com). The site is in the Atlantic Ocean along the eastern coast of North America, in portions of the Caribbean Sea, the Gulf of Mexico, at Navy pier-side locations and port transit channels, near civilian ports, and in bays, harbors, and inland waterways (e.g., the lower Chesapeake Bay). The study area covers approximately 2.6 million square nautical miles of ocean area and includes designated Navy operating areas and special use airspace. The explosives include projectiles (gunnery), bombs, ship shock (wave) charges that simulate near misses, diver demolition charges, torpedoes, and sonobuoys. Ecological impacts could include marine mammals, sea turtles, fish, birds, marine vegetation, marine invertebrates, marine habitats, public health and safety, submerged historic sites, and air quality. A risk assessment for this project might use tables similar to Tables 17.2 and 17.3 to evaluate different environmental impacts. Make a list of some environmental impacts the investigators might want to assess for this project.

TABLE 17.2 Sample risk matrix based on probability of event and severity of event.

Likelihood	Degree of Severity (Consequences)				
	Insignificant No significant environmental harm. Minor problem.	**Minor** Some harm to environment. Some disruption possible. Costs ≤$500k).	**Moderate** Damage to environment. Significant time/resources required. Costs of $1 million.	**Major** High environmental damage. Operations severely damaged. Costs of $10 million.	**Catastrophic** Toxic environmental damage. Business survival is at risk. Costs ≥$25 million.
Almost certain (>90% chance)	High	High	Extreme	Extreme	Extreme
Likely (between 50% and 90% chance)	Moderate	High	High	Extreme	Extreme
Moderate (between 10% and 50% chance)	Low	Moderate	High	Extreme	Extreme

Likelihood	Degree of Severity (Consequences)				
Unlikely (between 3% and 10% chance)	Low	Low	Moderate	High	Extreme
Rare (>3% chance)	Low	Low	Moderate	High	High

TABLE 17.3 Program risk management assessment scale for federal defense contractors.

Event Rating	Description	Technical Performance	Cost	Schedule
5 Severe	A risk event that will have a severe impact on critical outcome objectives	Performance unacceptable; does not meet a Key Performance Parameter (KPP) requirement	Program budget impacted by greater than $20 million	Key program event or milestone delayed by more than 3 months
4 Significant	A risk event that will bring one or more stated outcome objectives below acceptable levels	Performance unacceptable; significant changes required; fails a threshold requirement	Program budget impacted by greater than $10 million but less than $20 million	Increases critical path schedule by 2–3 months
3 Moderate	A risk event that will have bring one or more stated outcome objectives well below goals but above minimum acceptable levels	Performance below goal; moderate changes required; does not meet a threshold requirement	Program budget impacted by greater than $2 million but less than $10 million; does not require significant use of program cost and/or schedule reserves	Moderate (1–2 months) schedule slip
2 Minor	A risk event that brings one or more stated outcome objectives below goals but well above minimum acceptable levels	Performance below goal but within acceptable limits; no changes required; does not meet an objective requirement	Program budget impacted by less than $2 million; development or production cost goals exceeded by 1%–5%	Minor schedule slip; non-critical path activities late; impact to critical path up to 1-month slip
1 Minimal	A risk event with little or no impact on achieving outcome objectives	Required minor performance trades within the threshold objective range; no impact on program success	Program budget not affected; cost increase can be managed within the plan	Schedule not affected; schedule adjustments can be managed within plan; able to meet key milestones with no schedule float

Source: US Department of Defense (2014).

17.3 DECISION-MAKING TOOLS

Decision tools in EIA determine how a project moves forward. They can be used to compare impacts, techniques, processes, and other things. If the decision tool compares impacts, they must be clearly understood. There are many ways impacts can vary. Relevance is the measure for how detailed the EIA needs to be in describing impacts. As a reminder, a few of the ways impacts can vary are proximity, magnitude, importance, probability, classification, relation to regulations, quantifiable, stability, area of occurrence, relation to cost, how mitigatable it might be, and commitment (reversibility).

The Leopold method or Leopold matrix (Leopold et al., 1971) is perhaps the most well-known EIA matrix approach linking projects to physical, biological and social conditions of an environment. It involves a matrix system of project activities on a horizontal axis (100 rows) and environmental factors on a vertical axis (88 columns). The intersecting cells are identified with a magnitude value and an importance rating. Magnitude is a measurement of physical area and of severity. Importance refers to the ecological consequences or significance to the impact. There is a certain degree of subjectivity in assigning values of "probable importance." First suggested for planning by the US Geological Survey, the Leopold method can help to ensure that alternative actions are evaluated and considered. It can incorporate all phases of a project and can be adapted to specific ecosystems. It is useful in providing background information for decision-making.

Assignment

1. The Leopold method was proposed by Luna Leopold, son of famed conservationist Aldo Leopold in 1971. How has this method stood up to the test of time? What are its major strengths and weaknesses?
2. The Leopold method is used for EIA project planning in many countries. Locate a recent EIS, EIA report, or journal article involving the use of a Leopold matrix and report on its utility.

The rational planning model, from urban and transportation planning practices is useful for EIA decision-making (Dzurik, 2002). A first step in this model is identification of the problem and information gathering for solutions. Step 2 is the generation of solutions, and Step 3 is developing objective assessment criteria to apply to the potential solutions and alternatives. Step 4 is selection of the best solution, which is implemented in step 5 and monitored in Step 6. Step 7 is the iterative process of feedback from the previous steps into refinement of solutions and approaches.

Many decision-making tools provide some form of impact trade-off analysis. An analysis matrix might contain a qualitative approach for considering each alternative or each impact. The approach might include a qualitative approach or a combination of qualitative and quantitative.

Importance weighting techniques

Ranking
Nominal-group process

Rating—importance to a series of decision factors
Redefined importance scale
Multi-attribute utility measurement
Unranked pairwise comparison
Ranked pairwise comparison
Delphi study.

Ranking is a value of importance by comparison with other items or units. Importance ranks first to last. The nominal-group process is a particular form of ranking derived from studies of participation in program planning.

1. Nominal (silent, independent list of ideas by a panel);
2. Round-robin is a listing of ideas in serial discussion;
3. Group discussion of each recorded idea;
4. Independent voting on priority ideas and group decision based on mathematical rank ordering.

Multi-attribute decision-making techniques can get pretty complex The variables are ranked, rated, or scaled in some sort of weighted approach. Different impacts are compared with a weighted ranking factor. An index might be created in the form $I = \Sigma\ IW_i\ R_{ij}$, where n is the number of decision factors. A composite index sums up the jth factor, and Iw_i is the importance weight of ith decision factor. R_{ij} is a ranking, rating, or scale of the ith decision factor. All that will be embedded in a software program used to make the matrix.

Assignment

1. Locate and summarize the use of the rational planning model for an agency EIA decision.
2. In making decisions we may be in a situation where we need to "compare apples to oranges." Consider Apple River, a large company with holdings in 13 states. In Maine it has 100,000 acres and a plan to build and sell 50 vacation condominium homes, 900 lots, and a golf course. Apple River will manage timber on the undeveloped property. Its management is considering whether to construct a 100,000-square-foot ecotourism facility on 150 acres and another 10 miles of trails. Apple River must deal with approvals from regional planning, fish, and wildlife, and the state department of environmental protection. You, an environmental development consultant group, are hired by a regional planning agency to advise its staff on evaluating the project. One method used under NEPA is called *trade-off analysis*. It is part of the rational planning model.
3. What criteria should the preparers consider? How should they weight or value each criterion? How should they use the weighted criteria in a decision matrix or system? What is your opinion about this system?
4. Provide a table that labels environmental/social impact areas or categories, their measurements, and how the factor is weighed for importance.

"Value tree analysis" is an environmental policy and biological impact tool often used at the federal level (Figure 17.2). It is useful for organizational management and other decision-making

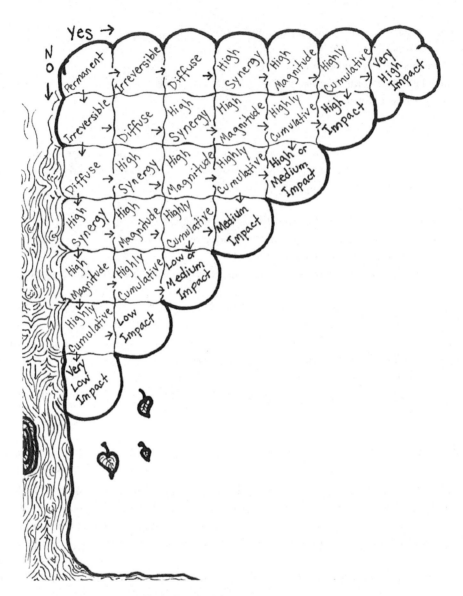

FIGURE 17.2 Value tree analysis for decisions.

processes (Pavlopoulos et al., 2010). Value tree analysis phases are problem structuring, preference calculation, recommended decisions, and sensitivity analysis. An assumption for the overall process is that there are three different parties to the process: the decision maker, a decision analyst, and more than one stakeholder. A decision maker (DM) is a person, an organization, or any other decision-making entity. In most cases the DM is also responsible for the decision and possible consequences. A decision analyst (DA) provides insight and advice to the DM in difficult decisions. The DA's task is to find the most appropriate decision alternative(s) with possible reasoning and will facilitate the decision-making process.

Problem structuring seeks answers to the following questions, which should become evident as the DM process unfolds. What are the objectives? What is the real problem? Who are the groups that fall into one or more of the three parties involved (DM, DA, stakeholders)? What information is available on these groups and on the problem?

Preference elicitation is the measurement and estimation of DM's preferences over a set of objectives. It asks the question, "What should the objectives be?" A recommended decision is developed. Sensitivity analysis explores how changes in the model might influence the decision recommendation. It asks, "What changes might be likely?" Depending on results of the sensitivity analysis, the problem might be restructured and preferences reassessed.

To see the application of value tree analysis, we can look at the case of hydrofracking the Marcellus Shale in Dimock, Pennsylvania. (For some background to this case, use Google or a similar resource to look up hydrofracking and Dimock.) The next step is to list all the organizations that might be involved in this issue—municipal, state, federal, non-profit, corporate, citizen—any entities. Then sort them into who might be a decision maker (DM), a decision analyst (DA), and stakeholders (entities that are involved or affected but are not DM or DA).

The next step is to write a problem statement, like this:

The problem is contaminated water with elevated methane and other dangerous substances like arsenic. But Dimock is the site of previous coal seam mining, which historically puts out even more methane than does mining for natural gas. So a question to be addressed is: Is the methane in Dimock from the coal or from hydrofracking or some combination?

After the problem statement is completed, define the decision context. For example: "The DM is the Pennsylvania Department of Environmental Protection (DEP) and the federal Agency for Toxic Substances and Disease Registry (ATSDR) and the EPA." You would identify the aparent or stated objectives for each decision maker, decision analyst, and stakeholder. For example:

DEP wants to protect water quality. ATSDR wants to proect human health by registering sites. EPA thinks the water is close to safe levels so that it is now treatable. Objectives: (1) provide residents with safe drinking water now, (2) determine what caused the contamination, (3) clean up the contaminants, and (4) prevent or reduce future contaminants.

The third step in the problem structuring phase is to identify decision alternatives. In this case we might say:

The EPA has lately seemed leery of controversy. Cabot Oil and Gas has denied responsibility for the contaminants. ATSDR is a tiny organization with little ability other than to collect data. DEP is based in Pennsylvania and thus beholden to the energy industry, as well as the general public.

We sort the objectives into a hierarchy of time and importance. For example: "Item 1 must be attended to immediately. Items proceed in order of importance and timeliness." We describe the attributes of the problem. For example we might say:

It is difficult to "prove" who is the culprit, but the contaminants and health issues are real. This is a difficult situation, and it has garnered national attention. Hydrofracking has the potential to be a great energy generation source, but it also has the potential for a huge environmental impact.

The above steps consitute an application of Phase A of value tree analysis. The next phase is Phase B (preference elicitation). The final phases are Phase C (recommended decision) and Phase D (sensitivity analysis).

Assignment

1. If we are going to solve the water quality problem in our hydrofracking operation, we need to describe the objectives of the deciders, so we can see how that includes what they might do. Choose one agency: EPA, DEP, or ATSDR. What are the major characteristics of or influences on this agency? These will inform how it might respond to hydrofracking.
2. Choose one stakeholder group and list its likely position in terms of the four objectives: (1) provide residents with safe drinking water now, (2) determine what caused the contamination, (3) clean up the contaminants, and (4) prevent or reduce future contaminants.
3. How sensitive is this issue? List the major factors that make this a sensitive issue. What course of action for each of the four objectives would you recommend for the agency you selected?

EIA decision-making can have public inputs at all stages of preparation and use. Public meetings and hearings should have media-sharing that allows on-site and remote viewing or input. Meetings should be held at accessible times and places. The ROD for the overall process will require documentation of public involvement in addition to an agency's specific requirements. These meetings can be used for a variety of purposes: gathering data, sharing data, promoting good relations, and making decisions. For orderliness and documentation, the meeting should use a signup sheet that calls for the attendee's name, organization or affiliation, contact information, and position regarding the action or project. For large or controversial projects, facilitators will use different sheets for different stances, as this can reduce conflict by allowing different signup tables or areas for people who are pro-project, anti-project, or neutral. Public meetings can also be organized by issues or by having issue-based kiosks in large facilities, so the people concerned about visual impacts can gather at those kiosks, for example.

Decision-making can be facilitated by setting up frameworks for analysis. Regulatory bodies will have to do this based on their jurisdiction. But issues can also be framed in terms of general importance such as sustainability or planning.

Assignment

Recall the four spheres or categories of involvement for federal projects that may have transborder impacts: direct physical impacts, trade issues, international environmental law, and compliance with federal policies and regulations. News media and public perceptions of sustainability and globalization contribute to project go or no-go decisions. The Keystone

Pipeline System route was proposed to run from the Western Canadian Sedimentary Basin in Alberta to refineries in Illinois and Texas and to oil tank farms and an oil pipeline distribution center in Cushing, Oklahoma (TransCanada Pipelines, 2021). Three phases or portions of the project are in operation, and the fourth was awaiting US government approval that appeared to be granted but then was withdrawn.

Phase I is the Keystone Pipeline, delivering oil from Hardisty, Alberta, 2,147 miles (3,456 km) to the junction at Steele City, Nebraska, and on to Wood River Refinery in Roxana, Illinois, and Patoka Oil Terminal Hub (tank farm) north of Patoka, Illinois, completed in June 2010.

Phase II is the Keystone-Cushing extension running 300 miles (480 km) from Steele City to storage and distribution facilities (tank farm) at Cushing, Oklahoma, completed in February 2011.

Phase III is the Gulf Coast Extension, running 487 miles (784 km) from Cushing to refineries at Port Arthur, Texas. It was completed in January 2014.

Phase IV, the proposed Keystone XL Pipeline, would essentially duplicate the Phase I pipeline between Hardisty, Alberta, and Steele City, Nebraska, with a shorter route and a larger-diameter pipe. It would run through Baker, Montana, where American-produced light crude oil from the Williston Basin (Bakken formation) of Montana and North Dakota would be added to the Keystone's current throughput of synthetic crude oil ("syncrude") and diluted bitumen ("dilbit") from the oil sands of Canada.

The first two phases have the capacity to deliver up to 590,000 barrels per day (94,000 m^3/d) of oil into the Midwest refineries. Phase III has the capacity to deliver up to 700,000 barrels per day (110,000 m^3/d) to the Texas refineries. By comparison, US oil production was about 9 million barrels per day (1,400,000 m^3/d) in early November 2014; in the 12 months preceding August 2014, the United States imported an average of about 7.5 million barrels of oil per day.

1. Outline the likely major aspects/implications overall for the entire Keystone XL Pipeline under the four carbon spheres of direct physical impacts, trade, international law, and compliance with federal policies and regulations.
2. What components of the project still remain under construction or under review?

17.4 NEPA AND FAST DECISION-MAKING

NEPA provides for Environmental Assessments and "alternative arrangements" for emergency response actions by federal agencies (CEQ, 2020). Insight can be gained from the Rapid Environmental Assessment (REA), developed in 2003. The REA is a qualitative data gathering and decision-making approach useful in situations where a decision must be made prior to or in absence of a full environmental assessment (EA) or EIS. Responses to emergencies, relief operations, wartime evacuation, and other situations may call for an REA. The UN High Commissioner for Refugees (UNHCR) and other UN offices are advocates for REA. Other applications include the World Bank. Climate change will likely continue to develop as a driving force for REA, as will more generalized humanitarian efforts for disaster response. Of necessity, REA tends to be more subjective than an EA or EIS but is useful to the process because, in addition to being a quick disaster relief tool, it can be used in the initial stages of impact analysis because of its ability to rapidly develop information, and because it is a driver towards standardization of data gathering tools such as indices and matrices (Kelly, 2005). A REA has three modules, an organization level assessment, a community level assessment and a "consolidation and analysis" module, leading up to an action step. The modules treat five topical areas:

1. The general context in which the disaster is taking place;
2. The identification of disaster-related factors that may have an immediate impact on the environment;
3. The identification of possible immediate environmental impacts of disaster agents;
4. The identification of unmet basic needs of disaster affected populations that could lead to an adverse impact on the environment;
5. The identification of negative environmental consequences of relief operations.

The REA identifies critical issues and provides a summary of conditions useful proceeding with a disaster relief action (Tables 17.4 and 17.5). The REA is designed to work even if people who do the assessment are not environmental impact specialists. A 2018 update addressed changes in humanitarian relief systems (Hauer, 2018). The use of REA has significant environmental justice implications.

Assignment

1. What other environmental impact factors could be considered for the REA to be used effectively for a NEPA EA on selecting emergency relocation sites for flood disasters?
2. Adapt the REA forms to make a checklist or matrix that you could use for deciding emergency relocation sites or projects in your state.

In making decisions, there are three general forms of feedback mechanisms: the EIA process, public/community outcry, and permits and licenses. The EIA process leaves it to each agency to design procedures for monitoring and reporting back on decisions and on impacts from projects. Public outcry may be considered as what occurs when the EIA process has been faulty or insufficient. Permits and licenses will have their own specific compliance procedures. If the EIA process is an "umbrella process," it may incorporate permits and licenses as part of its feedback assessment methodology.

Assignment

Mintzberg and Westley (2001) proposed three approaches to strategic decision-making: the *procedural rationality* ("think first"), the *insight and intuition* ("see first"), and *sense-making* ("do first"). Agency decisions on projects tend to be procedural, leading them to the procedural rational approach. Mintzberg and Westley argue that healthy organizations use all three approaches. Each has their use, with sense-making being useful in dealing with engagement of others and procedural rationality being good for consensus building and organizational goal achievement. The insight and intuition approach is good for creative adaptation to change. Do you think this argument applies to EIA? If so, how?

TABLE 17.4 Environmental Situation Rapid Environmental Assessment Response Form.

Natural Resources	Presence and Condition	Primary Uses	Information Source	Strength of Local Rules	Current Threat Level	Cultural Importance	Economic Importance	Total Score (sum of row)
Dominant tree species								
Dominant shrubs								
Dominant grass								
Wildlife								
Surface water								
Ground water								
Other water								
Soils								
Protected areas								
Ecologically sensitive areas								
Culturally important areas								
Other								
Priority issues				Comments on priority issues			Action and time frame	

Source: Based on Checklist 3, FRAME Toolkit (UNHCR, 2009).

TABLE 17.5 Environmental Impacts of Relief Activities Rapid Environmental Assessment Response Form.

Relief Activity	Environmental Standard or "Best practice"	Possible Environmental Impact (not exclusive)	Severity of impact	Permanence of impact	Probability of occurrence	Cultural importance	Urgency	Total score
Camp settlement location	≥15 km from sensitive areas, flood zones and slopes	Encroachment, disturbance, contamination, flooding, erosion						
Camp size	<20,000 people, ≥30 m² per person	Natural resources, health						
Ground cover and topsoil removal	Minimize disturbance	Shade, dust, erosion						
Road construction	Follow land contours, balance cut-and-fill	Drainage, dust, erosion, flooding, natural resources						
Pole for shelters	Alternative methods such as soil bricks	Deforestation excavation pits						
Other shelter material sources	Control access	Natural resource depletion, deforestation						
Water supply	Sustainable, management plan	Depletion, disturbance, contamination, population density						
Water treatment	Safe chemical use and disposal	Contamination						
Protection of water sources	Drained and protected from waste, livestock, and pollution	Mosquito breeding, pollution						
Latrines	≥30 m distance to water sources. Bottom of pit ≥2 m above water table.	Contamination of groundwater and water supply						
Site drainage	Maintain natural drainage patterns. Drain along contour lines.	Downstream pollution						
Wastewater drainage	Drain along contour lines, away from water sources. Washing and laundry downstream.	Pollution erosion						

			Comments	Action and time frame
Solid waste	Maximize re-use and recycling. Sufficient collection points. Lined landfills. Incinerate hazardous waste.	Pollution, disease vectors		
Transport of relief materials	Avoids damage to infrastructure.	Road and bridge damage		
Procurement	Prioritize recyclable and/or easily disposable materials. Reduce packaging.	Solid waste, pollution		
Food distribution and change in cooking practices	Promote fast-cooking foods. Promote energy-saving (presoaking, milling, use lids, double cooking, and improved stoves).	Deforestation: user conflicts, air pollution		
Firewood distribution	Minimal exchange for work. Assess household needs.	Deforestation: conflict with local authorities		
Expansion of farming	Sustainable practices, stakeholder land use plan	Habitat loss, biodiversity loss, deforestation, fallow period, invasive species, seed management		
Fertilizer, pesticide	Minimal	Contamination, resource extraction		
Livestock	Sustainable numbers, sanitary processing, protect watercourses	Habitat and biodiversity, animal disease, erosion, pollution, user conflicts		
Tree planting	Natural regeneration, native species	Conversion of farm land, monoculture, habitat loss		
Area wood harvest	Assessed and managed	Deforestation conflict with existing users		
Income generation	Environmentally friendly	Resource depletion, pollution, waste		
Priority issues				

Source: Based on Checklist 4, FRAME Toolkit (UNHCR, 2009).

17.5 USING EIA TO IMPROVE THE EIA PROCESS

The process of government environmental decision-making is improved using EIA. Other inputs to environmental decision-making, either through the EIA or separately, include technical analysis, public involvement, cost benefit, public policy, and political priorities (Sadler and McCabe, 2002). Improvements to any one or more of these inputs can improve overall decision-making processes and their results.

Noted public policy scholar Robert Bartlett pointed out the utility of impact assessment as a policy-making tool, labeling it "one of the major innovations in policy making and administration" (Bartlett, 1989: 1). Strategic Impact Assessment has grown into a major policy tool (Sadler, 2005). But every EIA has the ability to contribute in addition to SEAs. The EIA process has intentional feedback and iterative techniques for planning and decision-making. Essentially, there are four types of feedback loops:

1. The EIA process itself, with periodic check-in and reassessment at various points;
2. Public participation, public and community outcry;
3. Feedback built into various permits and licenses that may be associated with the project;
4. The monitoring that occurs after build-out or deployment of the project.

These four categories are not mutually exclusive. Public participation is built into the EIA process. Public participation can allow for a wide range of involvement. Many mitigation conditions include monitoring and reporting back, which contributes to the effectiveness of future EIAs (Sadler and McCabe, 2002).

Addressing cumulative impacts is a persistent issue in NEPA. Cumulative impacts must be considered, yet NEPA projects are typically reviewed on a case-by-case basis, which makes it inherently difficult to address accumulation. A strategic EIS might contemplate a program or series of future projects, all of which are related. They might be part of a master plan for infrastructure that stretches across states, or a series of independent public service projects of the same nature. Cumulative impacts can result from the accumulation of a series of impacts, from magnification of impacts, multiple impacts compounding each other, and synergistic relations in which impacts interact to create new impacts not originally contemplated (Noble, 2010).

17.6 CONCEPTS AND TERMS

Environmental Audit: Used to assess environmental impacts of an operating project. It examines site activities and is part of a periodic cycle for monitoring and review of projects.

Impact Significance Assessment (ISA): A tool for comparative analysis and decision-making in EIA.

Natural capital: An approach in which biodiversity and other natural resources are treated analogous to economic capital as depletable and subject to related qualitative and quantitative characteristics. Assessment scores of natural capital can be used in trade-off analysis and other decision-making approaches.

Rapid Environmental Assessment (REA): A United Nations tool for disaster response in cases where there is no time for a comprehensive environmental assessment report or equivalent EIS. May also be a first step in an eventual environmental assessment report or EIS.
Rapid Impact Assessment Matrix (RIAM): An analytical European Union tool for environmental impact assessment.
Strategic Environmental Assessment (SEA): The analysis of environmental impacts of plans, policies, and strategies.

17.7 SELECTED RESOURCES

Bass, Ronald E., A. I. Herson and K. E. Bogdan. (2001). *The NEPA Book: A Step-by-Step Guide on How to Comply with the National Environmental Policy Act*, 2nd ed. Point Arena, CA: Solano Press.
Council on Environmental Quality (CEQ). (2020). *Emergencies and the National Environmental Policy Act Guidance. CEQ Memorandum for Heads of Federal Departments and Agencies. Guidance Documents*. Washington, DC: Office of the President. https://ceq.doe.gov/docs/nepa-practice/emergencies-and-nepa-guidance-2020.pdf.
Council on Environmental Quality (CEQ). (2021). *CEQ Guidance Documents. NEPA.GOV.* Washington, DC: Office of the President. www.energy.gov/nepa/ceq-guidance-documents.
Environmental Law Institute. (2010). *NEPA Success Stories: Celebrating 40 Years of Transparency and Open Government*. Washington, DC: Environmental Law Institute.
Federal Permitting Improvement Steering Council (FPISC). (2017). Recommended Best Practices for Environmental Reviews and Authorizations for Infrastructure Projects. Issued in Accordance with Title 41 of the Fixing American's Surface Transportation Act (FAST-41) of 2015. https://permits.performance.gov.
Office of Management and Budget (OMB) and Council on Environmental Quality (CEQ). (2017). Guidance to Federal Agencies Regarding the Environmental Review and Authorization Process for Infrastructure Projects. www.permits.performance.gov.
Ortolono, Leonard. (1984). *Environmental Planning and Decision Making*. New York: Wiley.
Rikhtegar, Navid, N. Mansouri, A. Ahadi Oroumieh, A. Yazdani-Chamzini, E. K. Zavadskas and Simona Kildienė. (2014). Environmental Impact Assessment Based on Group Decision-Making Methods in Mining Projects. *Economic Research-Ekonomska Istraživanja*, 27(1), 378–392. doi: 10.1080/1331677X.2014.966971.
United Nations. (2002). *Environmental Impact Assessment Training Resource Manual*, 2nd ed. United Nations Environmental Programme. https://wedocs.unep.org/handle/20.500.11822/26503.
Vagiona, D. (2015). Environmental Performance Value of Projects: An Environmental Impact Assessment Tool. *International Journal of Sustainable Development and Planning*, 10(3), 315–330.

17.8 TOPIC REFERENCES

Bartlett, Robert V. (Ed.). (1989). *Policy Through Impact Assessment: Institutionalized Analysis as a Policy Strategy*. New York: Greenwood Press.
Canter, Larry W. (1996). *Environmental Impact Assessment*, 2nd ed. New York: McGraw-Hill.
Dzurik, Andrew A. (2002). *Water Resources Planning*, 3rd ed. Lanham, MD: Rowman & Littlefield.
Hauer, Moritz. (2018). *Guidelines for Rapid Environmental Impact Assessment in Disasters. Version 5. Cooperative for Assistance and Relief Everywhere (CARE)*. London: US Agency for International Development and United Nations Programme.

Kelly, Charles. (2005). *Guidelines for Rapid Environmental Impact Assessment in Disasters. Version 4.4. Cooperative for Assistance and Relief Everywhere (CARE)*. London: US Agency for International Development and United Nations Programme.

Leopold, Luna Bergere, Frank Eldridge Clarke, Bruce B. Hanshaw and James R. Balsley. (1971). *A Procedure for Evaluating Environmental Impact. US Geological Circular 645*. Washington, DC: US Geological Survey.

Mintzberg, H. and F. Westley. (2001). Decision Making: It's Not What You Think. *MIT Sloan Management Review*, 42(3), 89–93.

Noble, Bram F. (2010). *Introduction to Environmental Impact Assessment: A Guide to Principles and Practice*, 2nd ed. Oxford: Oxford University Press.

Pavlopoulos, Georgios, T. G. Soldatos, A. Barbosa-Silva and R. Schneider. (2010). A Reference Guide for Tree Analysis and Visualization. *BioData Mining*, 3(1). Published online 2010 Feb 22. doi: 10.1186/1756-0381-3-1.

PVT Land Company. (2020). PVT Integrated Solid Waste Management Facility Relocation, Waianae District, Oahu, Hawaii. Final Environmental Impact Statement. TMK: (1) 8-7-009:07. Hawaii Revised Statutes Chapter 343 Environmental Impact Statements. www.pvtland.com/wp-content/uploads/2020/07/PVT-FEIS_Vol-I-200129_new.pdf.

Sadler, Barry (Ed.). (2005). *Strategic Environmental Assessment at the Policy Level: Recent Progress, Current Status and Future Prospects*. Chlumec nad Cidlinou, Czech Republic: Regional Environmental Centre for Central and Eastern Europe.

Sadler, Barry and Mary McCabe (Eds.). (2002). *Environmental Impact Assessment Training Resources Manual*, 2nd ed. Geneva: United Nations Environmental Programme.

Steinemann, Anne. (2001). Improving Alternatives for Environmental Impact Assessment. *Environmental Impact Assessment Review*, 21, 3–21.

Tagliani, Paulo R. A. and Tatiana Walter. (2018). How to Assess the Significance of Environmental Impacts. *WIT Transactions on Ecology and the Environment*, 215, 47–55.

TransCanada Pipelines. (2021). Keystone XL Map. Terminated Route and Current Keystone Pipeline System. www.keystonexl.com/maps/.

United Nations High Commissioner for Refugees (UNHCR). (2009). Module III: Rapid Environmental Assessment. FRAME Toolkit: Framework for Assessing, Monitoring and Evaluating the Environment in Refugee-Related Operations. www.unhcr.org/4a9690239.pdf.

US Department of Defense. (2014). *Department of Defense Risk Management Guide for Defense Acquisition Programs*, 7th ed. Office of the Deputy Assistant Secretary of Defense for Systems Engineering. Washington, DC: Department of Defense.

US Environmental Protection Agency. (2015). *Peer Review Handbook*, 4th ed. Science and Technology Policy Council, US EPA. Washington, DC: EPA.

US General Accounting Office. (1999). *Federal Research: Peer Review Practices at Federal Agencies Vary*. GAO/RCED-99-99. Washington, DC: US Government Printing Office.

Appendices

Council on Environmental Quality Executive Office of the President

NATIONAL ENVIRONMENTAL POLICY ACT IMPLEMENTING REGULATIONS

40 CFR Parts 1500–1508
(May 20, 2022)

Inside Front Cover
(page intentionally left blank)

COUNCIL ON ENVIRONMENTAL QUALITY

The Council on Environmental Quality (CEQ) is housed within the Executive Office of the President. CEQ has offices within the Eisenhower Executive Office Building (EEOB) and within the Jackson Place townhouses on Lafayette Square.

CEQ provides this non-official copy for user reference. CEQ has taken steps to ensure the accuracy of this copy. Refer to the Code of Federal Regulations for the official copy of the Regulations and to the U.S. Code for the official copy of the Statutes.

Contact Information:

Council on Environmental Quality
730 Jackson Place, NW
Washington, DC 20503
(202) 395–5750

www.whitehouse.gov/ceq
www.nepa.gov/

Table of Contents

Title 40 Code of Federal Regulations
Chapter V—Council on Environmental Quality
Subchapter A—National Environmental Policy Act Implementing
Regulations 243

Part 1500—Purpose and policy   243
- § 1500.1   Purpose and policy 243
- § 1500.2   [Reserved] 244
- § 1500.3   NEPA compliance 244
- § 1500.4   Reducing paperwork 245
- § 1500.5   Reducing delay 246
- § 1500.6   Agency authority 247

Part 1501—NEPA and agency planning   247
- § 1501.1   NEPA thresholds 248
- § 1501.2   Apply NEPA early in the process 248
- § 1501.3   Determine the appropriate level of NEPA review 249
- § 1501.4   Categorical exclusions 249
- § 1501.5   Environmental assessments 250
- § 1501.6   Findings of no significant impact 250
- § 1501.7   Lead agencies 251
- § 1501.8   Cooperating agencies 252
- § 1501.9   Scoping 253
- § 1501.10 Time limits 255
- § 1501.11 Tiering 256
- § 1501.12 Incorporation by reference 256

Part 1502—Environmental impact statement   256
- § 1502.1   Purpose of environmental impact statement 257
- § 1502.2   Implementation 258
- § 1502.3   Statutory requirements for statements 258
- § 1502.4   Major Federal actions requiring the preparation of environmental impact statements 258
- § 1502.5   Timing 259
- § 1502.6   Interdisciplinary preparation 259
- § 1502.7   Page limits 260
- § 1502.8   Writing 260
- § 1502.9   Draft, final, and supplemental statements 260
- § 1502.10 Recommended format 261
- § 1502.11 Cover 261
- § 1502.12 Summary 262
- § 1502.13 Purpose and need 262
- § 1502.14 Alternatives including the proposed action 262

§ 1502.15   Affected environment 262
§ 1502.16   Environmental consequences 263
§ 1502.17   Summary of submitted alternatives, information, and analyses 264
§ 1502.18   List of preparers 264
§ 1502.19   Appendix 264
§ 1502.20   Publication of the environmental impact statement 265
§ 1502.21   Incomplete or unavailable information 265
§ 1502.22   Cost-benefit analysis 265
§ 1502.23   Methodology and scientific accuracy 266
§ 1502.24   Environmental review and consultation requirements 266

Part 1503—Commenting on environmental impact statements   266
§ 1503.1   Inviting comments and requesting information and analyses 267
§ 1503.2   Duty to comment 267
§ 1503.3   Specificity of comments and information 267
§ 1503.4   Response to comments 268

Part 1504—Pre-decisional referrals to the council of proposed federal actions determined to be environmentally unsatisfactory   269
§ 1504.1   Purpose 269
§ 1504.2   Criteria for referral 269
§ 1504.3   Procedure for referrals and response 270

Part 1505—NEPA and agency decision making   271
§ 1505.1   [Reserved] 272
§ 1505.2   Record of decision in cases requiring environmental impact statements 272
§ 1505.3   Implementing the decision 272

Part 1506—Other requirements of NEPA   273
§ 1506.1   Limitations on actions during NEPA process 273
§ 1506.2   Elimination of duplication with State, Tribal, and local procedures 274
§ 1506.3   Adoption 274
§ 1506.4   Combining documents 275
§ 1506.5   Agency responsibility for environmental documents 275
§ 1506.6   Public involvement 276
§ 1506.7   Further guidance 277
§ 1506.8   Proposals for legislation 277
§ 1506.9   Proposals for regulations 278
§ 1506.10   Filing requirements 278
§ 1506.11   Timing of agency action 278
§ 1506.12   Emergencies 279
§ 1506.13   Effective date 280

Part 1507—Agency compliance   280
§ 1507.1   Compliance 280
§ 1507.2   Agency capability to comply 280

§ 1507.3 Agency NEPA procedures 281
§ 1507.4 Agency NEPA program information 283

Part 1508—Definitions 284
§ 1508.1 Definitions 284
§ 1508.2 [Reserved] 288

The National Environmental Policy Act of 1969 289
42 U.S.C. 4321. Congressional declaration of purpose [Sec. 2]   289

Subchapter I—Policies and goals [Title I] 289
42 U.S.C. 4331. Congressional declaration of national environmental policy [Sec. 101]   289
42 U.S.C. 4332. Cooperation of agencies; reports; availability of information; recommendations; international and national coordination of efforts [Sec. 102]   290
42 U.S.C. 4333. Conformity of administrative procedures to national environmental policy [Sec. 103]   291
42 U.S.C. 4334. Other statutory obligations of agencies [Sec. 104]   292
42 U.S.C. 4335. Efforts supplemental to existing authorizations [Sec. 105]   292

Subchapter II—Council on Environmental Quality [Title II] 292
42 U.S.C. 4341. [Sec. 201] Omitted   292
42 U.S.C. 4342. Establishment; membership; Chairman; appointments [Sec. 202]   292
42 U.S.C. 4343. Employment of personnel, experts and consultants [Sec. 203]   293
42 U.S.C. 4344. Duties and functions [Sec. 204]   293
42 U.S.C. 4345. Consultation with Citizens' Advisory Committee on Environmental Quality and other representatives [Sec. 205]   294
42 U.S.C. 4346. Tenure and compensation of members [Sec. 206]   294
42 U.S.C. 4346a. Travel reimbursement by private organizations and Federal, State, and local governments [Sec. 207]   294
42 U.S.C. 4346b. Expenditures in support of international activities [Sec. 208]   295
42 U.S.C. 4347. Authorization of appropriations [Sec. 209]   295

The Environmental Quality Improvement Act of 1970 295
42 U.S.C. 4371. Congressional findings, declarations, and purposes [Sec. 202]   295
42 U.S.C. 4372. Office of Environmental Quality [Sec. 203]   296
42 U.S.C. 4373. Referral of Environmental Quality Reports to standing committees having jurisdiction [Sec. 204]   297
42 U.S.C. 4374. Authorization of appropriations [Sec. 205]   297
42 U.S.C. 4375. Office of Environmental Quality Management Fund [Sec. 206]   298

The Clean Air Act—Section 309 298
42 U.S.C. 7609. Policy review [Sec. 309]   298

Executive Order 11514—Protection and Enhancement of Environmental Quality, as Amended by Executive Order 11991 299

Title 40 Code of Federal Regulations
Chapter V—Council on Environmental Quality
Subchapter A—National Environmental Policy Act Implementing Regulations

Part 1500—Purpose and policy

Sec.
1500.1 Purpose and policy.
1500.2 [Reserved]
1500.3 NEPA compliance.
1500.4 Reducing paperwork.
1500.5 Reducing delay.
1500.6 Agency authority.

AUTHORITY: 42 U.S.C. 4321–4347; 42 U.S.C. 4371–4375; 42 U.S.C. 7609; E.O. 11514, 35 FR 4247, 3 CFR, 1966–1970, Comp., p. 902, as amended by E.O. 11991, 42 FR 26967, 3 CFR, 1977 Comp., p. 123; and E.O. 13807, 82 FR 40463, 3 CFR, 2017, Comp., p. 369.

SOURCE: 85 FR 43357, July 16, 2020

§ 1500.1 Purpose and policy

(a) The National Environmental Policy Act (NEPA) is a procedural statute intended to ensure Federal agencies consider the environmental impacts of their actions in the decision-making process. Section 101 of NEPA establishes the national environmental policy of the Federal Government to use all practicable means and measures to foster and promote the general welfare, create and maintain conditions under which man and nature can exist in productive harmony, and fulfill the social, economic, and other requirements of present and future generations of Americans. Section 102(2) of NEPA establishes the procedural requirements to carry out the policy stated in section 101 of NEPA. In particular, it requires Federal agencies to provide a detailed statement on proposals for major Federal actions significantly affecting the quality of the human environment. The purpose and function of NEPA is satisfied if Federal agencies have considered relevant environmental information, and the public has been informed regarding the decision-making process. NEPA does not mandate particular results or substantive outcomes. NEPA's purpose is not to generate paperwork or litigation, but to provide for informed decision making and foster excellent action.

(b) The regulations in this subchapter implement section 102(2) of NEPA. They provide direction to Federal agencies to determine what actions are subject to NEPA's procedural requirements and the level of NEPA review where applicable. The regulations in this subchapter are intended to ensure that relevant environmental information is identified and considered early in the process in order to ensure informed decision making by Federal agencies. The regulations in this subchapter are also intended to ensure that Federal agencies conduct environmental reviews in a coordinated, consistent, predictable and timely manner, and to reduce unnecessary burdens and delays. Finally, the regulations in this subchapter promote concurrent environmental reviews to ensure timely and efficient decision making.

§ 1500.2   [Reserved]

§ 1500.3   NEPA compliance

(a) *Mandate.* This subchapter is applicable to and binding on all Federal agencies for implementing the procedural provisions of the National Environmental Policy Act of 1969, as amended (Pub. L. 91–190, 42 U.S.C. 4321 *et seq.*) (NEPA or the Act), except where compliance would be inconsistent with other statutory requirements. The regulations in this subchapter are issued pursuant to NEPA; the Environmental Quality Improvement Act of 1970, as amended (Pub. L. 91–224, 42 U.S.C. 4371 *et seq.*); section 309 of the Clean Air Act, as amended (42 U.S.C. 7609); Executive Order 11514, Protection and Enhancement of Environmental Quality (March 5, 1970), as amended by Executive Order 11991, Relating to the Protection and Enhancement of Environmental Quality (May 24, 1977); and Executive Order 13807, Establishing Discipline and Accountability in the Environmental Review and Permitting Process for Infrastructure Projects (August 15, 2017). The regulations in this subchapter apply to the whole of section 102(2) of NEPA. The provisions of the Act and the regulations in this subchapter must be read together as a whole to comply with the law.

(b) *Exhaustion.*
   (1) To ensure informed decision making and reduce delays, agencies shall include a request for comments on potential alternatives and impacts, and identification of any relevant information, studies, or analyses of any kind concerning impacts affecting the quality of the human environment in the notice of intent to prepare an environmental impact statement (§ 1501.9(d)(7) of this chapter).
   (2) The draft and final environmental impact statements shall include a summary of all alternatives, information, and analyses submitted by State, Tribal, and local governments and other public commenters for consideration by the lead and cooperating agencies in developing the draft and final environmental impact statements (§ 1502.17 of this chapter).
   (3) For consideration by the lead and cooperating agencies, State, Tribal, and local governments and other public commenters must submit comments within the comment periods provided, and comments shall be as specific as possible (§§ 1503.1 and 1503.3 of this chapter). Comments or objections of any kind not submitted, including those based on submitted alternatives, information, and analyses, shall be forfeited as unexhausted.
   (4) Informed by the submitted alternatives, information, and analyses, including the summary in the final environmental impact statement (§ 1502.17 of this chapter) and the agency's response to comments in the final environmental impact statement (§ 1503.4 of this chapter), together with any other material in the record that he or she determines relevant, the decision maker shall certify in the record of decision that

the agency considered all of the alternatives, information, and analyses, and objections submitted by States, Tribal, and local governments and other public commenters for consideration by the lead and cooperating agencies in developing the environmental impact statement (§ 1505.2(b) of this chapter).
(c) *Review of NEPA compliance.* It is the Council's intention that judicial review of agency compliance with the regulations in this subchapter not occur before an agency has issued the record of decision or taken other final agency action. It is the Council's intention that any allegation of noncompliance with NEPA and the regulations in this subchapter should be resolved as expeditiously as possible. Consistent with their organic statutes, and as part of implementing the exhaustion provisions in paragraph (b) of this section, agencies may structure their procedures to include an appropriate bond or other security requirement.
(d) *Remedies.* Harm from the failure to comply with NEPA can be remedied by compliance with NEPA's procedural requirements as interpreted in the regulations in this subchapter. It is the Council's intention that the regulations in this subchapter create no presumption that violation of NEPA is a basis for injunctive relief or for a finding of irreparable harm. The regulations in this subchapter do not create a cause of action or right of action for violation of NEPA, which contains no such cause of action or right of action. It is the Council's intention that any actions to review, enjoin, stay, vacate, or otherwise alter an agency decision on the basis of an alleged NEPA violation be raised as soon as practicable after final agency action to avoid or minimize any costs to agencies, applicants, or any affected third parties. It is also the Council's intention that minor, non-substantive errors that have no effect on agency decision making shall be considered harmless and shall not invalidate an agency action.
(e) *Severability.* The sections of this subchapter are separate and severable from one another. If any section or portion therein is stayed or determined to be invalid, or the applicability of any section to any person or entity is held invalid, it is the Council's intention that the validity of the remainder of those parts shall not be affected, with the remaining sections to continue in effect.

§ 1500.4 Reducing paperwork

Agencies shall reduce excessive paperwork by:

(a) Using categorical exclusions to define categories of actions that normally do not have a significant effect on the human environment and therefore do not require preparation of an environmental impact statement (§ 1501.4 of this chapter).
(b) Using a finding of no significant impact when an action not otherwise excluded will not have a significant effect on the human environment and therefore does not require preparation of an environmental impact statement (§ 1501.6 of this chapter).
(c) Reducing the length of environmental documents by means such as meeting appropriate page limits (§§ 1501.5(f) and 1502.7 of this chapter).
(d) Preparing analytic and concise environmental impact statements (§ 1502.2 of this chapter).
(e) Discussing only briefly issues other than significant ones (§ 1502.2(b) of this chapter).
(f) Writing environmental impact statements in plain language (§ 1502.8 of this chapter).

(g) Following a clear format for environmental impact statements (§ 1502.10 of this chapter).
(h) Emphasizing the portions of the environmental impact statement that are useful to decision makers and the public (*e.g.*, §§ 1502.14 and 1502.15 of this chapter) and reducing emphasis on background material (§ 1502.1 of this chapter).
(i) Using the scoping process, not only to identify significant environmental issues deserving of study, but also to deemphasize insignificant issues, narrowing the scope of the environmental impact statement process accordingly (§ 1501.9 of this chapter).
(j) Summarizing the environmental impact statement (§ 1502.12 of this chapter).
(k) Using programmatic, policy, or plan environmental impact statements and tiering from statements of broad scope to those of narrower scope, to eliminate repetitive discussions of the same issues (§§ 1501.11 and 1502.4 of this chapter).
(l) Incorporating by reference (§ 1501.12 of this chapter).
(m) Integrating NEPA requirements with other environmental review and consultation requirements (§ 1502.24 of this chapter).
(n) Requiring comments to be as specific as possible (§ 1503.3 of this chapter).
(o) Attaching and publishing only changes to the draft environmental impact statement, rather than rewriting and publishing the entire statement when changes are minor (§ 1503.4(c) of this chapter).
(p) Eliminating duplication with State, Tribal, and local procedures, by providing for joint preparation of environmental documents where practicable (§ 1506.2 of this chapter), and with other Federal procedures, by providing that an agency may adopt appropriate environmental documents prepared by another agency (§ 1506.3 of this chapter).
(q) Combining environmental documents with other documents (§ 1506.4 of this chapter).

§ 1500.5 Reducing delay

Agencies shall reduce delay by:

(a) Using categorical exclusions to define categories of actions that normally do not have a significant effect on the human environment (§ 1501.4 of this chapter) and therefore do not require preparation of an environmental impact statement.
(b) Using a finding of no significant impact when an action not otherwise excluded will not have a significant effect on the human environment (§ 1501.6 of this chapter) and therefore does not require preparation of an environmental impact statement.
(c) Integrating the NEPA process into early planning (§ 1501.2 of this chapter).
(d) Engaging in interagency cooperation before or as the environmental assessment or environmental impact statement is prepared, rather than awaiting submission of comments on a completed document (§§ 1501.7 and 1501.8 of this chapter).
(e) Ensuring the swift and fair resolution of lead agency disputes (§ 1501.7 of this chapter).
(f) Using the scoping process for an early identification of what are and what are not the real issues (§ 1501.9 of this chapter).
(g) Meeting appropriate time limits for the environmental assessment and environmental impact statement processes (§ 1501.10 of this chapter).

(h) Preparing environmental impact statements early in the process (§ 1502.5 of this chapter).
(i) Integrating NEPA requirements with other environmental review and consultation requirements (§ 1502.24 of this chapter).
(j) Eliminating duplication with State, Tribal, and local procedures by providing for joint preparation of environmental documents where practicable (§ 1506.2 of this chapter) and with other Federal procedures by providing that agencies may jointly prepare or adopt appropriate environmental documents prepared by another agency (§ 1506.3 of this chapter).
(k) Combining environmental documents with other documents (§ 1506.4 of this chapter).
(l) Using accelerated procedures for proposals for legislation (§ 1506.8 of this chapter).

§ 1500.6   Agency authority

Each agency shall interpret the provisions of the Act as a supplement to its existing authority and as a mandate to view policies and missions in the light of the Act's national environmental objectives, to the extent consistent with its existing authority. Agencies shall review their policies, procedures, and regulations accordingly and revise them as necessary to ensure full compliance with the purposes and provisions of the Act as interpreted by the regulations in this subchapter. The phrase "to the fullest extent possible" in section 102 of NEPA means that each agency of the Federal Government shall comply with that section, consistent with § 1501.1 of this chapter. Nothing contained in the regulations in this subchapter is intended or should be construed to limit an agency's other authorities or legal responsibilities.

Part 1501—NEPA and agency planning

Sec.
1501.1   NEPA thresholds.
1501.2   Apply NEPA early in the process.
1501.3   Determine the appropriate level of NEPA review.
1501.4   Categorical exclusions.
1501.5   Environmental assessments.
1501.6   Findings of no significant impact.
1501.7   Lead agencies.
1501.8   Cooperating agencies.
1501.9   Scoping.
1501.10 Time limits.
1501.11 Tiering.
1501.12 Incorporation by reference.

AUTHORITY: 42 U.S.C. 4321–4347; 42 U.S.C. 4371–4375; 42 U.S.C. 7609; E.O. 11514, 35 FR 4247, 3 CFR, 1966–1970, Comp., p. 902, as amended by E.O. 11991, 42 FR 26967, 3 CFR, 1977 Comp., p. 123; and E.O. 13807, 82 FR 40463, 3 CFR, 2017, Comp., p. 369.

SOURCE: 85 FR 43357, July 16, 2020

§ 1501.1 NEPA thresholds

(a) In assessing whether NEPA applies or is otherwise fulfilled, Federal agencies should determine:
   (1) Whether the proposed activity or decision is expressly exempt from NEPA under another statute;
   (2) Whether compliance with NEPA would clearly and fundamentally conflict with the requirements of another statute;
   (3) Whether compliance with NEPA would be inconsistent with Congressional intent expressed in another statute;
   (4) Whether the proposed activity or decision is a major Federal action;
   (5) Whether the proposed activity or decision, in whole or in part, is a non-discretionary action for which the agency lacks authority to consider environmental effects as part of its decision-making process; and
   (6) Whether the proposed action is an action for which another statute's requirements serve the function of agency compliance with the Act.
(b) Federal agencies may make determinations under this section in their agency NEPA procedures (§ 1507.3(d) of this chapter) or on an individual basis, as appropriate.
   (1) Federal agencies may seek the Council's assistance in making an individual determination under this section.
   (2) An agency shall consult with other Federal agencies concerning their concurrence in statutory determinations made under this section where more than one Federal agency administers the statute.

§ 1501.2 Apply NEPA early in the process

(a) Agencies should integrate the NEPA process with other planning and authorization processes at the earliest reasonable time to ensure that agencies consider environmental impacts in their planning and decisions, to avoid delays later in the process, and to head off potential conflicts.
(b) Each agency shall:
   (1) Comply with the mandate of section 102(2)(A) of NEPA to utilize a systematic, interdisciplinary approach which will ensure the integrated use of the natural and social sciences and the environmental design arts in planning and in decision making which may have an impact on man's environment, as specified by § 1507.2(a) of this chapter.
   (2) Identify environmental effects and values in adequate detail so the decision maker can appropriately consider such effects and values alongside economic and technical analyses. Whenever practicable, agencies shall review and publish environmental documents and appropriate analyses at the same time as other planning documents.
   (3) Study, develop, and describe appropriate alternatives to recommended courses of action in any proposal that involves unresolved conflicts concerning alternative uses of available resources as provided by section 102(2)(E) of NEPA.
   (4) Provide for actions subject to NEPA that are planned by private applicants or other non-Federal entities before Federal involvement so that:

(i) Policies or designated staff are available to advise potential applicants of studies or other information foreseeably required for later Federal action.
(ii) The Federal agency consults early with appropriate State, Tribal, and local governments and with interested private persons and organizations when their involvement is reasonably foreseeable.
(iii) The Federal agency commences its NEPA process at the earliest reasonable time (§§ 1501.5(d) and 1502.5(b) of this chapter).

§ 1501.3 Determine the appropriate level of NEPA review

(a) In assessing the appropriate level of NEPA review, Federal agencies should determine whether the proposed action:
(1) Normally does not have significant effects and is categorically excluded (§ 1501.4);
(2) Is not likely to have significant effects or the significance of the effects is unknown and is therefore appropriate for an environmental assessment (§ 1501.5); or
(3) Is likely to have significant effects and is therefore appropriate for an environmental impact statement (part 1502 of this chapter).
(b) In considering whether the effects of the proposed action are significant, agencies shall analyze the potentially affected environment and degree of the effects of the action. Agencies should consider connected actions consistent with § 1501.9(e)(1).
(1) In considering the potentially affected environment, agencies should consider, as appropriate to the specific action, the affected area (national, regional, or local) and its resources, such as listed species and designated critical habitat under the Endangered Species Act. Significance varies with the setting of the proposed action. For instance, in the case of a site-specific action, significance would usually depend only upon the effects in the local area.
(2) In considering the degree of the effects, agencies should consider the following, as appropriate to the specific action:
(i) Both short- and long-term effects.
(ii) Both beneficial and adverse effects.
(iii) Effects on public health and safety.
(iv) Effects that would violate Federal, State, Tribal, or local law protecting the environment.

§ 1501.4 Categorical exclusions

(a) For efficiency, agencies shall identify in their agency NEPA procedures (§ 1507.3(e)(2)(ii) of this chapter) categories of actions that normally do not have a significant effect on the human environment, and therefore do not require preparation of an environmental assessment or environmental impact statement.
(b) If an agency determines that a categorical exclusion identified in its agency NEPA procedures covers a proposed action, the agency shall evaluate the action for extraordinary circumstances in which a normally excluded action may have a significant effect.

(1) If an extraordinary circumstance is present, the agency nevertheless may categorically exclude the proposed action if the agency determines that there are circumstances that lessen the impacts or other conditions sufficient to avoid significant effects.
(2) If the agency cannot categorically exclude the proposed action, the agency shall prepare an environmental assessment or environmental impact statement, as appropriate.

§ 1501.5 Environmental assessments

(a) An agency shall prepare an environmental assessment for a proposed action that is not likely to have significant effects or when the significance of the effects is unknown unless the agency finds that a categorical exclusion (§ 1501.4) is applicable or has decided to prepare an environmental impact statement.
(b) An agency may prepare an environmental assessment on any action in order to assist agency planning and decision making.
(c) An environmental assessment shall:
   (1) Briefly provide sufficient evidence and analysis for determining whether to prepare an environmental impact statement or a finding of no significant impact; and
   (2) Briefly discuss the purpose and need for the proposed action, alternatives as required by section 102(2)(E) of NEPA, and the environmental impacts of the proposed action and alternatives, and include a listing of agencies and persons consulted.
(d) For applications to the agency requiring an environmental assessment, the agency shall commence the environmental assessment as soon as practicable after receiving the application.
(e) Agencies shall involve the public, State, Tribal, and local governments, relevant agencies, and any applicants, to the extent practicable in preparing environmental assessments.
(f) The text of an environmental assessment shall be no more than 75 pages, not including appendices, unless a senior agency official approves in writing an assessment to exceed 75 pages and establishes a new page limit.
(g) Agencies may apply the following provisions to environmental assessments:
   (1) Section 1502.21 of this chapter—Incomplete or unavailable information;
   (2) Section 1502.23 of this chapter—Methodology and scientific accuracy; and
   (3) Section 1502.24 of this chapter—Environmental review and consultation requirements.

§ 1501.6 Findings of no significant impact

(a) An agency shall prepare a finding of no significant impact if the agency determines, based on the environmental assessment, not to prepare an environmental impact statement because the proposed action will not have significant effects.
   (1) The agency shall make the finding of no significant impact available to the affected public as specified in § 1506.6(b) of this chapter.
   (2) In the following circumstances, the agency shall make the finding of no significant impact available for public review for 30 days before the agency makes its final determination whether to prepare an environmental impact statement and before the action may begin:

(i) The proposed action is or is closely similar to one that normally requires the preparation of an environmental impact statement under the procedures adopted by the agency pursuant to § 1507.3 of this chapter; or

(ii) The nature of the proposed action is one without precedent.

(b) The finding of no significant impact shall include the environmental assessment or incorporate it by reference and shall note any other environmental documents related to it (§ 1501.9(f)(3)). If the assessment is included, the finding need not repeat any of the discussion in the assessment but may incorporate it by reference.

(c) The finding of no significant impact shall state the authority for any mitigation that the agency has adopted and any applicable monitoring or enforcement provisions. If the agency finds no significant impacts based on mitigation, the mitigated finding of no significant impact shall state any enforceable mitigation requirements or commitments that will be undertaken to avoid significant impacts.

§ 1501.7 Lead agencies

(a) A lead agency shall supervise the preparation of an environmental impact statement or a complex environmental assessment if more than one Federal agency either:

(1) Proposes or is involved in the same action; or

(2) Is involved in a group of actions directly related to each other because of their functional interdependence or geographical proximity.

(b) Federal, State, Tribal, or local agencies, including at least one Federal agency, may act as joint lead agencies to prepare an environmental impact statement or environmental assessment (§ 1506.2 of this chapter).

(c) If an action falls within the provisions of paragraph (a) of this section, the potential lead agencies shall determine, by letter or memorandum, which agency will be the lead agency and which will be cooperating agencies. The agencies shall resolve the lead agency question so as not to cause delay. If there is disagreement among the agencies, the following factors (which are listed in order of descending importance) shall determine lead agency designation:

(1) Magnitude of agency's involvement.

(2) Project approval or disapproval authority.

(3) Expertise concerning the action's environmental effects.

(4) Duration of agency's involvement.

(5) Sequence of agency's involvement.

(d) Any Federal agency, or any State, Tribal, or local agency or private person substantially affected by the absence of lead agency designation, may make a written request to the senior agency officials of the potential lead agencies that a lead agency be designated.

(e) If Federal agencies are unable to agree on which agency will be the lead agency or if the procedure described in paragraph (c) of this section has not resulted in a lead agency designation within 45 days, any of the agencies or persons concerned may file a request with the Council asking it to determine which Federal agency shall be the lead agency. A copy of the request shall be transmitted to each potential lead agency. The request shall consist of:

(1) A precise description of the nature and extent of the proposed action; and

(2) A detailed statement of why each potential lead agency should or should not be the lead agency under the criteria specified in paragraph (c) of this section.

(f) Any potential lead agency may file a response within 20 days after a request is filed with the Council. As soon as possible, but not later than 20 days after receiving the request and all responses to it, the Council shall determine which Federal agency will be the lead agency and which other Federal agencies will be cooperating agencies.

(g) To the extent practicable, if a proposal will require action by more than one Federal agency and the lead agency determines that it requires preparation of an environmental impact statement, the lead and cooperating agencies shall evaluate the proposal in a single environmental impact statement and issue a joint record of decision. To the extent practicable, if a proposal will require action by more than one Federal agency and the lead agency determines that it requires preparation of an environmental assessment, the lead and cooperating agencies should evaluate the proposal in a single environmental assessment and, where appropriate, issue a joint finding of no significant impact.

(h) With respect to cooperating agencies, the lead agency shall:
(1) Request the participation of each cooperating agency in the NEPA process at the earliest practicable time.
(2) Use the environmental analysis and proposals of cooperating agencies with jurisdiction by law or special expertise, to the maximum extent practicable.
(3) Meet with a cooperating agency at the latter's request.
(4) Determine the purpose and need, and alternatives in consultation with any cooperating agency.

(i) The lead agency shall develop a schedule, setting milestones for all environmental reviews and authorizations required for implementation of the action, in consultation with any applicant and all joint lead, cooperating, and participating agencies, as soon as practicable.

(j) If the lead agency anticipates that a milestone will be missed, it shall notify appropriate officials at the responsible agencies. As soon as practicable, the responsible agencies shall elevate the issue to the appropriate officials of the responsible agencies for timely resolution.

§ 1501.8 Cooperating agencies

(a) The purpose of this section is to emphasize agency cooperation early in the NEPA process. Upon request of the lead agency, any Federal agency with jurisdiction by law shall be a cooperating agency. In addition, upon request of the lead agency, any other Federal agency with special expertise with respect to any environmental issue may be a cooperating agency. A State, Tribal, or local agency of similar qualifications may become a cooperating agency by agreement with the lead agency. An agency may request that the lead agency designate it a cooperating agency, and a Federal agency may appeal a denial of its request to the Council, in accordance with § 1501.7(e).

(b) Each cooperating agency shall:
(1) Participate in the NEPA process at the earliest practicable time.
(2) Participate in the scoping process (described in § 1501.9).

(3) On request of the lead agency, assume responsibility for developing information and preparing environmental analyses, including portions of the environmental impact statement or environmental assessment concerning which the cooperating agency has special expertise.

(4) On request of the lead agency, make available staff support to enhance the lead agency's interdisciplinary capability.

(5) Normally use its own funds. To the extent available funds permit, the lead agency shall fund those major activities or analyses it requests from cooperating agencies. Potential lead agencies shall include such funding requirements in their budget requests.

(6) Consult with the lead agency in developing the schedule (§ 1501.7(i)), meet the schedule, and elevate, as soon as practicable, to the senior agency official of the lead agency any issues relating to purpose and need, alternatives, or other issues that may affect any agencies' ability to meet the schedule.

(7) Meet the lead agency's schedule for providing comments and limit its comments to those matters for which it has jurisdiction by law or special expertise with respect to any environmental issue consistent with § 1503.2 of this chapter.

(8) To the maximum extent practicable, jointly issue environmental documents with the lead agency.

(c) In response to a lead agency's request for assistance in preparing the environmental documents (described in paragraph (b)(3), (4), or (5) of this section), a cooperating agency may reply that other program commitments preclude any involvement or the degree of involvement requested in the action that is the subject of the environmental impact statement or environmental assessment. The cooperating agency shall submit a copy of this reply to the Council and the senior agency official of the lead agency.

§ 1501.9 Scoping

(a) *Generally*. Agencies shall use an early and open process to determine the scope of issues for analysis in an environmental impact statement, including identifying the significant issues and eliminating from further study non-significant issues. Scoping may begin as soon as practicable after the proposal for action is sufficiently developed for agency consideration. Scoping may include appropriate pre-application procedures or work conducted prior to publication of the notice of intent.

(b) *Invite cooperating and participating agencies*. As part of the scoping process, the lead agency shall invite the participation of likely affected Federal, State, Tribal, and local agencies and governments, the proponent of the action, and other likely affected or interested persons (including those who might not be in accord with the action), unless there is a limited exception under § 1507.3(f)(1) of this chapter.

(c) *Scoping outreach*. As part of the scoping process the lead agency may hold a scoping meeting or meetings, publish scoping information, or use other means to communicate with those persons or agencies who may be interested or affected, which the agency may integrate with any other early planning meeting. Such a scoping meeting will often be appropriate when the impacts of a particular action are confined to specific sites.

(d) *Notice of intent.* As soon as practicable after determining that a proposal is sufficiently developed to allow for meaningful public comment and requires an environmental impact statement, the lead agency shall publish a notice of intent to prepare an environmental impact statement in the *Federal Register*, except as provided in § 1507.3(f)(3) of this chapter. An agency also may publish notice in accordance with § 1506.6 of this chapter. The notice shall include, as appropriate:
  (1) The purpose and need for the proposed action;
  (2) A preliminary description of the proposed action and alternatives the environmental impact statement will consider;
  (3) A brief summary of expected impacts;
  (4) Anticipated permits and other authorizations;
  (5) A schedule for the decision-making process;
  (6) A description of the public scoping process, including any scoping meeting(s);
  (7) A request for identification of potential alternatives, information, and analyses relevant to the proposed action (*see* § 1502.17 of this chapter); and
  (8) Contact information for a person within the agency who can answer questions about the proposed action and the environmental impact statement.
(e) *Determination of scope.* As part of the scoping process, the lead agency shall determine the scope and the significant issues to be analyzed in depth in the environmental impact statement. To determine the scope of environmental impact statements, agencies shall consider:
  (1) Actions (other than unconnected single actions) that may be connected actions, which means that they are closely related and therefore should be discussed in the same impact statement. Actions are connected if they:
    (i) Automatically trigger other actions that may require environmental impact statements;
    (ii) Cannot or will not proceed unless other actions are taken previously or simultaneously; or
    (iii) Are interdependent parts of a larger action and depend on the larger action for their justification.
  (2) Alternatives, which include the no action alternative; other reasonable courses of action; and mitigation measures (not in the proposed action).
  (3) Impacts.
(f) *Additional scoping responsibilities.* As part of the scoping process, the lead agency shall:
  (1) Identify and eliminate from detailed study the issues that are not significant or have been covered by prior environmental review(s) (§ 1506.3 of this chapter), narrowing the discussion of these issues in the statement to a brief presentation of why they will not have a significant effect on the human environment or providing a reference to their coverage elsewhere.
  (2) Allocate assignments for preparation of the environmental impact statement among the lead and cooperating agencies, with the lead agency retaining responsibility for the statement.
  (3) Indicate any public environmental assessments and other environmental impact statements that are being or will be prepared and are related to but are not part of the scope of the impact statement under consideration.

(4) Identify other environmental review, authorization, and consultation requirements so the lead and cooperating agencies may prepare other required analyses and studies concurrently and integrated with the environmental impact statement, as provided in § 1502.24 of this chapter.
(5) Indicate the relationship between the timing of the preparation of environmental analyses and the agencies' tentative planning and decision-making schedule.
(g) *Revisions.* An agency shall revise the determinations made under paragraphs (b), (c), (e), and (f) of this section if substantial changes are made later in the proposed action, or if significant new circumstances or information arise which bear on the proposal or its impacts.

§ 1501.10 Time limits

(a) To ensure that agencies conduct NEPA reviews as efficiently and expeditiously as practicable, Federal agencies should set time limits appropriate to individual actions or types of actions (consistent with the time intervals required by § 1506.11 of this chapter).
(b) To ensure timely decision making, agencies shall complete:
  (1) Environmental assessments within 1 year unless a senior agency official of the lead agency approves a longer period in writing and establishes a new time limit. One year is measured from the date of agency decision to prepare an environmental assessment to the publication of an environmental assessment or a finding of no significant impact.
  (2) Environmental impact statements within 2 years unless a senior agency official of the lead agency approves a longer period in writing and establishes a new time limit. Two years is measured from the date of the issuance of the notice of intent to the date a record of decision is signed.
(c) The senior agency official may consider the following factors in determining time limits:
  (1) Potential for environmental harm.
  (2) Size of the proposed action.
  (3) State of the art of analytic techniques.
  (4) Degree of public need for the proposed action, including the consequences of delay.
  (5) Number of persons and agencies affected.
  (6) Availability of relevant information.
  (7) Other time limits imposed on the agency by law, regulations, or Executive order.
(d) The senior agency official may set overall time limits or limits for each constituent part of the NEPA process, which may include:
  (1) Decision on whether to prepare an environmental impact statement (if not already decided).
  (2) Determination of the scope of the environmental impact statement.
  (3) Preparation of the draft environmental impact statement.
  (4) Review of any comments on the draft environmental impact statement from the public and agencies.
  (5) Preparation of the final environmental impact statement.
  (6) Review of any comments on the final environmental impact statement.
  (7) Decision on the action based in part on the environmental impact statement.

(e) The agency may designate a person (such as the project manager or a person in the agency's office with NEPA responsibilities) to expedite the NEPA process.
(f) State, Tribal, or local agencies or members of the public may request a Federal agency to set time limits.

§ 1501.11 Tiering

(a) Agencies should tier their environmental impact statements and environmental assessments when it would eliminate repetitive discussions of the same issues, focus on the actual issues ripe for decision, and exclude from consideration issues already decided or not yet ripe at each level of environmental review. Tiering may also be appropriate for different stages of actions.
(b) When an agency has prepared an environmental impact statement or environmental assessment for a program or policy and then prepares a subsequent statement or assessment on an action included within the entire program or policy (such as a project- or site-specific action), the tiered document needs only to summarize and incorporate by reference the issues discussed in the broader document. The tiered document shall concentrate on the issues specific to the subsequent action. The tiered document shall state where the earlier document is available.
(c) Tiering is appropriate when the sequence from an environmental impact statement or environmental assessment is:
  (1) From a programmatic, plan, or policy environmental impact statement or environmental assessment to a program, plan, or policy statement or assessment of lesser or narrower scope or to a site-specific statement or assessment.
  (2) From an environmental impact statement or environmental assessment on a specific action at an early stage (such as need and site selection) to a supplement (which is preferred) or a subsequent statement or assessment at a later stage (such as environmental mitigation). Tiering in such cases is appropriate when it helps the lead agency to focus on the issues that are ripe for decision and exclude from consideration issues already decided or not yet ripe.

§ 1501.12 Incorporation by reference

Agencies shall incorporate material, such as planning studies, analyses, or other relevant information, into environmental documents by reference when the effect will be to cut down on bulk without impeding agency and public review of the action. Agencies shall cite the incorporated material in the document and briefly describe its content. Agencies may not incorporate material by reference unless it is reasonably available for inspection by potentially interested persons within the time allowed for comment. Agencies shall not incorporate by reference material based on proprietary data that is not available for review and comment.

Part 1502—Environmental impact statement

Sec.
1502.1   Purpose of environmental impact statement.
1502.2   Implementation.

1502.3   Statutory requirements for statements.
1502.4   Major Federal actions requiring the preparation of environmental impact statements.
1502.5   Timing.
1502.6   Interdisciplinary preparation.
1502.7   Page limits.
1502.8   Writing.
1502.9   Draft, final, and supplemental statements.
1502.10 Recommended format.
1502.11 Cover.
1502.12 Summary.
1502.13 Purpose and need.
1502.14 Alternatives including the proposed action.
1502.15 Affected environment.
1502.16 Environmental consequences.
1502.17 Summary of submitted alternatives, information, and analyses.
1502.18 List of preparers.
1502.19 Appendix.
1502.20 Publication of the environmental impact statement.
1502.21 Incomplete or unavailable information.
1502.22 Cost-benefit analysis.
1502.23 Methodology and scientific accuracy.
1502.24 Environmental review and consultation requirements.

AUTHORITY: 42 U.S.C. 4321–4347; 42 U.S.C. 4371–4375; 42 U.S.C. 7609; and E.O. 11514, 35 FR 4247, 3 CFR, 1966–1970, Comp., p. 902, as amended by E.O. 11991, 42 FR 26967, 3 CFR, 1977 Comp., p. 123.

SOURCE: 85 FR 43357, July 16, 2020

§ 1502.1   Purpose of environmental impact statement

The primary purpose of an environmental impact statement prepared pursuant to section 102(2)(C) of NEPA is to ensure agencies consider the environmental impacts of their actions in decision making. It shall provide full and fair discussion of significant environmental impacts and shall inform decision makers and the public of reasonable alternatives that would avoid or minimize adverse impacts or enhance the quality of the human environment. Agencies shall focus on significant environmental issues and alternatives and shall reduce paperwork and the accumulation of extraneous background
    data. Statements shall be concise, clear, and to the point, and shall be supported by evidence that the agency has made the necessary environmental analyses. An environmental impact statement is a document that informs Federal agency decision making and the public.

§ 1502.2 Implementation

(a) Environmental impact statements shall not be encyclopedic.
(b) Environmental impact statements shall discuss impacts in proportion to their significance. There shall be only brief discussion of other than significant issues. As in a finding of no significant impact, there should be only enough discussion to show why more study is not warranted.
(c) Environmental impact statements shall be analytic, concise, and no longer than necessary to comply with NEPA and with the regulations in this subchapter. Length should be proportional to potential environmental effects and project size.
(d) Environmental impact statements shall state how alternatives considered in it and decisions based on it will or will not achieve the requirements of sections 101 and 102(1) of NEPA as interpreted in the regulations in this subchapter and other environmental laws and policies.
(e) The range of alternatives discussed in environmental impact statements shall encompass those to be considered by the decision maker.
(f) Agencies shall not commit resources prejudicing selection of alternatives before making a final decision (*see also* § 1506.1 of this chapter).
(g) Environmental impact statements shall serve as the means of assessing the environmental impact of proposed agency actions, rather than justifying decisions already made.

§ 1502.3 Statutory requirements for statements

As required by section 102(2)(C) of NEPA, environmental impact statements are to be included in every Federal agency recommendation or report on proposals for legislation and other major Federal actions significantly affecting the quality of the human environment.

§ 1502.4 Major Federal actions requiring the preparation of environmental impact statements

(a) Agencies shall define the proposal that is the subject of an environmental impact statement based on the statutory authorities for the proposed action. Agencies shall use the criteria for scope (§ 1501.9(e) of this chapter) to determine which proposal(s) shall be the subject of a particular statement. Agencies shall evaluate in a single environmental impact statement proposals or parts of proposals that are related to each other closely enough to be, in effect, a single course of action.
(b) Environmental impact statements may be prepared for programmatic Federal actions, such as the adoption of new agency programs. When agencies prepare such statements, they should be relevant to the program decision and timed to coincide with meaningful points in agency planning and decision making.
 (1) When preparing statements on programmatic actions (including proposals by more than one agency), agencies may find it useful to evaluate the proposal(s) in one of the following ways:
  (i) Geographically, including actions occurring in the same general location, such as body of water, region, or metropolitan area.

(ii) Generically, including actions that have relevant similarities, such as common timing, impacts, alternatives, methods of implementation, media, or subject matter.
(iii) By stage of technological development including Federal or federally assisted research, development or demonstration programs for new technologies that, if applied, could significantly affect the quality of the human environment. Statements on such programs should be available before the program has reached a stage of investment or commitment to implementation likely to determine subsequent development or restrict later alternatives.

(2) Agencies shall as appropriate employ scoping (§ 1501.9 of this chapter), tiering (§ 1501.11 of this chapter), and other methods listed in §§ 1500.4 and 1500.5 of this chapter to relate programmatic and narrow actions and to avoid duplication and delay. Agencies may tier their environmental analyses to defer detailed analysis of environmental impacts of specific program elements until such program elements are ripe for final agency action.

§ 1502.5 Timing

An agency should commence preparation of an environmental impact statement as close as practicable to the time the agency is developing or receives a proposal so that preparation can be completed in time for the final statement to be included in any recommendation or report on the proposal. The statement shall be prepared early enough so that it can serve as an important practical contribution to the decision-making process and will not be used to rationalize or justify decisions already made (§§ 1501.2 of this chapter and 1502.2).

For instance:

(a) For projects directly undertaken by Federal agencies, the agency shall prepare the environmental impact statement at the feasibility analysis (go/no-go) stage and may supplement it at a later stage, if necessary.
(b) For applications to the agency requiring an environmental impact statement, the agency shall commence the statement as soon as practicable after receiving the application. Federal agencies should work with potential applicants and applicable State, Tribal, and local agencies and governments prior to receipt of the application.
(c) For adjudication, the final environmental impact statement shall normally precede the final staff recommendation and that portion of the public hearing related to the impact study. In appropriate circumstances, the statement may follow preliminary hearings designed to gather information for use in the statements.
(d) For informal rulemaking, the draft environmental impact statement shall normally accompany the proposed rule.

§ 1502.6 Interdisciplinary preparation

Agencies shall prepare environmental impact statements using an interdisciplinary approach that will ensure the integrated use of the natural and social sciences and the environmental design arts (section 102(2)(A) of NEPA). The disciplines of the preparers shall be appropriate to the scope and issues identified in the scoping process (§ 1501.9 of this chapter).

§ 1502.7 Page limits

The text of final environmental impact statements (paragraphs (a)(4) through (6) of § 1502.10) shall be 150 pages or fewer and, for proposals of unusual scope or complexity, shall be 300 pages or fewer unless a senior agency official of the lead agency approves in writing a statement to exceed 300 pages and establishes a new page limit.

§ 1502.8 Writing

Agencies shall write environmental impact statements in plain language and may use appropriate graphics so that decision makers and the public can readily understand such statements. Agencies should employ writers of clear prose or editors to write, review, or edit statements, which shall be based upon the analysis and supporting data from the natural and social sciences and the environmental design arts.

§ 1502.9 Draft, final, and supplemental statements

(a) *Generally.* Except for proposals for legislation as provided in § 1506.8 of this chapter, agencies shall prepare environmental impact statements in two stages and, where necessary, supplement them, as provided in paragraph (d)(1) of this section.

(b) *Draft environmental impact statements.* Agencies shall prepare draft environmental impact statements in accordance with the scope decided upon in the scoping process (§ 1501.9 of this chapter). The lead agency shall work with the cooperating agencies and shall obtain comments as required in part 1503 of this chapter. To the fullest extent practicable, the draft statement must meet the requirements established for final statements in section 102(2)(C) of NEPA as interpreted in the regulations in this subchapter. If a draft statement is so inadequate as to preclude meaningful analysis, the agency shall prepare and publish a supplemental draft of the appropriate portion. At appropriate points in the draft statement, the agency shall discuss all major points of view on the environmental impacts of the alternatives including the proposed action.

(c) *Final environmental impact statements.* Final environmental impact statements shall address comments as required in part 1503 of this chapter. At appropriate points in the final statement, the agency shall discuss any responsible opposing view that was not adequately discussed in the draft statement and shall indicate the agency's response to the issues raised.

(d) *Supplemental environmental impact statements.* Agencies:
 (1) Shall prepare supplements to either draft or final environmental impact statements if a major Federal action remains to occur, and:
  (i) The agency makes substantial changes to the proposed action that are relevant to environmental concerns; or
  (ii) There are significant new circumstances or information relevant to environmental concerns and bearing on the proposed action or its impacts.
 (2) May also prepare supplements when the agency determines that the purposes of the Act will be furthered by doing so.

(3) Shall prepare, publish, and file a supplement to a statement (exclusive of scoping (§ 1501.9 of this chapter)) as a draft and final statement, as is appropriate to the stage of the statement involved, unless the Council approves alternative procedures (§ 1506.12 of this chapter).

(4) May find that changes to the proposed action or new circumstances or information relevant to environmental concerns are not significant and therefore do not require a supplement. The agency should document the finding consistent with its agency NEPA procedures (§ 1507.3 of this chapter), or, if necessary, in a finding of no significant impact supported by an environmental assessment.

§ 1502.10   Recommended format

(a) Agencies shall use a format for environmental impact statements that will encourage good analysis and clear presentation of the alternatives including the proposed action. Agencies should use the following standard format for environmental impact statements unless the agency determines that there is a more effective format for communication:
  (1) Cover.
  (2) Summary.
  (3) Table of contents.
  (4) Purpose of and need for action.
  (5) Alternatives including the proposed action (sections 102(2)(C)(iii) and 102(2)(E) of NEPA).
  (6) Affected environment and environmental consequences (especially sections 102(2)(C) (i), (ii), (iv), and (v) of NEPA).
  (7) Submitted alternatives, information, and analyses.
  (8) List of preparers.
  (9) Appendices (if any).
(b) If an agency uses a different format, it shall include paragraphs (a)(1) through (8) of this section, as further described in §§ 1502.11 through 1502.19, in any appropriate format.

§ 1502.11   Cover

The cover shall not exceed one page and include:

(a) A list of the responsible agencies, including the lead agency and any cooperating agencies.
(b) The title of the proposed action that is the subject of the statement (and, if appropriate, the titles of related cooperating agency actions), together with the State(s) and county(ies) (or other jurisdiction(s), if applicable) where the action is located.
(c) The name, address, and telephone number of the person at the agency who can supply further information.
(d) A designation of the statement as a draft, final, or draft or final supplement.
(e) A one-paragraph abstract of the statement.
(f) The date by which the agency must receive comments (computed in cooperation with EPA under § 1506.11 of this chapter).

(g) For the final environmental impact statement, the estimated total cost to prepare both the draft and final environmental impact statement, including the costs of agency full-time equivalent (FTE) personnel hours, contractor costs, and other direct costs. If practicable and noted where not practicable, agencies also should include costs incurred by cooperating and participating agencies, applicants, and contractors.

§ 1502.12 Summary

Each environmental impact statement shall contain a summary that adequately and accurately summarizes the statement. The summary shall stress the major conclusions, areas of disputed issues raised by agencies and the public, and the issues to be resolved (including the choice among alternatives). The summary normally will not exceed 15 pages.

§ 1502.13 Purpose and need

The statement shall briefly specify the underlying purpose and need to which the agency is responding in proposing the alternatives including the proposed action.

§ 1502.14 Alternatives including the proposed action

The alternatives section should present the environmental impacts of the proposed action and the alternatives in comparative form based on the information and analysis presented in the sections on the affected environment (§ 1502.15) and the environmental consequences (§ 1502.16). In this section, agencies shall:
(a) Evaluate reasonable alternatives to the proposed action, and, for alternatives that the agency eliminated from detailed study, briefly discuss the reasons for their elimination.
(b) Discuss each alternative considered in detail, including the proposed action, so that reviewers may evaluate their comparative merits.
(c) Include the no action alternative.
(d) Identify the agency's preferred alternative or alternatives, if one or more exists, in the draft statement and identify such alternative in the final statement unless another law prohibits the expression of such a preference.
(e) Include appropriate mitigation measures not already included in the proposed action or alternatives.
(f) Limit their consideration to a reasonable number of alternatives.

§ 1502.15 Affected environment

The environmental impact statement shall succinctly describe the environment of the area(s) to be affected or created by the alternatives under consideration, including the reasonably foreseeable environmental trends and planned actions in the area(s). The environmental impact statement may combine the description with evaluation of the environmental consequences (§ 1502.16), and it shall be no longer than is necessary to understand the effects of the alternatives.

Data and analyses in a statement shall be commensurate with the importance of the impact, with less important material summarized, consolidated, or simply referenced. Agencies shall avoid useless bulk in statements and shall concentrate effort and attention on important issues. Verbose descriptions of the affected environment are themselves no measure of the adequacy of an environmental impact statement.

§ 1502.16 Environmental consequences

(a) The environmental consequences section forms the scientific and analytic basis for the comparisons under § 1502.14. It shall consolidate the discussions of those elements required by sections 102(2)(C)(i), (ii), (iv), and (v) of NEPA that are within the scope of the statement and as much of section 102(2)(C)(iii) of NEPA as is necessary to support the comparisons. This section should not duplicate discussions in § 1502.14. The discussion shall include:
   (1) The environmental impacts of the proposed action and reasonable alternatives to the proposed action and the significance of those impacts. The comparison of the proposed action and reasonable alternatives shall be based on this discussion of the impacts.
   (2) Any adverse environmental effects that cannot be avoided should the proposal be implemented.
   (3) The relationship between short-term uses of man's environment and the maintenance and enhancement of long-term productivity.
   (4) Any irreversible or irretrievable commitments of resources that would be involved in the proposal should it be implemented.
   (5) Possible conflicts between the proposed action and the objectives of Federal, regional, State, Tribal, and local land use plans, policies and controls for the area concerned. (§ 1506.2(d) of this chapter)
   (6) Energy requirements and conservation potential of various alternatives and mitigation measures.
   (7) Natural or depletable resource requirements and conservation potential of various alternatives and mitigation measures.
   (8) Urban quality, historic and cultural resources, and the design of the built environment, including the reuse and conservation potential of various alternatives and mitigation measures.
   (9) Means to mitigate adverse environmental impacts (if not fully covered under § 1502.14(e)).
   (10) Where applicable, economic and technical considerations, including the economic benefits of the proposed action.
(b) Economic or social effects by themselves do not require preparation of an environmental impact statement. However, when the agency determines that economic or social and natural or physical environmental effects are interrelated, the environmental impact statement shall discuss and give appropriate consideration to these effects on the human environment.

§ 1502.17 Summary of submitted alternatives, information, and analyses

(a) The draft environmental impact statement shall include a summary that identifies all alternatives, information, and analyses submitted by State, Tribal, and local governments and other public commenters during the scoping process for consideration by the lead and cooperating agencies in developing the environmental impact statement.
  (1) The agency shall append to the draft environmental impact statement or otherwise publish all comments (or summaries thereof where the response has been exceptionally voluminous) received during the scoping process that identified alternatives, information, and analyses for the agency's consideration.
  (2) Consistent with § 1503.1(a)(3) of this chapter, the lead agency shall invite comment on the summary identifying all submitted alternatives, information, and analyses in the draft environmental impact statement.
(b) The final environmental impact statement shall include a summary that identifies all alternatives, information, and analyses submitted by State, Tribal, and local governments and other public commenters for consideration by the lead and cooperating agencies in developing the final environmental impact statement.

§ 1502.18 List of preparers

The environmental impact statement shall list the names, together with their qualifications (expertise, experience, professional disciplines), of the persons who were primarily responsible for preparing the environmental impact statement or significant background papers, including basic components of the statement. Where possible, the environmental impact statement shall identify the persons who are responsible for a particular analysis, including analyses in background papers. Normally the list will not exceed two pages.

§ 1502.19 Appendix

If an agency prepares an appendix, the agency shall publish it with the environmental impact statement, and it shall consist of:
(a) Material prepared in connection with an environmental impact statement (as distinct from material that is not so prepared and is incorporated by reference (§ 1501.12 of this chapter)).
(b) Material substantiating any analysis fundamental to the impact statement.
(c) Material relevant to the decision to be made.
(d) For draft environmental impact statements, all comments (or summaries thereof where the response has been exceptionally voluminous) received during the scoping process that identified alternatives, information, and analyses for the agency's consideration.
(e) For final environmental impact statements, the comment summaries and responses consistent with § 1503.4 of this chapter.

§ 1502.20 Publication of the environmental impact statement

Agencies shall publish the entire draft and final environmental impact statements and unchanged statements as provided in § 1503.4(c) of this chapter. The agency shall transmit the entire statement electronically (or in paper copy, if so requested due to economic or other hardship) to:
(a) Any Federal agency that has jurisdiction by law or special expertise with respect to any environmental impact involved and any appropriate Federal, State, Tribal, or local agency authorized to develop and enforce environmental standards.
(b) The applicant, if any.
(c) Any person, organization, or agency requesting the entire environmental impact statement.
(d) In the case of a final environmental impact statement, any person, organization, or agency that submitted substantive comments on the draft.

§ 1502.21 Incomplete or unavailable information

(a) When an agency is evaluating reasonably foreseeable significant adverse effects on the human environment in an environmental impact statement, and there is incomplete or unavailable information, the agency shall make clear that such information is lacking.
(b) If the incomplete but available information relevant to reasonably foreseeable significant adverse impacts is essential to a reasoned choice among alternatives, and the overall costs of obtaining it are not unreasonable, the agency shall include the information in the environmental impact statement.
(c) If the information relevant to reasonably foreseeable significant adverse impacts cannot be obtained because the overall costs of obtaining it are unreasonable or the means to obtain it are not known, the agency shall include within the environmental impact statement:
   (1) A statement that such information is incomplete or unavailable;
   (2) A statement of the relevance of the incomplete or unavailable information to evaluating reasonably foreseeable significant adverse impacts on the human environment;
   (3) A summary of existing credible scientific evidence that is relevant to evaluating the reasonably foreseeable significant adverse impacts on the human environment; and
   (4) The agency's evaluation of such impacts based upon theoretical approaches or research methods generally accepted in the scientific community.
(d) For the purposes of this section, "reasonably foreseeable" includes impacts that have catastrophic consequences, even if their probability of occurrence is low, provided that the analysis of the impacts is supported by credible scientific evidence, is not based on pure conjecture, and is within the rule of reason.

§ 1502.22 Cost-benefit analysis

If the agency is considering a cost-benefit analysis for the proposed action relevant to the choice among alternatives with different environmental effects, the agency shall incorporate the cost-benefit analysis by reference or append it to the statement as an aid in evaluating the environmental consequences. In such cases, to assess the adequacy of compliance with section 102(2)(B) of NEPA (ensuring appropriate consideration of unquantified environmental amenities and values in decision making, along with economical and technical considerations), the statement shall discuss the relationship between that analysis and any analyses of unquantified environmental impacts, values, and amenities. For purposes of complying with the Act, agencies need not display the weighing of the merits and drawbacks of the various alternatives in a monetary cost-benefit analysis and should not do so when there are important qualitative considerations. However, an environmental impact statement should at least indicate those considerations, including factors not related to environmental quality, that are likely to be relevant and important to a decision.

§ 1502.23   Methodology and scientific accuracy

Agencies shall ensure the professional integrity, including scientific integrity, of the discussions and analyses in environmental documents. Agencies shall make use of reliable existing data and resources. Agencies may make use of any reliable data sources, such as remotely gathered information or statistical models. They shall identify any methodologies used and shall make explicit reference to the scientific and other sources relied upon for conclusions in the statement. Agencies may place discussion of methodology in an appendix. Agencies are not required to undertake new scientific and technical research to inform their analyses. Nothing in this section is intended to prohibit agencies from compliance with the requirements of other statutes pertaining to scientific and technical research.

§ 1502.24   Environmental review and consultation requirements

(a) To the fullest extent possible, agencies shall prepare draft environmental impact statements concurrent and integrated with environmental impact analyses and related surveys and studies required by all other Federal environmental review laws and Executive orders applicable to the proposed action, including the Fish and Wildlife Coordination Act (16 U.S.C. 661 et seq.), the National Historic Preservation Act of 1966 (54 U.S.C. 300101 et seq.), and the Endangered Species Act of 1973 (16 U.S.C. 1531 et seq.).
(b) The draft environmental impact statement shall list all Federal permits, licenses, and other authorizations that must be obtained in implementing the proposal. If it is uncertain whether a Federal permit, license, or other authorization is necessary, the draft environmental impact statement shall so indicate.

Part 1503—Commenting on environmental impact statements

Sec.
1503.1   Inviting comments and requesting information and analyses.
1503.2   Duty to comment.
1503.3   Specificity of comments and information.
1503.4   Response to comments.

AUTHORITY: 42 U.S.C. 4321–4347; 42 U.S.C. 4371–4375; 42 U.S.C. 7609; E.O. 11514, 35 FR 4247, 3 CFR, 1966–1970, Comp., p. 902, as amended by E.O. 11991, 42 FR 26967, 3 CFR, 1977 Comp., p. 123; E.O. 13807, 82 FR 40463, 3 CFR, 2017, Comp., p. 369.

SOURCE: 85 FR 43357, July 16, 2020

§ 1503.1 Inviting comments and requesting information and analyses

(a) After preparing a draft environmental impact statement and before preparing a final environmental impact statement the agency shall:
  (1) Obtain the comments of any Federal agency that has jurisdiction by law or special expertise with respect to any environmental impact involved or is authorized to develop and enforce environmental standards.
  (2) Request the comments of:
    (i) Appropriate State, Tribal, and local agencies that are authorized to develop and enforce environmental standards;
    (ii) State, Tribal, or local governments that may be affected by the proposed action;
    (iii) Any agency that has requested it receive statements on actions of the kind proposed;
    (iv) The applicant, if any; and
    (v) The public, affirmatively soliciting comments in a manner designed to inform those persons or organizations who may be interested in or affected by the proposed action.
  (3) Invite comment specifically on the submitted alternatives, information, and analyses and the summary thereof (§ 1502.17 of this chapter).
(b) An agency may request comments on a final environmental impact statement before the final decision and set a deadline for providing such comments. Other agencies or persons may make comments consistent with the time periods under § 1506.11 of this chapter.
(c) An agency shall provide for electronic submission of public comments, with reasonable measures to ensure the comment process is accessible to affected persons.

§ 1503.2 Duty to comment

Cooperating agencies and agencies that are authorized to develop and enforce environmental standards shall comment on statements within their jurisdiction, expertise, or authority within the time period specified for comment in § 1506.11 of this chapter. A Federal agency may reply that it has no comment. If a cooperating agency is satisfied that the environmental impact statement adequately reflects its views, it should reply that it has no comment.

§ 1503.3 Specificity of comments and information

(a) To promote informed decision making, comments on an environmental impact statement or on a proposed action shall be as specific as possible, may address either the adequacy of the statement or the merits of the alternatives discussed or both, and shall provide as much detail as necessary to meaningfully participate and fully inform the agency of the commenter's position. Comments should explain why the issues raised are important to the consideration of potential environmental impacts and alternatives to the proposed action, as well as economic and employment impacts, and other impacts affecting the quality of the human environment. Comments should reference the corresponding section or page number of the draft environmental impact statement, propose specific changes to those parts of the statement, where possible, and include or describe the data sources and methodologies supporting the proposed changes.

(b) Comments on the submitted alternatives, information, and analyses and summary thereof (§ 1502.17 of this chapter) should be as specific as possible. Comments and objections of any kind shall be raised within the comment period on the draft environmental impact statement provided by the agency, consistent with § 1506.11 of this chapter. If the agency requests comments on the final environmental impact statement before the final decision, consistent with § 1503.1(b), comments and objections of any kind shall be raised within the comment period provided by the agency. Comments and objections of any kind not provided within the comment period(s) shall be considered unexhausted and forfeited, consistent with § 1500.3(b) of this chapter.

(c) When a participating agency criticizes a lead agency's predictive methodology, the participating agency should describe the alternative methodology that it prefers and why.

(d) A cooperating agency shall specify in its comments whether it needs additional information to fulfill other applicable environmental reviews or consultation requirements and what information it needs. In particular, it shall specify any additional information it needs to comment adequately on the draft statement's analysis of significant site-specific effects associated with the granting or approving by that cooperating agency of necessary Federal permits, licenses, or authorizations.

(e) When a cooperating agency with jurisdiction by law specifies mitigation measures it considers necessary to allow the agency to grant or approve applicable permit, license, or related requirements or concurrences, the cooperating agency shall cite to its applicable statutory authority.

§ 1503.4 Response to comments

(a) An agency preparing a final environmental impact statement shall consider substantive comments timely submitted during the public comment period. The agency may respond to individual comments or groups of comments. In the final environmental impact statement, the agency may respond by:
  (1) Modifying alternatives including the proposed action.
  (2) Developing and evaluating alternatives not previously given serious consideration by the agency.
  (3) Supplementing, improving, or modifying its analyses.
  (4) Making factual corrections.
  (5) Explaining why the comments do not warrant further agency response, recognizing that agencies are not required to respond to each comment.

(b) An agency shall append or otherwise publish all substantive comments received on the draft statement (or summaries thereof where the response has been exceptionally voluminous).

(c) If changes in response to comments are minor and are confined to the responses described in paragraphs (a)(4) and (5) of this section, an agency may write any changes on errata sheets and attach the responses to the statement instead of rewriting the draft statement. In such cases, only the comments, the responses, and the changes and not the final statement need be published (§ 1502.20 of this chapter). The agency shall file the entire document with a new cover sheet with the Environmental Protection Agency as the final statement (§ 1506.10 of this chapter).

Part 1504—Pre-decisional referrals to the council of proposed federal actions determined to be environmentally unsatisfactory

Sec.
1504.1   Purpose.
1504.2   Criteria for referral.
1504.3   Procedure for referrals and response.

AUTHORITY: 42 U.S.C. 4321–4347; 42 U.S.C. 4371–4375; 42 U.S.C. 7609; E.O. 11514, 35 FR 4247, 3 CFR, 1966–1970, Comp., p. 902, as amended by E.O. 11991, 42 FR 26967, 3 CFR, 1977 Comp., p. 123; E.O. 13807, 82 FR 40463, 3 CFR, 2017, Comp., p. 369.

SOURCE: 85 FR 43357, July 16, 2020

§ 1504.1   Purpose

(a) This part establishes procedures for referring to the Council Federal interagency disagreements concerning proposed major Federal actions that might cause unsatisfactory environmental effects. It provides means for early resolution of such disagreements.
(b) Section 309 of the Clean Air Act (42 U.S.C. 7609) directs the Administrator of the Environmental Protection Agency to review and comment publicly on the environmental impacts of Federal activities, including actions for which agencies prepare environmental impact statements. If, after this review, the Administrator determines that the matter is "unsatisfactory from the standpoint of public health or welfare or environmental quality," section 309 directs that the matter be referred to the Council (hereafter "environmental referrals").
(c) Under section 102(2)(C) of NEPA (42 U.S.C. 4332(2)(C)), other Federal agencies may prepare similar reviews of environmental impact statements, including judgments on the acceptability of anticipated environmental impacts. These reviews must be made available to the President, the Council, and the public.

§ 1504.2   Criteria for referral

Environmental referrals should be made to the Council only after concerted, timely (as early as practicable in the process), but unsuccessful attempts to resolve differences with the lead agency. In determining what environmental objections to the matter are appropriate to refer to the Council, an agency should weigh potential adverse environmental impacts, considering:

(a) Possible violation of national environmental standards or policies;
(b) Severity;
(c) Geographical scope;
(d) Duration;
(e) Importance as precedents;
(f) Availability of environmentally preferable alternatives; and
(g) Economic and technical considerations, including the economic costs of delaying or impeding the decision making of the agencies involved in the action.

§ 1504.3 Procedure for referrals and response

(a) A Federal agency making the referral to the Council shall:
   (1) Notify the lead agency at the earliest possible time that it intends to refer a matter to the Council unless a satisfactory agreement is reached;
   (2) Include such a notification whenever practicable in the referring agency's comments on the environmental assessment or draft environmental impact statement;
   (3) Identify any essential information that is lacking and request that the lead agency make it available at the earliest possible time; and
   (4) Send copies of the referring agency's views to the Council.
(b) The referring agency shall deliver its referral to the Council no later than 25 days after the lead agency has made the final environmental impact statement available to the Environmental Protection Agency, participating agencies, and the public, and in the case of an environmental assessment, no later than 25 days after the lead agency makes it available. Except when the lead agency grants an extension of this period, the Council will not accept a referral after that date.
(c) The referral shall consist of:
   (1) A copy of the letter signed by the head of the referring agency and delivered to the lead agency informing the lead agency of the referral and the reasons for it; and
   (2) A statement supported by factual evidence leading to the conclusion that the matter is unsatisfactory from the standpoint of public health or welfare or environmental quality. The statement shall:
      (i) Identify any disputed material facts and incorporate (by reference if appropriate) agreed upon facts;
      (ii) Identify any existing environmental requirements or policies that would be violated by the matter;
      (iii) Present the reasons for the referral;
      (iv) Contain a finding by the agency whether the issue raised is of national importance because of the threat to national environmental resources or policies or for some other reason;
      (v) Review the steps taken by the referring agency to bring its concerns to the attention of the lead agency at the earliest possible time; and
      (vi) Give the referring agency's recommendations as to what mitigation alternative, further study, or other course of action (including abandonment of the matter) are necessary to remedy the situation.

(d) No later than 25 days after the referral to the Council, the lead agency may deliver a response to the Council and the referring agency. If the lead agency requests more time and gives assurance that the matter will not go forward in the interim, the Council may grant an extension. The response shall:
   (1) Address fully the issues raised in the referral;
   (2) Be supported by evidence and explanations, as appropriate; and
   (3) Give the lead agency's response to the referring agency's recommendations.
(e) Applicants may provide views in writing to the Council no later than the response.
(f) No later than 25 days after receipt of both the referral and any response or upon being informed that there will be no response (unless the lead agency agrees to a longer time), the Council may take one or more of the following actions:
   (1) Conclude that the process of referral and response has successfully resolved the problem.
   (2) Initiate discussions with the agencies with the objective of mediation with referring and lead agencies.
   (3) Obtain additional views and information.
   (4) Determine that the issue is not one of national importance and request the referring and lead agencies to pursue their decision process.
   (5) Determine that the referring and lead agencies should further negotiate the issue, and the issue is not appropriate for Council consideration until one or more heads of agencies report to the Council that the agencies' disagreements are irreconcilable.
   (6) Publish its findings and recommendations (including, where appropriate, a finding that the submitted evidence does not support the position of an agency).
   (7) When appropriate, submit the referral and the response together with the Council's recommendation to the President for action.
(g) The Council shall take no longer than 60 days to complete the actions specified in paragraph (f)(2), (3), or (5) of this section.
(h) The referral process is not intended to create any private rights of action or to be judicially reviewable because any voluntary resolutions by the agency parties do not represent final agency action and instead are only provisional and dependent on later consistent action by the action agencies.

Part 1505—NEPA and agency decision making

Sec.
1505.1 [Reserved]
1505.2 Record of decision in cases requiring environmental impact statements.
1505.3 Implementing the decision.

AUTHORITY: 42 U.S.C. 4321–4347; 42 U.S.C. 4371–4375; 42 U.S.C. 7609; E.O. 11514, 35 FR 4247, 3 CFR, 1966–1970, Comp., p. 902, as amended by E.O. 11991, 42 FR 26967, 3 CFR, 1977 Comp., p. 123; and E.O. 13807, 82 FR 40463, 3 CFR, 2017, Comp., p. 369.

SOURCE: 85 FR 43357, July 16, 2020

§ 1505.1   [Reserved]

§ 1505.2   **Record of decision in cases requiring environmental impact statements**

(a) At the time of its decision (§ 1506.11 of this chapter) or, if appropriate, its recommendation to Congress, each agency shall prepare and timely publish a concise public record of decision or joint record of decision. The record, which each agency may integrate into any other record it prepares, shall:
   (1) State the decision.
   (2) Identify alternatives considered by the agency in reaching its decision, specifying the alternative or alternatives considered environmentally preferable. An agency may discuss preferences among alternatives based on relevant factors including economic and technical considerations and agency statutory missions. An agency shall identify and discuss all such factors, including any essential considerations of national policy, that the agency balanced in making its decision and state how those considerations entered into its decision.
   (3) State whether the agency has adopted all practicable means to avoid or minimize environmental harm from the alternative selected, and if not, why the agency did not. The agency shall adopt and summarize, where applicable, a monitoring and enforcement program for any enforceable mitigation requirements or commitments.
(b) Informed by the summary of the submitted alternatives, information, and analyses in the final environmental impact statement (§ 1502.17(b) of this chapter), together with any other material in the record that he or she determines to be relevant, the decision maker shall certify in the record of decision that the agency has considered all of the alternatives, information, analyses, and objections submitted by State, Tribal, and local governments and public commenters for consideration by the lead and cooperating agencies in developing the environmental impact statement. Agency environmental impact statements certified in accordance with this section are entitled to a presumption that the agency has considered the submitted alternatives, information, and analyses, including the summary thereof, in the final environmental impact statement

(§ 1502.17(b)). § 1505.3   **Implementing the decision**

Agencies may provide for monitoring to assure that their decisions are carried out and should do so in important cases. Mitigation (§ 1505.2(a)(3)) and other conditions established in the environmental impact statement or during its review and committed as part of the decision shall be implemented by the lead agency or other appropriate consenting agency. The lead agency shall:
(a) Include appropriate conditions in grants, permits, or other approvals.
(b) Condition funding of actions on mitigation.
(c) Upon request, inform cooperating or participating agencies on progress in carrying out mitigation measures that they have proposed and were adopted by the agency making the decision.
(d) Upon request, publish the results of relevant monitoring.

Part 1506—Other requirements of NEPA

Sec.
1506.1 Limitations on actions during NEPA process.
1506.2 Elimination of duplication with State, Tribal, and local procedures.
1506.3 Adoption.
1506.4 Combining documents.
1506.5 Agency responsibility for environmental documents.
1506.6 Public involvement.
1506.7 Further guidance.
1506.8 Proposals for legislation.
1506.9 Proposals for regulations.
1506.10 Filing requirements.
1506.11 Timing of agency action.
1506.12 Emergencies.
1506.13 Effective date.

AUTHORITY: 42 U.S.C. 4321–4347; 42 U.S.C. 4371–4375; 42 U.S.C. 7609; E.O. 11514, 35 FR 4247, 3 CFR, 1966–1970, Comp., p. 902, as amended by E.O. 11991, 42 FR 26967, 3 CFR, 1977 Comp., p. 123; and E.O. 13807, 82 FR 40463, 3 CFR, 2017, Comp., p. 369.

SOURCE: 85 FR 43357, July 16, 2020

§ 1506.1 Limitations on actions during NEPA process

(a) Except as provided in paragraphs (b) and (c) of this section, until an agency issues a finding of no significant impact, as provided in § 1501.6 of this chapter, or record of decision, as provided in § 1505.2 of this chapter, no action concerning the proposal may be taken that would:
  (1) Have an adverse environmental impact; or
  (2) Limit the choice of reasonable alternatives.
(b) If any agency is considering an application from a non-Federal entity and is aware that the applicant is about to take an action within the agency's jurisdiction that would meet either of the criteria in paragraph (a) of this section, then the agency shall promptly notify the applicant that the agency will take appropriate action to ensure that the objectives and procedures of NEPA are achieved. This section does not preclude development by applicants of plans or designs or performance of other activities necessary to support an application for Federal, State, Tribal, or local permits or assistance. An agency considering a proposed action for Federal funding may authorize such activities, including, but not limited to, acquisition of interests in land (*e.g.*, fee simple, rights-of-way, and conservation easements), purchase of long lead-time equipment, and purchase options made by applicants.
(c) While work on a required programmatic environmental review is in progress and the action is not covered by an existing programmatic review, agencies shall not undertake in

the interim any major Federal action covered by the program that may significantly affect the quality of the human environment unless such action:
(1) Is justified independently of the program;
(2) Is itself accompanied by an adequate environmental review; and
(3) Will not prejudice the ultimate decision on the program. Interim action prejudices the ultimate decision on the program when it tends to determine subsequent development or limit alternatives.

§ 1506.2 Elimination of duplication with State, Tribal, and local procedures

(a) Federal agencies are authorized to cooperate with State, Tribal, and local agencies that are responsible for preparing environmental documents, including those prepared pursuant to section 102(2)(D) of NEPA.
(b) To the fullest extent practicable unless specifically prohibited by law, agencies shall cooperate with State, Tribal, and local agencies to reduce duplication between NEPA and State, Tribal, and local requirements, including through use of studies, analysis, and decisions developed by State, Tribal, or local agencies. Except for cases covered by paragraph (a) of this section, such cooperation shall include, to the fullest extent practicable:
(1) Joint planning processes.
(2) Joint environmental research and studies.
(3) Joint public hearings (except where otherwise provided by statute).
(4) Joint environmental assessments.
(c) To the fullest extent practicable unless specifically prohibited by law, agencies shall cooperate with State, Tribal, and local agencies to reduce duplication between NEPA and comparable State, Tribal, and local requirements. Such cooperation shall include, to the fullest extent practicable, joint environmental impact statements. In such cases, one or more Federal agencies and one or more State, Tribal, or local agencies shall be joint lead agencies. Where State or Tribal laws or local ordinances have environmental impact statement or similar requirements in addition to but not in conflict with those in NEPA, Federal agencies may cooperate in fulfilling these requirements, as well as those of Federal laws, so that one document will comply with all applicable laws.
(d) To better integrate environmental impact statements into State, Tribal, or local planning processes, environmental impact statements shall discuss any inconsistency of a proposed action with any approved State, Tribal, or local plan or law (whether or not federally sanctioned). Where an inconsistency exists, the statement should describe the extent to which the agency would reconcile its proposed action with the plan or law. While the statement should discuss any inconsistencies, NEPA does not require reconciliation.

§ 1506.3 Adoption

(a) *Generally.* An agency may adopt a Federal draft or final environmental impact statement, environmental assessment, or portion thereof, or categorical exclusion determination provided that the statement, assessment, portion thereof, or determination meets the standards for an adequate statement, assessment, or determination under the regulations in this subchapter.

(b) *Environmental impact statements.*
   (1) If the actions covered by the original environmental impact statement and the proposed action are substantially the same, the adopting agency shall republish it as a final statement consistent with § 1506.10. If the actions are not substantially the same, the adopting agency shall treat the statement as a draft and republish it, consistent with § 1506.10.
   (2) Notwithstanding paragraph (b)(1) of this section, a cooperating agency may adopt in its record of decision without republishing the environmental impact statement of a lead agency when, after an independent review of the statement, the cooperating agency concludes that its comments and suggestions have been satisfied.
(c) *Environmental assessments.* If the actions covered by the original environmental assessment and the proposed action are substantially the same, the adopting agency may adopt the environmental assessment in its finding of no significant impact and provide notice consistent with § 1501.6 of this chapter.
(d) *Categorical exclusions.* An agency may adopt another agency's determination that a categorical exclusion applies to a proposed action if the action covered by the original categorical exclusion determination and the adopting agency's proposed action are substantially the same. The agency shall document the adoption.
(e) *Identification of certain circumstances.* The adopting agency shall specify if one of the following circumstances is present:

   (1) The agency is adopting an assessment or statement that is not final within the agency that prepared it.
   (2) The action assessed in the assessment or statement is the subject of a referral under part 1504 of this chapter.
   (3) The assessment or statement's adequacy is the subject of a judicial action that is not final.

§ 1506.4   Combining documents

Agencies should combine, to the fullest extent practicable, any environmental document with any other agency document to reduce duplication and paperwork.

§ 1506.5   Agency responsibility for environmental documents

(a) *Responsibility.* The agency is responsible for the accuracy, scope (§ 1501.9(e) of this chapter), and content of environmental documents prepared by the agency or by an applicant or contractor under the supervision of the agency.
(b) *Information.* An agency may require an applicant to submit environmental information for possible use by the agency in preparing an environmental document. An agency also may direct an applicant or authorize a contractor to prepare an environmental document under the supervision of the agency.
   (1) The agency should assist the applicant by outlining the types of information required or, for the preparation of environmental documents, shall provide guidance to the applicant or contractor and participate in their preparation.

(2) The agency shall independently evaluate the information submitted or the environmental document and shall be responsible for its accuracy, scope, and contents.
(3) The agency shall include in the environmental document the names and qualifications of the persons preparing environmental documents, and conducting the independent evaluation of any information submitted or environmental documents prepared by an applicant or contractor, such as in the list of preparers for environmental impact statements (§ 1502.18 of this chapter). It is the intent of this paragraph (b)(3) that acceptable work not be redone, but that it be verified by the agency.
(4) Contractors or applicants preparing environmental assessments or environmental impact statements shall submit a disclosure statement to the lead agency that specifies any financial or other interest in the outcome of the action. Such statement need not include privileged or confidential trade secrets or other confidential business information.
(5) Nothing in this section is intended to prohibit any agency from requesting any person, including the applicant, to submit information to it or to prohibit any person from submitting information to any agency for use in preparing environmental documents.

§ 1506.6 Public involvement

Agencies shall:

(a) Make diligent efforts to involve the public in preparing and implementing their NEPA procedures (§ 1507.3 of this chapter).
(b) Provide public notice of NEPA-related hearings, public meetings, and other opportunities for public involvement, and the availability of environmental documents so as to inform those persons and agencies who may be interested or affected by their proposed actions. When selecting appropriate methods for providing public notice, agencies shall consider the ability of affected persons and agencies to access electronic media.
  (1) In all cases, the agency shall notify those who have requested notice on an individual action.
  (2) In the case of an action with effects of national concern, notice shall include publication in the *Federal Register*. An agency may notify organizations that have requested regular notice.
  (3) In the case of an action with effects primarily of local concern, the notice may include:
      (i) Notice to State, Tribal, and local agencies that may be interested or affected by the proposed action.
      (ii) Notice to interested or affected State, Tribal, and local governments.
      (iii) Following the affected State or Tribe's public notice procedures for comparable actions.
      (iv) Publication in local newspapers (in papers of general circulation rather than legal papers).
      (v) Notice through other local media.
      (vi) Notice to potentially interested community organizations including small business associations.

(vii) Publication in newsletters that may be expected to reach potentially interested persons.
(viii) Direct mailing to owners and occupants of nearby or affected property.
(ix) Posting of notice on and off site in the area where the action is to be located.
(x) Notice through electronic media (*e.g.*, a project or agency website, email, or social media).

(c) Hold or sponsor public hearings, public meetings, or other opportunities for public involvement whenever appropriate or in accordance with statutory requirements applicable to the agency. Agencies may conduct public hearings and public meetings by means of electronic communication except where another format is required by law. When selecting appropriate methods for public involvement, agencies shall consider the ability of affected entities to access electronic media.
(d) Solicit appropriate information from the public.
(e) Explain in its procedures where interested persons can get information or status reports on environmental impact statements and other elements of the NEPA process.
(f) Make environmental impact statements, the comments received, and any underlying documents available to the public pursuant to the provisions of the Freedom of Information Act, as amended (5 U.S.C. 552).

§ 1506.7 Further guidance

(a) The Council may provide further guidance concerning NEPA and its procedures consistent with Executive Order 13807, Establishing Discipline and Accountability in the Environmental Review and Permitting Process for Infrastructure Projects (August 5, 2017), Executive Order 13891, Promoting the Rule of Law Through Improved Agency Guidance Documents (October 9, 2019), and any other applicable Executive orders.
(b) To the extent that Council guidance issued prior to September 14, 2020 is in conflict with this subchapter, the provisions of this subchapter apply.

§ 1506.8 Proposals for legislation

(a) When developing legislation, agencies shall integrate the NEPA process for proposals for legislation significantly affecting the quality of the human environment with the legislative process of the Congress. Technical drafting assistance does not by itself constitute a legislative proposal. Only the agency that has primary responsibility for the subject matter involved will prepare a legislative environmental impact statement.
(b) A legislative environmental impact statement is the detailed statement required by law to be included in an agency's recommendation or report on a legislative proposal to Congress. A legislative environmental impact statement shall be considered part of the formal transmittal of a legislative proposal to Congress; however, it may be transmitted to Congress up to 30 days later in order to allow time for completion of an accurate statement that can serve as the basis for public and Congressional debate. The statement must be available in time for Congressional hearings and deliberations.

(c) Preparation of a legislative environmental impact statement shall conform to the requirements of the regulations in this subchapter, except as follows:
  (1) There need not be a scoping process.
  (2) Agencies shall prepare the legislative statement in the same manner as a draft environmental impact statement and need not prepare a final statement unless any of the following conditions exist. In such cases, the agency shall prepare and publish the statements consistent with §§ 1503.1 of this chapter and 1506.11:
    (i) A Congressional committee with jurisdiction over the proposal has a rule requiring both draft and final environmental impact statements.
    (ii) The proposal results from a study process required by statute (such as those required by the Wild and Scenic Rivers Act (16 U.S.C. 1271 *et seq.*)).
    (iii) Legislative approval is sought for Federal or federally assisted construction or other projects that the agency recommends be located at specific geographic locations. For proposals requiring an environmental impact statement for the acquisition of space by the General Services Administration, a draft statement shall accompany the Prospectus or the 11(b) Report of Building Project Surveys to the Congress, and a final statement shall be completed before site acquisition.
    (iv) The agency decides to prepare draft and final statements.
(d) Comments on the legislative statement shall be given to the lead agency, which shall forward them along with its own responses to the Congressional committees with jurisdiction.

§ 1506.9 Proposals for regulations

Where the proposed action is the promulgation of a rule or regulation, procedures and documentation pursuant to other statutory or Executive order requirements may satisfy one or more requirements of this subchapter. When a procedure or document satisfies one or more requirements of this subchapter, the agency may substitute it for the corresponding requirements in this subchapter and need not carry out duplicative procedures or documentation. Agencies shall identify which corresponding requirements in this subchapter are satisfied and consult with the Council to confirm such determinations.

§ 1506.10 Filing requirements

(a) Agencies shall file environmental impact statements together with comments and responses with the Environmental Protection Agency (EPA), Office of Federal Activities, consistent with EPA's procedures.
(b) Agencies shall file statements with the EPA no earlier than they are also transmitted to participating agencies and made available to the public. EPA may issue guidelines to agencies to implement its responsibilities under this section and § 1506.11.

§ 1506.11 Timing of agency action

(a) The Environmental Protection Agency shall publish a notice in the *Federal Register* each week of the environmental impact statements filed since its prior notice. The minimum time periods set forth in this section are calculated from the date of publication of this notice.

(b) Unless otherwise provided by law, including statutory provisions for combining a final environmental impact statement and record of decision, Federal agencies may not make or issue a record of decision under § 1505.2 of this chapter for the proposed action until the later of the following dates:
   (1) 90 days after publication of the notice described in paragraph (a) of this section for a draft environmental impact statement.
   (2) 30 days after publication of the notice described in paragraph (a) of this section for a final environmental impact statement.
(c) An agency may make an exception to the rule on timing set forth in paragraph (b) of this section for a proposed action in the following circumstances:
   (1) Some agencies have a formally established appeal process after publication of the final environmental impact statement that allows other agencies or the public to take appeals on a decision and make their views known. In such cases where a real opportunity exists to alter the decision, the agency may make and record the decision at the same time it publishes the environmental impact statement. This means that the period for appeal of the decision and the 30-day period set forth in paragraph (b)(2) of this section may run concurrently. In such cases, the environmental impact statement shall explain the timing and the public's right of appeal and provide notification consistent with § 1506.10; or
   (2) An agency engaged in rulemaking under the Administrative Procedure Act or other statute for the purpose of protecting the public health or safety may waive the time period in paragraph (b)(2) of this section, publish a decision on the final rule simultaneously with publication of the notice of the availability of the final environmental impact statement, and provide notification consistent with § 1506.10, as described in paragraph (a) of this section.
(d) If an agency files the final environmental impact statement within 90 days of the filing of the draft environmental impact statement with the Environmental Protection Agency, the decision-making period and the 90-day period may run concurrently. However, subject to paragraph (e) of this section, agencies shall allow at least 45 days for comments on draft statements.
(e) The lead agency may extend the minimum periods in paragraph (b) of this section and provide notification consistent with § 1506.10. Upon a showing by the lead agency of compelling reasons of national policy, the Environmental Protection Agency may reduce the minimum periods and, upon a showing by any other Federal agency of compelling reasons of national policy, also may extend the minimum periods, but only after consultation with the lead agency. The lead agency may modify the minimum periods when necessary to comply with other specific statutory requirements. (§ 1507.3(f)(2) of this chapter) Failure to file timely comments shall not be a sufficient reason for extending a period. If the lead agency does not concur with the extension of time, EPA may not extend it for more than 30 days. When the Environmental Protection Agency reduces or extends any period of time it shall notify the Council.

§ 1506.12   Emergencies

Where emergency circumstances make it necessary to take an action with significant environmental impact without observing the provisions of the regulations in this subchapter, the Federal agency taking the action should consult with the Council about alternative arrangements for compliance with section 102(2)(C) of NEPA. Agencies and the Council will limit such arrangements to actions necessary to control the immediate impacts of the emergency. Other actions remain subject to NEPA review.

§ 1506.13 Effective date

The regulations in this subchapter apply to any NEPA process begun after September 14, 2020. An agency may apply the regulations in this subchapter to ongoing activities and environmental documents begun before September 14, 2020.

Part 1507—Agency compliance

Sec.
1507.1   Compliance.
1507.2   Agency capability to comply.
1507.3   Agency NEPA procedures.
1507.4   Agency NEPA program information.

AUTHORITY: 42 U.S.C. 4321–4347; 42 U.S.C. 4371–4375; 42 U.S.C. 7609; and E.O. 11514, 35 FR 4247, 3 CFR, 1966–1970, Comp., p. 902, as amended by E.O. 11991, 42 FR 26967, 3 CFR, 1977 Comp., p. 123.

SOURCE: 85 FR 43357, July 16, 2020, unless otherwise noted.

§ 1507.1 Compliance

All agencies of the Federal Government shall comply with the regulations in this subchapter.

§ 1507.2 Agency capability to comply

Each agency shall be capable (in terms of personnel and other resources) of complying with the requirements of NEPA and the regulations in this subchapter. Such compliance may include use of the resources of other agencies, applicants, and other participants in the NEPA process, but the agency using the resources shall itself have sufficient capability to evaluate what others do for it and account for the contributions of others. Agencies shall:

(a) Fulfill the requirements of section 102(2)(A) of NEPA to utilize a systematic, interdisciplinary approach that will ensure the integrated use of the natural and social sciences and the environmental design arts in planning and in decision making that may have an impact on the human environment. Agencies shall designate a senior agency official to be responsible for overall review of agency NEPA compliance, including resolving implementation issues.

(b) Identify methods and procedures required by section 102(2)(B) of NEPA to ensure that presently unquantified environmental amenities and values may be given appropriate consideration.

(c) Prepare adequate environmental impact statements pursuant to section 102(2)(C) of NEPA and cooperate on the development of statements in the areas where the agency has jurisdiction by law or special expertise or is authorized to develop and enforce environmental standards.
(d) Study, develop, and describe alternatives to recommended courses of action in any proposal that involves unresolved conflicts concerning alternative uses of available resources, consistent with section 102(2)(E) of NEPA.
(e) Comply with the requirements of section 102(2)(H) of NEPA that the agency initiate and utilize ecological information in the planning and development of resource-oriented projects.
(f) Fulfill the requirements of sections 102(2)(F), 102(2)(G), and 102(2)(I), of NEPA, Executive Order 11514, Protection and Enhancement of Environmental Quality, section 2, as amended by Executive Order 11991, Relating to Protection and Enhancement of Environmental Quality, and Executive Order 13807, Establishing Discipline and Accountability in the Environmental Review and Permitting for Infrastructure Projects.

§ 1507.3 Agency NEPA procedures

(a) The Council has determined that the categorical exclusions contained in agency NEPA procedures as of September 14, 2020 are consistent with this subchapter.
(b) No more than 36 months after September 14, 2020, or 9 months after the establishment of an agency, whichever comes later, each agency shall develop or revise, as necessary, proposed procedures to implement the regulations in this subchapter. When the agency is a department, it may be efficient for major subunits (with the consent of the department) to adopt their own procedures.
  (1) Each agency shall consult with the Council while developing or revising its proposed procedures and before publishing them in the *Federal Register* for comment. Agencies with similar programs should consult with each other and the Council to coordinate their procedures, especially for programs requesting similar information from applicants.
  (2) Agencies shall provide an opportunity for public review and review by the Council for conformity with the Act and the regulations in this subchapter before adopting their final procedures. The Council shall complete its review within 30 days of the receipt of the proposed final procedures. Once in effect, the agency shall publish its NEPA procedures and ensure that they are readily available to the public.
(c) Agencies shall adopt, as necessary, agency NEPA procedures to improve agency efficiency and ensure that agencies make decisions in accordance with the Act's procedural requirements. Such procedures shall include:
  (1) Designating the major decision points for the agency's principal programs likely to have a significant effect on the human environment and assuring that the NEPA process begins at the earliest reasonable time, consistent with § 1501.2 of this chapter, and aligns with the corresponding decision points.
  (2) Requiring that relevant environmental documents, comments, and responses be part of the record in formal rulemaking or adjudicatory proceedings.

(3) Requiring that relevant environmental documents, comments, and responses accompany the proposal through existing agency review processes so that decision makers use the statement in making decisions.

(4) Requiring that the alternatives considered by the decision maker are encompassed by the range of alternatives discussed in the relevant environmental documents and that the decision maker consider the alternatives described in the environmental documents. If another decision document accompanies the relevant environmental documents to the decision maker, agencies are encouraged to make available to the public before the decision is made any part of that document that relates to the comparison of alternatives.

(5) Requiring the combination of environmental documents with other agency documents. Agencies may designate and rely on one or more procedures or documents under other statutes or Executive orders as satisfying some or all of the requirements in this subchapter, and substitute such procedures and documentation to reduce duplication. When an agency substitutes one or more procedures or documents for the requirements in this subchapter, the agency shall identify the respective requirements that are satisfied.

(d) Agency procedures should identify those activities or decisions that are not subject to NEPA, including:
(1) Activities or decisions expressly exempt from NEPA under another statute;
(2) Activities or decisions where compliance with NEPA would clearly and fundamentally conflict with the requirements of another statute;
(3) Activities or decisions where compliance with NEPA would be inconsistent with Congressional intent expressed in another statute;
(4) Activities or decisions that are non-major Federal actions;
(5) Activities or decisions that are non-discretionary actions, in whole or in part, for which the agency lacks authority to consider environmental effects as part of its decision-making process; and
(6) Actions where the agency has determined that another statute's requirements serve the function of agency compliance with the Act.

(e) Agency procedures shall comply with the regulations in this subchapter except where compliance would be inconsistent with statutory requirements and shall include:
(1) Those procedures required by §§ 1501.2(b)(4) (assistance to applicants) and 1506.6(e) of this chapter (status information).
(2) Specific criteria for and identification of those typical classes of action:
  (i) Which normally do require environmental impact statements.
  (ii) Which normally do not require either an environmental impact statement or an environmental assessment and do not have a significant effect on the human environment (categorical exclusions (§ 1501.4 of this chapter)). Any procedures under this section shall provide for extraordinary circumstances in which a normally excluded action may have a significant environmental effect. Agency NEPA procedures shall identify when documentation of a categorical exclusion determination is required.

(iii) Which normally require environmental assessments but not necessarily environmental impact statements.
(3) Procedures for introducing a supplement to an environmental assessment or environmental impact statement into its formal administrative record, if such a record exists.
(f) Agency procedures may:
(1) Include specific criteria for providing limited exceptions to the provisions of the regulations in this subchapter for classified proposals. These are proposed actions that are specifically authorized under criteria established by an Executive order or statute to be kept secret in the interest of national defense or foreign policy and are in fact properly classified pursuant to such Executive order or statute. Agencies may safeguard and restrict from public dissemination environmental assessments and environmental impact statements that address classified proposals in accordance with agencies' own regulations applicable to classified information. Agencies should organize these documents so that classified portions are included as annexes, so that the agencies can make the unclassified portions available to the public.
(2) Provide for periods of time other than those presented in § 1506.11 of this chapter when necessary to comply with other specific statutory requirements, including requirements of lead or cooperating agencies.
(3) Provide that, where there is a lengthy period between the agency's decision to prepare an environmental impact statement and the time of actual preparation, the agency may publish the notice of intent required by § 1501.9(d) of this chapter at a reasonable time in advance of preparation of the draft statement. Agency procedures shall provide for publication of supplemental notices to inform the public of a pause in its preparation of an environmental impact statement and for any agency decision to withdraw its notice of intent to prepare an environmental impact statement.
(4) Adopt procedures to combine its environmental assessment process with its scoping process.
(5) Establish a process that allows the agency to use a categorical exclusion listed in another agency's NEPA procedures after consulting with that agency to ensure the use of the categorical exclusion is appropriate. The process should ensure documentation of the consultation and identify to the public those categorical exclusions the agency may use for its proposed actions. Then, the agency may apply the categorical exclusion to its proposed actions.

[85 FR 43373, July 16, 2020, as amended at 86 FR 34158, June 29, 2021]

§ 1507.4 Agency NEPA program information

(a) To allow agencies and the public to efficiently and effectively access information about NEPA reviews, agencies shall provide for agency websites or other means to make available environmental documents, relevant notices, and other relevant information for use by agencies, applicants, and interested persons. Such means of publication may include:
(1) Agency planning and environmental documents that guide agency management and provide for public involvement in agency planning processes;

(2) A directory of pending and final environmental documents;
(3) Agency policy documents, orders, terminology, and explanatory materials regarding agency decision-making processes;
(4) Agency planning program information, plans, and planning tools; and
(5) A database searchable by geographic information, document status, document type, and project type.
(b) Agencies shall provide for efficient and effective interagency coordination of their environmental program websites, including use of shared databases or application programming interface, in their implementation of NEPA and related authorities.

Part 1508—Definitions

Sec.
1508.1   Definitions.
1508.2   [Reserved]

AUTHORITY: 42 U.S.C. 4321–4347; 42 U.S.C. 4371–4375; 42 U.S.C. 7609; and E.O. 11514, 35 FR 4247, 3 CFR, 1966–1970, Comp., p. 902, as amended by E.O. 11991, 42 FR 26967, 3 CFR, 1977 Comp., p. 123.

SOURCE: 85 FR 43357, July 16, 2020

§ 1508.1   Definitions

The following definitions apply to the regulations in this subchapter. Federal agencies shall use these terms uniformly throughout the Federal Government.
(a) *Act* or *NEPA* means the National Environmental Policy Act, as amended (42 U.S.C. 4321, et seq.).
(b) *Affecting* means will or may have an effect on.
(c) *Authorization* means any license, permit, approval, finding, determination, or other administrative decision issued by an agency that is required or authorized under Federal law in order to implement a proposed action.
(d) *Categorical exclusion* means a category of actions that the agency has determined, in its agency NEPA procedures (§ 1507.3 of this chapter), normally do not have a significant effect on the human environment.
(e) *Cooperating agency* means any Federal agency (and a State, Tribal, or local agency with agreement of the lead agency) other than a lead agency that has jurisdiction by law or special expertise with respect to any environmental impact involved in a proposal (or a reasonable alternative) for legislation or other major Federal action that may significantly affect the quality of the human environment.
(f) *Council* means the Council on Environmental Quality established by title II of the Act.
(g) *Effects* or *impacts* means changes to the human environment from the proposed action or alternatives that are reasonably foreseeable and include the following:
(1) Direct effects, which are caused by the action and occur at the same time and place.

(2) Indirect effects, which are caused by the action and are later in time or farther removed in distance, but are still reasonably foreseeable. Indirect effects may include growth inducing effects and other effects related to induced changes in the pattern of land use, population density or growth rate, and related effects on air and water and other natural systems, including ecosystems.

(3) Cumulative effects, which are effects on the environment that result from the incremental effects of the action when added to the effects of other past, present, and reasonably foreseeable actions regardless of what agency (Federal or non-Federal) or person undertakes such other actions. Cumulative effects can result from individually minor but collectively significant actions taking place over a period of time.

(4) Effects include ecological (such as the effects on natural resources and on the components, structures, and functioning of affected ecosystems), aesthetic, historic, cultural, economic, social, or health, whether direct, indirect, or cumulative. Effects may also include those resulting from actions which may have both beneficial and detrimental effects, even if on balance the agency believes that the effects will be beneficial.

(h) *Environmental assessment* means a concise public document prepared by a Federal agency to aid an agency's compliance with the Act and support its determination of whether to prepare an environmental impact statement or a finding of no significant impact, as provided in § 1501.6 of this chapter.

(i) *Environmental document* means an environmental assessment, environmental impact statement, finding of no significant impact, or notice of intent.

(j) *Environmental impact statement* means a detailed written statement as required by section 102(2)(C) of NEPA.

(k) *Federal agency* means all agencies of the Federal Government. It does not mean the Congress, the Judiciary, or the President, including the performance of staff functions for the President in his Executive Office. For the purposes of the regulations in this subchapter, Federal agency also includes States, units of general local government, and Tribal governments assuming NEPA responsibilities from a Federal agency pursuant to statute.

(l) *Finding of no significant impact* means a document by a Federal agency briefly presenting the reasons why an action, not otherwise categorically excluded (§ 1501.4 of this chapter), will not have a significant effect on the human environment and for which an environmental impact statement therefore will not be prepared.

(m) *Human environment* means comprehensively the natural and physical environment and the relationship of present and future generations of Americans with that environment. (*See also* the definition of "effects" in paragraph (g) of this section.)

(n) *Jurisdiction by law* means agency authority to approve, veto, or finance all or part of the proposal.

(o) *Lead agency* means the agency or agencies, in the case of joint lead agencies, preparing or having taken primary responsibility for preparing the environmental impact statement.

(p) *Legislation* means a bill or legislative proposal to Congress developed by a Federal agency, but does not include requests for appropriations or legislation recommended by the President.

(q) *Major Federal action* or *action* means an activity or decision subject to Federal control and responsibility subject to the following:
   (1) Major Federal action does not include the following activities or decisions:
      (i) Extraterritorial activities or decisions, which means agency activities or decisions with effects located entirely outside of the jurisdiction of the United States;
      (ii) Activities or decisions that are non-discretionary and made in accordance with the agency's statutory authority;
      (iii) Activities or decisions that do not result in final agency action under the Administrative Procedure Act or other statute that also includes a finality requirement;
      (iv) Judicial or administrative civil or criminal enforcement actions;
      (v) Funding assistance solely in the form of general revenue sharing funds with no Federal agency control over the subsequent use of such funds;
      (vi) Non-Federal projects with minimal Federal funding or minimal Federal involvement where the agency does not exercise sufficient control and responsibility over the outcome of the project; and
      (vii) Loans, loan guarantees, or other forms of financial assistance where the Federal agency does not exercise sufficient control and responsibility over the effects of such assistance (for example, action does not include farm ownership and operating loan guarantees by the Farm Service Agency pursuant to 7 U.S.C. 1925 and 1941 through 1949 and business loan guarantees by the Small Business Administration pursuant to 15 U.S.C. 636(a), 636(m), and 695 through 697g).
   (2) Major Federal actions may include new and continuing activities, including projects and programs entirely or partly financed, assisted, conducted, regulated, or approved by Federal agencies; new or revised agency rules, regulations, plans, policies, or procedures; and legislative proposals (§ 1506.8 of this chapter).
   (3) Major Federal actions tend to fall within one of the following categories:
      (i) Adoption of official policy, such as rules, regulations, and interpretations adopted under the Administrative Procedure Act, 5 U.S.C. 551 *et seq.* or other statutes; implementation of treaties and international conventions or agreements, including those implemented pursuant to statute or regulation; formal documents establishing an agency's policies which will result in or substantially alter agency programs.
      (ii) Adoption of formal plans, such as official documents prepared or approved by Federal agencies, which prescribe alternative uses of Federal resources, upon which future agency actions will be based.
      (iii) Adoption of programs, such as a group of concerted actions to implement a specific policy or plan; systematic and connected agency decisions allocating agency resources to implement a specific statutory program or executive directive.
      (iv) Approval of specific projects, such as construction or management activities located in a defined geographic area. Projects include actions approved by permit or other regulatory decision as well as Federal and federally assisted activities.
(r) *Matter* includes for purposes of part 1504 of this chapter:
   (1) With respect to the Environmental Protection Agency, any proposed legislation, project, action or regulation as those terms are used in section 309(a) of the Clean Air Act (42 U.S.C. 7609).

(2) With respect to all other agencies, any proposed major Federal action to which section 102(2)(C) of NEPA applies.
(s) *Mitigation* means measures that avoid, minimize, or compensate for effects caused by a proposed action or alternatives as described in an environmental document or record of decision and that have a nexus to those effects. While NEPA requires consideration of mitigation, it does not mandate the form or adoption of any mitigation. Mitigation includes:
  (1) Avoiding the impact altogether by not taking a certain action or parts of an action.
  (2) Minimizing impacts by limiting the degree or magnitude of the action and its implementation.
  (3) Rectifying the impact by repairing, rehabilitating, or restoring the affected environment.
  (4) Reducing or eliminating the impact over time by preservation and maintenance operations during the life of the action.
  (5) Compensating for the impact by replacing or providing substitute resources or environments.
(t) *NEPA process* means all measures necessary for compliance with the requirements of section 2 and title I of NEPA.
(u) *Notice of intent* means a public notice that an agency will prepare and consider an environmental impact statement.
(v) *Page* means 500 words and does not include explanatory maps, diagrams, graphs, tables, and other means of graphically displaying quantitative or geospatial information.
(w) *Participating agency* means a Federal, State, Tribal, or local agency participating in an environmental review or authorization of an action.
(x) *Proposal* means a proposed action at a stage when an agency has a goal, is actively preparing to make a decision on one or more alternative means of accomplishing that goal, and can meaningfully evaluate its effects. A proposal may exist in fact as well as by agency declaration that one exists.
(y) *Publish* and *publication* mean methods found by the agency to efficiently and effectively make environmental documents and information available for review by interested persons, including electronic publication, and adopted by agency NEPA procedures pursuant to § 1507.3 of this chapter.
(z) *Reasonable alternatives* means a reasonable range of alternatives that are technically and economically feasible, and meet the purpose and need for the proposed action.
  (aa) *Reasonably foreseeable* means sufficiently likely to occur such that a person of ordinary prudence would take it into account in reaching a decision.
  (bb) *Referring agency* means the Federal agency that has referred any matter to the Council after a determination that the matter is unsatisfactory from the standpoint of public health or welfare or environmental quality.
  (cc) *Scope* consists of the range of actions, alternatives, and impacts to be considered in an environmental impact statement. The scope of an individual statement may depend on its relationships to other statements (§ 1501.11 of this chapter).
  (dd) *Senior agency official* means an official of assistant secretary rank or higher (or equivalent) that is designated for overall agency NEPA compliance, including resolving implementation issues.

(ee) *Special expertise* means statutory responsibility, agency mission, or related program experience.

(ff) *Tiering* refers to the coverage of general matters in broader environmental impact statements or environmental assessments (such as national program or policy statements) with subsequent narrower statements or environmental analyses (such as regional or basin-wide program statements or ultimately site-specific statements) incorporating by reference the general discussions and concentrating solely on the issues specific to the statement subsequently prepared.

§ 1508.2   [Reserved]

THE NATIONAL ENVIRONMENTAL POLICY ACT OF 1969

42 U.S.C. 4321. Congressional declaration of purpose [Sec. 2]

The purposes of this chapter are: To declare a national policy which will encourage productive and enjoyable harmony between man and his environment; to promote efforts which will prevent or eliminate damage to the environment and biosphere and stimulate the health and welfare of man; to enrich the understanding of the ecological systems and natural resources important to the Nation; and to establish a Council on Environmental Quality.

(Pub. L. 91–190, § 2, Jan. 1, 1970, 83 Stat. 852)

Subchapter I—Policies and goals [Title I]

42 U.S.C. 4331. Congressional declaration of national environmental policy [Sec. 101]

(a) The Congress, recognizing the profound impact of man's activity on the interrelations of all components of the natural environment, particularly the profound influences of population growth, high-density urbanization, industrial expansion, resource exploitation, and new and expanding technological advances and recognizing further the critical importance of restoring and maintaining environmental quality to the overall welfare and development of man, declares that it is the continuing policy of the Federal Government, in cooperation with state and local governments, and other concerned public and private organizations, to use all practicable means and measures, including financial and technical assistance, in a manner calculated to foster and promote the general welfare, to create and maintain conditions under which man and nature can exist in productive harmony, and fulfill the social, economic, and other requirements of present and future generations of Americans.

(b) In order to carry out the policy set forth in this chapter, it is the continuing responsibility of the Federal Government to use all practicable means, consistent with other essential considerations of national policy, to improve and coordinate Federal plans, functions, programs, and resources to the end that the Nation may—

(1) fulfill the responsibilities of each generation as trustee of the environment for succeeding generations;
(2) assure for all Americans safe, healthful, productive, and esthetically and culturally pleasing surroundings;
(3) attain the widest range of beneficial uses of the environment without degradation, risk to health or safety, or other undesirable and unintended consequences;
(4) preserve important historic, cultural, and natural aspects of our national heritage, and maintain, wherever possible, an environment which supports diversity, and variety of individual choice;
(5) achieve a balance between population and resource use which will permit high standards of living and a wide sharing of life's amenities; and

(6) enhance the quality of renewable resources and approach the maximum attainable recycling of depletable resources.

(c) The Congress recognizes that each person should enjoy a healthful environment and that each person has a responsibility to contribute to the preservation and enhancement of the environment.

(Pub. L. 91–190, title I, § 101, Jan. 1, 1970, 83 Stat. 852)

42 U.S.C. 4332. Cooperation of agencies; reports; availability of information; recommendations; international and national coordination of efforts [Sec. 102]

The Congress authorizes and directs that, to the fullest extent possible: (1) the policies, regulations, and public laws of the United States shall be interpreted and administered in accordance with the policies set forth in this chapter and (2) all agencies of the Federal Government shall—

(A) utilize a systematic, interdisciplinary approach which will insure the integrated use of the natural and social sciences and the environmental design arts in planning and in decision making which may have an impact on man's environment;

(B) identify and develop methods and procedures, in consultation with the Council on Environmental Quality established by subchapter II of this chapter, which will insure that presently unquantified environmental amenities and values may be given appropriate consideration in decision-making along with economic and technical considerations;

(C) include in every recommendation or report on proposals for legislation and other major Federal actions significantly affecting the quality of the human environment, a detailed statement by the responsible official on—
  (i) the environmental impact of the proposed action,
  (ii) any adverse environmental effects which cannot be avoided should the proposal be implemented,
  (iii) alternatives to the proposed action,
  (iv) the relationship between local short-term uses of man's environment and the maintenance and enhancement of long-term productivity, and
  (v) any irreversible and irretrievable commitments of resources which would be involved in the proposed action should it be implemented.
Prior to making any detailed statement, the responsible Federal official shall consult with and obtain the comments of any Federal agency which has jurisdiction by law or special expertise with respect to any environmental impact involved. Copies of such statement and the comments and views of the appropriate Federal, State, and local agencies, which are authorized to develop and enforce environmental standards, shall be made available to the President, the Council on Environmental Quality and to the public as provided by section 552 of title 5, and shall accompany the proposal through the existing agency review processes;

(D) Any detailed statement required under subparagraph (C) after January 1, 1970, for any major Federal action funded under a program of grants to States shall not be deemed to be legally insufficient solely by reason of having been prepared by a state agency or official, if:

(i) the State agency or official has statewide jurisdiction and has the responsibility for such action,

(ii) the responsible Federal official furnishes guidance and participates in such preparation,

(iii) the responsible Federal official independently evaluates such statement prior to its approval and adoption, and

(iv) after January 1, 1976, the responsible Federal official provides early notification to, and solicits the views of, any other state or any Federal land management entity of any action or any alternative thereto which may have significant impacts upon such state or affected Federal land management entity and, if there is any disagreement on such impacts, prepares a written assessment of such impacts and views for incorporation into such detailed statement.

The procedures in this subparagraph shall not relieve the Federal official of his responsibilities for the scope, objectivity, and content of the entire statement or of any other responsibility under this Act; and further, this subparagraph does not affect the legal sufficiency of statements prepared by State agencies with less than statewide jurisdiction.

(E) study, develop, and describe appropriate alternatives to recommended courses of action in any proposal which involves unresolved conflicts concerning alternative uses of available resources;

(F) recognize the worldwide and long-range character of environmental problems and, where consistent with the foreign policy of the United States, lend appropriate support to initiatives, resolutions, and programs designed to maximize international cooperation in anticipating and preventing a decline in the quality of mankind's world environment;

(G) make available to States, counties, municipalities, institutions, and individuals, advice and information useful in restoring, maintaining, and enhancing the quality of the environment;

(H) initiate and utilize ecological information in the planning and development of resource-oriented projects; and

(I) assist the Council on Environmental Quality established by subchapter II of this chapter.

(Pub. L. 91–190, title I, § 102, Jan. 1, 1970, 83 Stat. 853; Pub. L. 94–83, Aug. 9, 1975, 89 Stat. 424)

42 U.S.C. 4333. Conformity of administrative procedures to national environmental policy [Sec. 103]

All agencies of the Federal Government shall review their present statutory authority, administrative regulations, and current policies and procedures for the purpose of determining whether there are any deficiencies or inconsistencies therein which prohibit full compliance with the purposes and provisions of this chapter and shall propose to the President not later than July 1, 1971, such measures as may be necessary to bring their authority and policies into conformity with the intent, purposes, and procedures set forth in this chapter.

(Pub. L. 91–190, title I, § 103, Jan. 1, 1970, 83 Stat. 854)

42 U.S.C. 4334. Other statutory obligations of agencies [Sec. 104]

Nothing in section 4332 [Sec. 102] or 4333 [Sec. 103] shall in any way affect the specific statutory obligations of any Federal agency (1) to comply with criteria or standards of environmental quality, (2) to coordinate or consult with any other Federal or State agency, or (3) to act, or refrain from acting contingent upon the recommendations or certification of any other Federal or State agency.

(Pub. L. 91–190, title I, § 104, Jan. 1, 1970, 83 Stat. 854)

42 U.S.C. 4335. Efforts supplemental to existing authorizations [Sec. 105]

The policies and goals set forth in this chapter are supplementary to those set forth in existing authorizations of Federal agencies.

(Pub. L. 91–190, title I, § 105, Jan. 1, 1970, 83 Stat. 854)

Subchapter II—Council on Environmental Quality [Title II]

42 U.S.C. 4341. [Sec. 201] Omitted

Section 201 which required the President to transmit to Congress annually an Environmental Quality Report, was terminated by Congress, effective May 15, 2000, pursuant to section 3003 of Pub. L. 104–66, as amended, set out as a note under section 1113 of Title 31, Money and Finance.

(Pub. L. 91–190, title II, § 201, Jan. 1, 1970, 83 Stat. 854; Pub. L. 104–66, title III, § 3003, Dec. 21, 1995 of as amended, 31 U.S.C. 1113)

42 U.S.C. 4342. Establishment; membership; Chairman; appointments [Sec. 202]

There is created in the Executive Office of the President a Council on Environmental Quality (hereinafter referred to as the "Council"). The Council shall be composed of three members who shall be appointed by the President to serve at his pleasure, by and with the advice and consent of the Senate. The President shall designate one of the members of the Council to serve as Chairman. Each member shall be a person who, as a result of his training, experience, and attainments, is exceptionally well qualified to analyze and interpret environmental trends and information of all kinds; to appraise programs and activities of the Federal Government in the light of the policy set forth in subchapter I of this chapter; to be conscious of and responsive to the scientific, economic, social, esthetic, and cultural needs and interests of the Nation; and to formulate and recommend national policies to promote the improvement of the quality of the environment.

(Pub. L. 91–190, title II, § 202, Jan. 1, 1970, 83 Stat. 854)

Provisions stating that notwithstanding this section, the Council was to consist of one member, appointed by the President, by and with the advice and consent of the Senate, serving as chairman and exercising all powers, functions, and duties of the Council, were contained in the Department of the Interior, Environment, and Related Agencies Appropriations Act, 2006, Pub. L. 109–54, title III, Aug. 2, 2005, 119 Stat. 543, and were repeated in provisions of subsequent appropriations acts which are not set out in the Code.

42 U.S.C. 4343. Employment of personnel, experts and consultants [Sec. 203]

(a) The Council may employ such officers and employees as may be necessary to carry out its functions under this chapter. In addition, the Council may employ and fix the compensation of such experts and consultants as may be necessary for the carrying out of its functions under this Act, in accordance with section 3109 of title 5, (but without regard to the last sentence thereof).

(b) Notwithstanding section 1342 of Title 31, the Council may accept and employ voluntary and uncompensated services in furtherance of the purposes of the Council.

(Pub. L. 91–190, title II, § 203, Jan. 1, 1970, 83 Stat. 855; Pub. L. 94–52, § 2, July 3, 1975, 89 Stat. 258)

42 U.S.C. 4344. Duties and functions [Sec. 204]

It shall be the duty and function of the Council—

(1) to assist and advise the President in the preparation of the Environmental Quality Report required by section 4341[Sec. 201] of this title;[1]
(2) to gather timely and authoritative information concerning the conditions and trends in the quality of the environment both current and prospective, to analyze and interpret such information for the purpose of determining whether such conditions and trends are interfering, or are likely to interfere, with the achievement of the policy set forth in subchapter I of this chapter, and to compile and submit to the President studies relating to such conditions and trends;
(3) to review and appraise the various programs and activities of the Federal Government in the light of the policy set forth in subchapter I of this chapter for the purpose of determining the extent to which such programs and activities are contributing to the achievement of such policy, and to make recommendations to the President with respect thereto;
(4) to develop and recommend to the President national policies to foster and promote the improvement of environmental quality to meet the conservation, social, economic, health, and other requirements and goals of the Nation;
(5) to conduct investigations, studies, surveys, research, and analyses relating to ecological systems and environmental quality;

(6) to document and define changes in the natural environment, including the plant and animal systems, and to accumulate necessary data and other information for a continuing analysis of these changes or trends and an interpretation of their underlying causes;

(7) to report at least once each year to the President on the state and condition of the environment; and

(8) to make and furnish such studies, reports thereon, and recommendations with respect to matters of policy and legislation as the President may request.

(Pub. L. 91–190, title II, § 204, Jan. 1, 1970, 83 Stat. 855)

42 U.S.C. 4345. Consultation with Citizens' Advisory Committee on Environmental Quality and other representatives [Sec. 205]

In exercising its powers, functions, and duties under this Act, the Council shall—

(1) consult with the Citizens' Advisory Committee on Environmental Quality established by Executive Order numbered 11472, dated May 29, 1969, and with such representatives of science, industry, agriculture, labor, conservation organizations, State and local governments and other groups, as it deems advisable; and

(2) utilize, to the fullest extent possible, the services, facilities and information (including statistical information) of public and private agencies and organizations, and individuals, in order that duplication of effort and expense may be avoided, thus assuring that the Council's activities will not unnecessarily overlap or conflict with similar activities authorized by law and performed by established agencies.

(Pub. L. 91–190, title II, § 205, Jan. 1, 1970, 83 Stat. 855)

42 U.S.C. 4346. Tenure and compensation of members [Sec. 206]

Members of the Council shall serve full time and the Chairman of the Council shall be compensated at the rate provided for Level II of the Executive Schedule Pay Rates (5 U.S.C. 5313). The other members of the Council shall be compensated at the rate provided for Level IV o[f] the Executive Schedule Pay Rates (5 U.S.C. 5315).

(Pub. L. 91–190, title II, § 206, Jan. 1, 1970, 83 Stat. 856)

42 U.S.C. 4346a. Travel reimbursement by private organizations and Federal, State, and local governments [Sec. 207]

The Council may accept reimbursements from any private nonprofit organization or from any department, agency, or instrumentality of the Federal Government, any State, or local government, for the reasonable travel expenses incurred by an officer or employee of the Council in connection with his attendance at any conference, seminar, or similar meeting conducted for the benefit of the Council.

(Pub. L. 91–190, title II, § 207, as added Pub. L. 94–52, § 3, July 3, 1975, 89 Stat. 258)

42 U.S.C. 4346b. Expenditures in support of international activities [Sec. 208]

The Council may make expenditures in support of its international activities, including expenditures for: (1) international travel; (2) activities in implementation of international agreements; and (3) the support of international exchange programs in the United States and in foreign countries.

(Pub. L. 91–190, title II, § 208, as added Pub. L. 94–52, § 3, July 3, 1975, 89 Stat. 258)

42 U.S.C. 4347. Authorization of appropriations [Sec. 209]

There are authorized to be appropriated to carry out the provisions of this chapter not to exceed $300,000 for fiscal year 1970, $700,000 for fiscal year 1971, and $1,000,000 for each fiscal year thereafter.

(Pub. L. 91–190, title II, § 209, formerly § 207, Jan. 1, 1970, 83 Stat. 856, renumbered § 209, Pub. L. 94–52, § 3, July 3, 1975, 89 Stat. 258)

THE ENVIRONMENTAL QUALITY IMPROVEMENT ACT OF 1970

42 U.S.C. 4371. Congressional findings, declarations, and purposes [Sec. 202]

(a) The Congress finds –
  (1) that man has caused changes in the environment;
  (2) that many of these changes may affect the relationship between man and his environment; and
  (3) that population increases and urban concentration contribute directly to pollution and the degradation of our environment.
(b) (1) The Congress declares that there is a national policy for the environment which provides for the enhancement of environmental quality. This policy is evidenced by statutes heretofore enacted relating to the prevention, abatement, and control of environmental pollution, water and land resources, transportation, and economic and regional development.
  (2) The primary responsibility for implementing this policy rests with State and local government.
  (3) The Federal Government encourages and supports implementation of this policy through appropriate regional organizations established under existing law.

(c) The purposes of this chapter are—
   (1) to assure that each Federal department and agency conducting or supporting public works activities which affect the environment shall implement the policies established under existing law; and
   (2) to authorize an Office of Environmental Quality, which, notwithstanding any other provision of law, shall provide the professional and administrative staff for the Council on Environmental Quality established by Public Law 91–190.

(Pub. L. 91–224, title II, § 202, Apr. 3, 1970, 84 Stat. 114)

42 U.S.C. 4372. Office of Environmental Quality [Sec. 203]

(a) Establishment; Director; Deputy Director

There is established in the Executive Office of the President an office to be known as the Office of Environmental Quality (hereafter in this chapter referred to as the "Office"). The Chairman of the Council on Environmental Quality established by Public Law 91–190 shall be the Director of the Office. There shall be in the Office a Deputy Director who shall be appointed by the President, by and with the advice and consent of the Senate.

(b) *Compensation of Deputy Director*

The compensation of the Deputy Director shall be fixed by the President at a rate not in excess of the annual rate of compensation payable to the Deputy Director of the Office of Management and Budget.

(c) *Employment of personnel, experts, and consultants; compensation*

The Director is authorized to employ such officers and employees (including experts and consultants) as may be necessary to enable the Office to carry out its functions; under this chapter and Public Law 91–190, except that he may employ no more than ten specialists and other experts without regard to the provisions of title 5, governing appointments in the competitive service, and pay such specialists and experts without regard to the provisions of chapter 51 and subchapter III of chapter 53 of such title relating to classification and General Schedule pay rates, but no such specialist or expert shall be paid at a rate in excess of the maximum rate for GS-18 of the General Schedule under section 5332 of Title 5.

(d) *Duties and Functions of Director*

In carrying out his functions the Director shall assist and advise the President on policies and programs of the Federal Government affecting environmental quality by—
   (1) providing the professional and administrative staff and support for the Council on Environmental Quality established by Public Law 91–190;
   (2) assisting the Federal agencies and departments in appraising the effectiveness of existing and proposed facilities, programs, policies, and activities of the Federal Government, and those specific major projects designated by the President which do not require individual project authorization by Congress, which affect environmental quality;
   (3) reviewing the adequacy of existing systems for monitoring and predicting environmental changes in order to achieve effective coverage and efficient use of research facilities and other resources;

(4) promoting the advancement of scientific knowledge of the effects of actions and technology on the environment and encourage[ing] the development of the means to prevent or reduce adverse effects that endanger the health and well-being of man;

(5) assisting in coordinating among the Federal departments and agencies those programs and activities which affect, protect, and improve environmental quality;

(6) assisting the Federal departments and agencies in the development and inter-relationship of environmental quality criteria and standards established throughout the Federal Government;

(7) collecting, collating, analyzing, and interpreting data and information on environmental quality, ecological research, and evaluation.

(e) Authority of Director to contract

The Director is authorized to contract with public or private agencies, institutions, and organizations and with individuals without regard to section 3324(a) and (b) of title 31 and section 5 of title 41 in carrying out his functions.

(Pub. L. 91–224, title II, § 203, Apr. 3, 1970, 84 Stat. 114; 1970 Reorg. Plan No. 2, § 102, eff. July 1, 1970, 35 F.R. 7959, 84 Stat. 2085)

42 U.S.C. 4373. Referral of Environmental Quality Reports to standing committees having jurisdiction [Sec. 204]

Each Environmental Quality Report required by Public Law 91–190 [NEPA] shall, upon transmittal to Congress, be referred to each standing committee having jurisdiction over any part of the subject matter of the Report.[2]

(Pub. L. 91–224, title II, § 204, Apr. 3, 1970, 84 Stat. 115)

42 U.S.C. 4374. Authorization of appropriations [Sec. 205]

There are hereby authorized to be appropriated for the operations of the Office of Environmental Quality and the Council on Environmental Quality not to exceed the following sums for the following fiscal years which sums are in addition to those contained in Public Law 91–190 [NEPA]:

(a) $2,126,000 for the fiscal year ending September 30, 1979.
(b) $3,000,000 for the fiscal years ending September 30, 1980, and September 30, 1981.
(c) $44,000 for the fiscal years ending September 30, 1982, 1983, and 1984.
(d) $480,000 for each of the fiscal years ending September 30, 1985 and 1986.

(Pub. L. 91–224, title II, § 205, Apr. 3, 1970, 84 Stat. 115; Pub. L. 93–36, May 18, 1973, 87 Stat. 72; Pub. L. 94–52, § 1, July 3, 1975, 89 Stat. 258; Pub. L. 94–298, May 29, 1976, 90 Stat. 587; Pub. L. 95–300, June 26, 1978, 92 Stat. 342; Pub. L. 97–350, § 1, Oct. 18, 1982, 96 Stat. 1661; Pub. L. 98–581, § 1, Oct. 30, 1984, 98 Stat. 3093)

42 U.S.C. 4375. Office of Environmental Quality Management Fund [Sec. 206]

(a) Establishment; financing of study contracts and Federal interagency environmental projects

There is established an Office of Environmental Quality Management Fund (hereinafter referred to as the "Fund") to receive advance payments from other agencies or accounts that may be used solely to finance—

(1) study contracts that are jointly sponsored by the Office and one or more other Federal agencies; and

(2) Federal interagency environmental projects (including task forces) in which the Office participates.

(b) Study contract or project initiative

Any study contract or project that is to be financed under subsection (a) of this section may be initiated only with the approval of the Director.

(c) Regulations

The Director shall promulgate regulations setting forth policies and procedures for operation of the Fund.

(Pub. L. 91–224, title II, § 206, as added Pub. L. 98–581, § 2, Oct. 30, 1984, 98 Stat. 3093)

THE CLEAN AIR ACT—SECTION 309

42 U.S.C. 7609. Policy review [Sec. 309]

(a) Environmental impact

The Administrator shall review and comment in writing on the environmental impact of any matter relating to duties and responsibilities granted pursuant to this chapter or other provisions of the authority of the Administration, contained in any (1) legislation proposed by any Federal department or agency, (2) newly authorized Federal projects for construction and any major Federal agency action (other than a project for construction) to which section 4332(2)(C) of the title applies, and (3) proposed regulations published by any department or agency of the Federal Government. Such written comment shall be made public at the conclusion of any such review.

(b) Unsatisfactory legislation, action, or regulation

In the event the Administrator determines that any such legislation, action, or regulation is unsatisfactory from the standpoint of public health or welfare or environmental quality, he shall publish his determination and the matter shall be referred to the Council on Environmental Quality.

(July 14, 1955, ch. 360, title III, § 309, as added Pub. L. 91–604, § 12(a), Dec. 31, 1970, 84 Stat. 1709)

EXECUTIVE ORDER 11514—PROTECTION AND ENHANCEMENT OF ENVIRONMENTAL QUALITY, AS AMENDED BY EXECUTIVE ORDER 11991

SOURCE: The provisions of Executive Order 11514 of Mar. 5, 1970, appear at 35 FR 4247, 3 CFR, 1966–1970, Comp., p. 902, unless otherwise noted.

By virtue of the authority vested in me as President of the United States and in furtherance of the purpose and policy of the National Environmental Policy Act of 1969 (Public Law No. 91–190, approved January 1, 1970), it is ordered as follows:

Section 1. *Policy.* The Federal Government shall provide leadership in protecting and enhancing the quality of the Nation's environment to sustain and enrich human life. Federal agencies shall initiate measures needed to direct their policies, plans and programs so as to meet national environmental goals. The Council on Environmental Quality, through the Chairman, shall advise and assist the President in leading this national effort.

Sec. 2. *Responsibilities of Federal agencies.* Consonant with Title I of the National Environmental Policy Act of 1969, hereafter referred to as the "Act", the heads of Federal agencies shall:

(a) Monitor, evaluate, and control on a continuing basis their agencies' activities so as to protect and enhance the quality of the environment. Such activities shall include those directed to controlling pollution and enhancing the environment and those designed to accomplish other program objectives which may affect the quality of the environment. Agencies shall develop programs and measures to protect and enhance environmental quality and shall assess progress in meeting the specific objectives of such activities. Heads of agencies shall consult with appropriate Federal, State and local agencies in carrying out their activities as they affect the quality of the environment.

(b) Develop procedures to ensure the fullest practicable provision of timely public information and understanding of Federal plans and programs with environmental impact in order to obtain the views of interested parties. These procedures shall include, whenever appropriate, provision for public hearings, and shall provide the public with relevant information, including information on alternative courses of action. Federal agencies shall also encourage State and local agencies to adopt similar procedures for informing the public concerning their activities affecting the quality of the environment.

(c) Insure that information regarding existing or potential environmental problems and control methods developed as part of research, development, demonstration, test, or evaluation activities is made available to Federal agencies, states, counties, municipalities, institutions, and other entities, as appropriate.

(d) Review their agencies' statutory authority, administrative regulations, policies, and procedures, including those relating to loans, grants, contracts, leases, licenses, or permits, in order to identify any deficiencies or inconsistencies therein which prohibit or limit full compliance with the purposes and provisions of the Act. A report on this review and the corrective actions taken or planned, including such measures to be proposed to the

President as may be necessary to bring their authority and policies into conformance with the intent, purposes, and procedures of the Act, shall be provided to the Council on Environmental Quality not later than September 1, 1970.

(e) Engage in exchange of data and research results, and cooperate with agencies of other governments to foster the purposes of the Act.

(f) Proceed, in coordination with other agencies, with actions required by section 102 of the Act.

(g) In carrying out their responsibilities under the Act and this Order, comply with the regulations issued by the Council except where such compliance would be inconsistent with statutory requirements.

(Sec. 2 amended by Executive Order 11991 of May 24, 1977, 42 FR 26967, 3 CFR, 1977 Comp., p. 123)

Sec. 3. *Responsibilities of Council on Environmental Quality.* The Council on Environmental Quality shall:

(a) Evaluate existing and proposed policies and activities of the Federal Government directed to the control of pollution and the enhancement of the environment and to the accomplishment of other objectives which affect the quality of the environment. This shall include continuing review of procedures employed in the development and enforcement of Federal standards affecting environmental quality. Based upon such evaluations the Council shall, where appropriate, recommend to the President policies and programs to achieve more effective protection and enhancement of environmental quality and shall, where appropriate, seek resolution of significant environmental issues.

(b) Recommend to the President and to the agencies priorities among programs designed for the control of pollution and for the enhancement of the environment.

(c) Determine the need for new policies and programs for dealing with environmental problems not being adequately addressed.

(d) Conduct, as it determines to be appropriate, public hearings or conferences on issues of environmental significance.

(e) Promote the development and use of indices and monitoring systems (1) to assess environmental conditions and trends, (2) to predict the environmental impact of proposed public and private actions, and (3) to determine the effectiveness of programs for protecting and enhancing environmental quality.

(f) Coordinate Federal programs related to environmental quality.

(g) Advise and assist the President and the agencies in achieving international cooperation for dealing with environmental problems, under the foreign policy guidance of the Secretary of State.

(h) Issue regulations to Federal agencies for the implementation of the procedural provisions of the Act (42 U.S.C. 4332(2)). Such regulations shall be developed after consultation with affected agencies and after such public hearings as may be appropriate. They will be designed to make the environmental impact statement process more useful to decisionmakers and the public; and to reduce paperwork and the accumulation of extraneous background data, in order to emphasize the need to focus on real environmental issues and

alternatives. They will require impact statements to be concise, clear, and to the point, and supported by evidence that agencies have made the necessary environmental analyses. The Council shall include in its regulations procedures (1) for the early preparation of environmental impact statements, and (2) for the referral to the Council of conflicts between agencies concerning the implementation of the National Environmental Policy Act of 1969, as amended, and Section 309 of the Clean Air Act, as amended, for the Council's recommendation as to their prompt resolution.
(i) Issue such other instructions to agencies, and request such reports and other information from them, as may be required to carry out the Council's responsibilities under the Act.
(j) Assist the President in preparing the annual Environmental Quality Report provided for in section 201 of the Act.
(k) Foster investigations, studies, surveys, research, and analyses relating to (i) ecological systems and environmental quality, (ii) the impact of new and changing technologies thereon, and (iii) means of preventing or reducing adverse effects from such technologies.

(Sec. 3 amended by Executive Order 11991 of May 24, 1977, 42 FR 26967, 3 CFR, 1977 Comp., p. 123)

Sec. 4. *Amendments of E.O. 11472.*

[Selection 4 amends Executive Order 11472 of May 29, 1969, Chapter 40.]

Notes

1 CEQ notes that Congress amended 42 U.S.C. 4341 to remove the Environmental Quality Report requirement.

PREPARATION OF STUDENTS AS ENVIRONMENTAL PRACTITIONERS

Robert M. Sanford

Environmental practitioners now are more likely to come from interdisciplinary education programs than in the previous half century since the rise of the environmental professions. But regardless of the nature of their training, one of the most common questions and concerns lies in the nature of deciphering what to expect when leaving the "cocoon" of academia and entering the world of environmental practice; a world of jobs, competition, clients, and regulatory compliance. Bringing the experiences of practitioners into the classroom is one way to address this concern.

As a government regulator who moonlighted by teaching graduate courses in environmental impact assessment, I was always interested in the pragmatic aspects of environmental science. Pleased at the success of one of my recent students in the fast-paced arena of environmental consulting, I asked that student to address the next crop of EIA students: "Please write a 'letter from the trenches' summarizing the advice and wisdom you've gained from your new job." The student replied, asking me to keep his identity anonymous. After six years of continued success in preparing environmental impact statements, this same professional was asked, "What would you add or change in your earlier letter?" and we published the results in 1999 (Sanford, 1999). Now it has been well over two decades since I became a full-time member of academia, but I remain interested in the applied nature of science and in preparing students for the workforce. Almost 20 years later, I asked my former student the same question as before. The letters, edited, are reproduced below. The advice from the early two letters is still applicable, and all three letters convey a work ethic useful to entering professionals (and the rest of us) whether contemplating the private sector or government service.

Letter 1: December 5, 1992
Dear EIA Class:

In 1991, I completed my coursework in the resource management and administration (MS) degree program at Antioch [Antioch University New England]. My academic concentrations were in waste management and water resources. I am writing you to discuss some things that I have learned "from the trenches" of environmental impact assessment.
I am a research analyst at an engineering consulting firm where I conduct environmental assessments (EAs) and environmental impact assessments (EIAs). I have learned a number of things during my employment over the past half-year. Some of the things I have found to be most important are summarized below.

- The scope of work is very important to understand. The scope of work should be fully understood by the contracted firm. A personal meeting with the client should be held to discuss any uncertainties. Try to get yourself included in such meetings so that they are not left to senior staff alone; you are the one who will be "in the field."

- Use the scope of work as an outline, and provide your work in the requested format noted in the scope of work. The scope of work is a summary of what services would be provided by the contracted firm. Accordingly, this should serve as a basic outline for the work completed.
- Make and follow an outline, which should also be used in cost proposal development. This outline should serve as an initial table of contents to help the preparers stay organized during project work. It should also serve as a task sheet for estimating costs for a cost proposal.
- Attempt to get due dates for the different deliverables for your project and plan backwards for your internal due dates. It is sometimes ambiguous when certain drafts (deliverables) are due. This is especially true for subcontracting work, where due dates are filtered through the prime contractor. Every effort should be made to get these dates and develop a timeline for the work effort. If no time-line is available, develop internal timelines that are conservative. Don't be late!
- Write the method of your work as you progress and not afterward. It takes more time, but it is well worth it for accuracy. It is well worth the effort to keep a legal pad especially for methodology. Should anyone (including yourself) have a question about your methods, it will be readily available.
- Know well where to find and how to use your data resources for each project. Remember that as a consultant, you are a problem-solver and should be well versed on your available resources and how to use them. Seek every opportunity to learn and develop skills in obtaining data . . . any data. All knowledge is good!
- Data are rarely in the form requested and if they are, the requested format is rarely perfect. This is a bitter truth. Never have I been given a completely "clean" data set that is ready for analysis. Media form, software format, content, and structure are typical problems encountered in data. Plan ahead for problems with data manipulation . . . both an art and science.
- Be realistic with what can be provided with certain costs. Strive for excellence but not perfection. The effort expended in this marginal difference is not typically justified.
- Realize that after the contract is signed, you and your firm are responsible for all things in the corresponding scope of work. Negotiate carefully. As noted above, the contracted firm must fully understand its responsibilities outlined in the contract and corresponding scope of work. You cannot be too rigid in this phase of the project.
- Learn what level of detail is needed for each task to be completed. Do not discuss insignificant items in great detail while only marginally discussing important items. Detail should correspond to significance.
- Your discussion of impacts must be accompanied by a preceding section of corresponding baseline conditions to the level of detail of the impacts discussion. Enough said.
- Don't be afraid to use boilerplate information to avoid wasting time. In certain instances, information from one portion of the same or different projects may be relevant for another portion of the same project or for another. Do not reinvent the wheel. Utilize your information and always note your sources.
- Don't be afraid of qualitative rather than quantitative assessment where appropriate. In some instances, qualitative assessment is the only feasible method. This is perfectly acceptable. As a consultant, you have been hired for your skills and professional judgment. However, always fully explain your qualitative conclusions.

- The baseline section must contain the information to substantiate findings of no significant impacts. If no significant impacts are found in the EA or Environmental Impact Statement (EIS), this conclusion must be fully developed in the baseline and impacts sections. A simple statement of no significant impacts is not enough!
- The EA or EIS is not complete after all sections are complete—editing, tables, table of contents, page numbering, inserts, copying, binding, mailing, and cover letters must also be completed. Plan ahead! Realize the logistics in getting multiple copies (sometimes 100 or more) edited, copied, bound and mailed. Conform to the organizational systems of your key players, your typists. Helping these people will make the hectic due day much more smooth. Stay calm on the "due day," and remember that few projects are sent without the overnight Federal Express Label.
- Volunteer for special projects to expand your skill base. The most important skill in a multidiscipline consulting environment is flexibility. Use your ability to learn how to learn—the *raison d'être* for a liberal arts degree. Develop your skills at reading technical manuals. This will greatly enhance your flexibility.
- Develop a subject specialty within environmental assessment; i.e., socioeconomic, physical environment, or natural environment. In order for you to be included on a project team, you must have a specialized skill that will allow you to take responsibility for a section of the project. Specialization in a certain area will contribute to your usefulness and the firm's dependence on you! But keep your generalist orientation too; your "eye on the big picture."
- Attempt to specialize in a type of project in order for the firm to become dependent on your skills. Certain types of projects may recur. For example, highway expansion projects and military movement projects are common types.
- Do not underestimate your skills. Many people are winging it and you may be more experienced! While only a rookie, you do have a significantly wide knowledge base from which to draw. Use it.
- Be confident of your knowledge and do not be afraid to ask your project team members questions. Your project team is a team. You cannot learn unless you first understand, and the team is the fastest way to do this.
- Know the differences between EAs and EIAs. Learn the level of detail expected for the agency for which you are completing the work.
- Develop memorandum-writing skills for coordination letters. Concise memo writing skills will save time in coordination between clients, different departments, and stake holding agencies.
- Realize the existence of bureaucratic dislocation, disorganization, and inefficiency before beginning to deal with federal, state, or local governments. Government employees have to deal with this too, and may be just as frustrated. Things go smoother if you understand the system.
- Make sure that you always keep a priority list. Sometimes your plate is empty, and sometimes it's overflowing. Always know what is more important and act accordingly. If you don't know, your supervisor will certainly tell you.
- Personal meetings are much more effective than telephone conversations. This is especially true in scope of work and contract negotiations. It's too easy to hang up a phone! People will work much harder towards a compromise if they pay for flight, food and lodging for a meeting.

- Master group interaction. Most work is completed through cooperative efforts in one way or another. Learn to share praise and responsibility.
- Understand the unofficial and unspoken powers of the firm's typists and supply orderers. The typists rule on "due day." Befriend these special people and your life will be much more pleasant. Also, always conform to their organizational systems, no matter how bizarre they may appear! For instance, if any draft document that is to be typed must be stamped in red ink before being typed by the typist, so be it! The supply orderers also play a similarly powerful role.
- Learn to seek and accept criticism with a smile. Remember, the people who will be giving this criticism typically are more experienced and may know best. Technical writing is different from scholarly writing. Be concise and roll with the punches. The writing will come with practice.
- Learn to budget your time during the day. Set daily goals. Don't split hairs if you do not need to. Go on to other things. Setting daily goals will help you to stay on line for longer time lines. The tendency to procrastinate or to work slowly may become overwhelming without these progress checks along the way.
- Do not forget that in the consulting business (and for many government agencies), every minute of your day is billable to some project or overhead account . . . productive! Enough said.
- Sacrifice social acquainting during the first few weeks of employment in order to establish your lasting picture of diligence. While difficult and seemingly fake, the first few weeks will give your supervisors a lasting impression of your performance.
- Understand to whom you report and accept responsibility only through this person. Changes in scope or other important changes should be given through the one person to whom you are responsible. It is important that directives be accepted from only one person. This will avoid conflicting directives. But do not exclude constructive input and advice from others.
- Become proficient in MS/IBM DOS based computers. The government tends to concentrate in MS/IBM DOS format data. MS Office, WordPerfect, Quattro Pros are several software packages in which project work is frequently requested for the government.
- Be prepared to learn the firm's software in the off-hours. Staff development time is frowned upon although usually offered. Staff development is an overhead account at my firm, and it is by definition non-productive in the monetary sense. Avoid it.
- Purchase a compatible home computer for your word processing and table formation. It will be well worth the money to work at home. Just a suggestion for a relaxing work environment for after hours work.
- Understand that consultants are constantly sought for advice and sometimes have abundant confidence with large egos. Be careful with criticism while new at your job. Use your best judgment.
- Realize that the consulting business is interesting and profitable while the contracts are current. Few things last forever. Job security is dependent on contracts. Last one hired, first one fired.
- Always be outwardly positive about progress. Amazing things can be accomplished in a few extra hours of work. Stay cool. Nervousness shows and never helps. You will be amazed at your productivity under pressure. You will appear much more professional if you avoid

"sighs," and other signs of stress at work. It is the same logic as, "Never say that you are tired at work."
- Tackle the most time-consuming portion of the project first by completing it or by planning on how to complete this section. Plan ahead and get organized!
- Complete fieldwork as early as impracticable. More questions will be answered at an earlier time and your time will be used more wisely in the end. If this can be arranged, it is most productive. However, it is not always possible.

I hope that these comments "from the trenches" are useful. Good luck with your career, and consider consulting. It's a very interesting career. With new projects every week your job can be a lot of fun.

Letter 2: February 15, 1998
Dear EIA Class:

I've now been in the consulting business for over six years doing interdisciplinary impact analyses. I no longer work for the company in which I worked when I prepared this memo in 1992, but after rereading it, the tips still apply. A few additional comments may be helpful, however:

- Get involved in societies, organizations, and the like. It looks great on a resume, and extra people skill development never hurts. It's also very important for networking.
- Network and market. Worthy of its own tip, networking will help you during your transition or pull you through a project when you need guidance but don't want to ask the boss! Do it; maintain it up on a Rolodex or computer card file; and utilize it when you can. Regarding marketing, you never know when your sub-consultant will be your prime. Keep your eyes open. Read the papers, Commerce and Business Daily (a compilation of bid notices for federal contract services of $25,000 or more), and other.
- Write articles, papers, and otherwise toot your own horn. Everyone works hard. Your hard work will likely be unnoticed at times. Don't try to get "atta-boys/girls" for every accomplishment, but do have your work noticed. Write email that time-stamps your work if you work at night. There are other subtle ways to do this. Writing papers for publication or for presentation at conferences is a great way to get recognition, get to go to the conference, and sometimes honorary bonuses! Also, it's great for your resume.
- Keep a detailed list of every project on which you have worked and your responsibilities. Additionally, try to get a copy of each report that you've been involved with preparing. It's great to bring some of these materials, especially the list and a shining example of your work, to interviews.
- Go the extra mile . . . on your own time. For instance, if you believe that a certain form used at work is obsolete, make a new one, and suggest its use. Be industrious. Take command, and be invaluable.
- Help out other divisions when you can. Perhaps another division in your company may do very different work than you, but that doesn't mean that they couldn't use your help in a pinch. Try to identify these times, and suggest help. Such "team" attitude can only help when managers discuss among themselves personnel actions.

- Know the internet protocols, email protocols, FTP protocols, and in general, your computer. Computer maintenance (loading new programs, uninstalling programs, cleaning hard drives, and the like) is becoming assumed knowledge for employees. Office Management Information System Coordinators are taxed enough with maintaining networks. They don't want someone asking them how to reload a computer program. And, you certainly don't want them complaining about such menial requests from you to anyone important—on purpose or otherwise!
- Shop around for the best vendor if you are spending company or project funds. This is obvious, but it is important that you document your search. This may even be required per company policy.
- Analyze outsourcing versus in-house photocopying. Photocopying can be expensive if leasing machines in-house. Often, vendors will work out deals that make spending the per page and in-house labor look very wasteful. Work the savings up on paper and keep it to justify your decision.
- Document everything. Document all telephone conversations and meetings where plans or decisions are made or data are collected; have someone else review work before it is submitted and have them document this review. Don't forget to check spreadsheet formulas and other calculations when reviewing the submittals. Often such spreadsheets are considered a "black box" after they are developed, with users assuming they are correct.
- Check your work. This also deserves its own tip. This is most important to avoid costly and embarrassing errors.
- Stay away from gossip and socializing too much. Not to be so straight-laced about this, but it's just a good professional decision. You can still be friendly without such things, and you'll stay away from many problems that arise from it. And there are problems.

I have found that students appreciate and relate to the straightforward advice suggested above. However, my former student and I are the first to acknowledge that it is based on common sense and reflects typical management principles (e.g., Total Quality Management). Nevertheless, the pragmatic application of basic business practices provides reassurance during the transition from student to environmental professional.

Letter 3:
Dear EIA class,

It's 2017, and I've now been in the environmental impact consulting business for 25 years, six years with my first company as a research analyst, 12 years with another company as a project planner and project manager, and 8 years with my present company as a senior project manager and officer. While dated, my 1992 and 1998 "pearls of wisdom" still apply. However, I thought it would be useful to review some of the key things that are attractive from the employer's standpoint and which I can contribute my greatest career growth and rewards in the interdisciplinary environmental impact assessment consulting field:

- *Again, check your work.* You are responsible for your work product—your sentences, tables, and graphics. We don't make machines; we make arguments and documents. Do it as best as you can.

- *It's often not possible to be the smartest person in the room, but you can be the hardest working and best organized.* No one wants team members who must be reminded or redirected. Do this yourself; take initiative.
- *Prepare for and run a good meeting.* While a well-run meeting doesn't always get praise, a poorly run meeting nearly always gets criticisms. Meetings are expensive. Typically your high-dollar staff attend, and they have both productive attendance time and unproductive travel, waiting, and record-development time. Use it wisely.
- *Appreciate that different communication techniques are needed for different situations.* Email doesn't work for everything. While old school, face to face meetings are critical for some things like negotiations and persuasion.
- *Remember to stay professional in both spoken and written communications.* By now, I'm sure you know of embarrassing email stories personally. Such recklessness can have profound adverse business effects. In general—no negative email.
- *Companies don't care about your degree or grade point as much as they care about your energy, initiative and what skills you have and are working on.* We want to know what you can do for us; we don't care about what you have done for yourself.
- *Get a credential.* It will be expected as you progress in your career. The ABCEP CEP credential is a good fit, as may be the AICP credential. The PMI PMP credential is also useful if you gravitate towards project management.
- *Raise your hand.* Do so to help; do so to contribute; do so to grow.
- *Master Excel/Spreadsheets.* It's the most useful software tool to most. Be the go-to person when it comes to a special application need.
- *Grow your network and always stay tuned in to new contract opportunities.* You will remain on the standard salary curve of years experience versus salary for your role—unless you bring more to the table. More could be project management skills, administrative talents, and of course—business development. Rain-makers are largely insulated from workforce contractions. Be multi-dimensional.
- *Appreciate the potential strategic importance and seek the added value in all you do.* Small actions can mean a lot to your employer and client. Exceeding expectations should be your goal.
- *Organize a personal plan to meet your annual goals.* Your company will likely develop SMART goals for you—Specific, Measurable, Attainable, Relevant, and Time-bound. Don't lose track. Exceed or at least meet them. Acknowledge and address your growth needs and weaknesses; we all have them.
- *Understand that the grass is not always greener elsewhere.* It's a job. It's not vacation. Collaborate to improve the shortcomings of your work situation. Don't allow yourself to hop from one company to the next. This will be questioned, and you will be flagged as a "flight-risk." Loyalty is earned over time, and it is important in lean times.

REFERENCE

Sanford, R. M. (1999). Preparation of Students for Professional Practices: A View from the Trenches. *Environmental Practice*, 1(3), 121–124.

Index

Note: Numbers in **bold** indicate a table. Numbers in *italics* indicate a figure.

AASHTO *see* American Association of State Highway and Transportation Officials
abiotic factors: definition of 111
abstracting services 34
accretion: definition of 100
acid rock drainage: definition of 89
acre-foot: definition of 100
action: Environmental Action Plan (EAP) 6; "fairness" in 26; mitigation 15; ongoing 3; state or private 16
action areas 110
action (federal or state) *see* categorical exclusion; environmental assessment; environmental impact assessment; environmental impact statement; Finding of No Significant Impact; Record of Decision; scoping
ad hoc method 56
aesthetics and visual impact analysis 161–169
affected environment: Brant and Schultz climate change recommendations regarding 123–124; CEQ regulations including rectification or restoration of 110; definition of 24, 42; as step in sample EIA job plan **41**; step in EIA preparation 52; Title 40 Code of regulation § 1501.3 249; § 1502.15 262
African American settlement 56
Agency for Toxic Substances and Disease Registry (ATSDR) 225
agency planning and NEPA as stipulated in Title 40 Code of Federal Regulations, Part 1501 247–256
Air Quality Index (AQI) 117
Air Quality Management Plan (AQMP) 120
air quality models 120–121
air pollutants *see* criteria pollutants; Hazardous Air Pollutants; residence time
air toxics: definition of 125
alluvium: definition of 101
alternatives: as EIS component 24; definition of 42; *see also* Environmentally Preferable Alternative; project alternatives
ambient noise 153: definition of 158
amelioration: definition of 78
American Association of Civil Engineers 69
American Association of State Highway and Transportation Officials (AASHTO) 69; definition of 202; "Green Book" 203
American Institute of Architects 128
Americans *see* Native Americans
American Society of Heating, Refrigeration, and Air Conditioning Engineers (ASHRAE) 141; definition of 147
Americans with Disabilities Act (ADA) 202
annual fuel efficiency (AFUE): definition of 147
anthropogenic trace gases 119
AQI *see* Air Quality Index
aquifer 75, 96; definition of 101
aquifer classes 99
archeological site: definition of 133
archeology and historic preservation 127–135
area of potential effect (APE) 104, 127, 129, 161; affected environment and (definition of) 24
Army Corps of Engineers 27
Arnstein, Sherry R. 59
arterials and freeways: definition of 202

asbestos 36, 37, 125
ASHRAE *see* American Society of Heating, Refrigeration, and Air Conditioning Engineers
associational noise: definition of 158
ATSDR *see* Agency for Toxic Substances and Disease Registry
availability *see* Notice of Availability
average daily traffic (ADT): definition of 202
average travel speed 199
Aviation Environmental Design Tool (AEDT): definition of 158

baseline information 8; definition of 17
BEQI *see* Bicycle Environmental Quality Index
best available technology (or techniques) (BAT) 158; definition of 158
best management practices (BMPs) 95, 100; definition of 42
best on-site management practices (BMPs) **111**
BI *see* Biotic Index
Bicycle Environmental Quality Index (BEQI) 69
bids 35, 38
biodiversity: definition of 111; impacts to 106; Strategic Plan for Biodiversity 32; *see also* natural capital
biogeochemical cycles 103; definition of 111
biological activity 81
biological change 76
biological conditions, existing 104
biological impacts 104, 107, **109**; definition of 101; mitigation measures **111**; *see also* impacts
biological opinion (US FWS) 110
biological oxygen demand (BOD) **96**
biological species and habitats 103–113
biological water pollutant **97**
biotic community 106, 113
Biotic Index (BI) 69
BMPs *see* best management practices
BOD *see* biological oxygen demand
borehole: definition of 89
Brant, Leslie, and Courtney Shultz 123–124
building envelope: definition of 147

CAA *see* Clean Air Act
California checklist for sites in unstable areas (sample) **87**
California Environmental Quality Act (CEQA) 46; checklist of biological impacts **109**;
state-equivalent "little NEPA" law **29**; weaknesses identified in NEPA by 109–110
Callahan Mining Corp. 87, *88*
Canter, Larry 15, 19, 80; approaches to screening described by 47; checklist for describing the environmental setting 55; comparison of NEPA with equivalents in other countries 30; EIA preparation, steps as described by 52; EIA/EIS, ideal components of 77; EIS, discussion of what it should contain 19, 24, 77; EIA, six step approach to 80; environmental impacts, types of 72; fresh water, approaches to 92; issues for EIA effectiveness pointed out by 15; projects with significant earth resource issues described by **84**
capacity: carrying (species habitat) **105**, 111, 112; definition of 111, 191; overflow parking 106; school 188; visitor accommodation 95
capacity analysis 193
capital costs 189; definition of 191
carbon dioxide 118
carrying capacity (species habitat) **105**, 112; definition of 111
Carson, Rachel 2
categorical exclusion (CX or CE) 47, 48, definition of 17, 42, 62, 213
categorical exclusion as stipulated in Title 40 Code of Federal Regulations: § 1500.4 245, 246, 247; § 1501.4 249–250; § 1506.3 274
categorical exclusion determination (CATEX or CE or CX) 22
CEQ *see* Council on Environmental Quality
CEQA *see* California Environmental Quality Act
CEQR *see* New York City Environmental Quality Review
certified local government: definition of 133
charismatic megafauna: definition of 111
checklist 39, 55; data collection and preservation involving 56; impact assessment using 76; MAP 178; social impact 177
checklists: applications involving transportation review or permits (sample) **194**; assessing an environmental setting 105; avian impact assessment **108**; biological impacts from the California Environmental Quality Act for your state **109**; EIA energy component based on LEED green building certification **142**; EIA report completion **206**; energy policy and program impact assessment

143; energy-related emission reduction categories based on LEED green building certification **145**; environmental justice and social impacts of an EA report **175**; erosion control plans and details (sample) **86**; landscape checklist for energy efficient sites (sample) **144**; police department **185**; sewer and water **187**; sites in unstable areas, California (sample) **87**; soil engineering report (sample) **85**; steps for evaluating the environmental justice component of an EA report **175**; transportation construction plans (sample) **195**; water resource categories of impact for project phases in an EIA **99**; wetland functions and values for an EIA on a site that contains wetlands **94**
checks: mitigation 164; progress 305
checks and balances xiv, 25, 26
Chemical Mass Balance (CMB) model 121
chlorofluorocarbons 119
chroma **167**; definition of 168
CIA *see* Climate Impact Assessment
circle of poison 31
citizen control 59
citizen groups 174
citizen participation in EIA process 11; eight rungs on ladder of 59
Citizens' Advisory Committee on Environmental Quality 294
Clean Air Act (CAA) 28, 115, 206; carbon dioxide as pollutant 121 (assignment); section 309 244, 269, 286, 298, 301
Clean Air Act Amendments of 1990 118
Clean Water Act (CWA) 28, 206; definition of 101
climate change 14, 30, 103; air quality and 115–125; waterfront development and **124**
climate impact assessment (CIA): definition of 125
climate warming 32
climax community: definition of 111
clime: definition of 111
coastal stabilization 95
Coastal Plain Oil and Gas Leasing Program (EO 1399) 140
Coastal Zone Management Act of 1972 (CZMA) 87, 128
coastal zones 108
collector: definition of 202; subcollector, definition of 203
community: definition of 111

conservation bank: definition of 112
conservation easement: definition of 112
Constitution *see* United States Constitution
cooperating agency 12; cover sheet and 24; definition of 17, 213
cooperating agency as stipulated in Title 40 Code of Federal Regulations: § 1500.3 NEPA compliance 244; § 1501.8 cooperating agencies 253–254; § 1508.1 definition of 284
Council on Environmental Quality (CEQ) xiv, 239; definition of 17; environmental justice and 171; mitigation, as defined by 110; page limits specified by 6; responsibilities of 300–301
criteria pollutants: air 116–117; definition of 125
cultural resources: definition of 133; NEPA and 128, **129**
cumulative effects 124, 175; definition of 42
cumulative impacts *see* impacts, types of
cyanidation: definition of 89
CZMA *see* Coastal Zone Management Act of 1972

DEIS *see* Draft Environmental Impact Statement
deforestation 30, **230**, **231**
Delphi method 56, 75, 223 (assignment)
demand-side management 146 (assignment); definition of 147
density (traffic): definition of 202
deposition **83**; definition of 101
description of project 77; *see also* project description
descriptions of variables 71
desertification 30
design hourly volume (DHV): definition of 202
Detweiler, R. 206, 210 (assignment)
disbenefit: definition of 179
disclosure: NEPA requirements regarding 58–59; Section 304 of the National Historic Preservation Act (16 USC § 470w-3) restrictions on 133
disclosure statement **186**, 276
dispersion modeling 120–121
dissolved oxygen (DO) 70, **97**; definition of 101
district: cultural 166; definition of 134
disturbance: definition of 112
diversity: definition of 112; *see also* biodiversity
DO *see* dissolved oxygen
dominant species: definition of 112
Draft Environmental Impact Statement (DEIS): agency guidelines pertaining to 47; commenting

agencies' criteria applied to 106; definition of 17, 42, 62; distribution of 11; Environmental Assessment (EA) and 23; FEIS and 43; Lake Champlain Lampricide example 211, 212; review of draft of 19

earthquake fault 87
earthquakes 85, 153
earth resources: definition of 89
EC *see* environmental concerns
Eccleston, Charles 30
ecological diversity: definition of 112
ecosystem 1–3; birds and 108; definition of 112
ecosystem health 109
ecosystem change 76
ecosystem concept of trade and resource management 33
ecosystem diversity 107
ecosystems services 15, 30, 103, 177; definition of 42, 179
ecotone: definition of 112
ecotourism 177, 223
effect: definition of 134
effectiveness of EIA *see* EIA
emissivity: definition of 147; low **145**
endangered species 3, 27, 30, 31, **73**; biological mitigation measures **111**; biological resources and 103; federal and state lists of 104–109; water resources and **99**; wetland **94**
Endangered Species Act (ESA) 26 (assignment), 28, 1107 (assignment), 124; definition of 112; Green Book guidance on 206; Title 40 Code of Federal Regulation/ NEPA § 1501.3 249, § 1502.24 266
Endangered Species List 108 (assignment); definition of 113
energy 138–148; conserving 125
energy audit 141
energy efficient vehicles 122
energy growth 3
Energy Independence and Security Act of 2007 141
Energy Policy Act 141
energy policy checklist **143**
ENERGY STAR 141; definition of 147
Environmental Assessment (EA, ES) 2; assessing environmental impacts using 77, 105–106; CATEX lists that do not require 23; definition of 10, 17, 42, 77, 213; environmental justice report, checklist for evaluating **175**; Finding of No Significance (FONSI) 10; job plan and 39; mitigated 13; NEPA and 227; as "Phase 1" of site analysis 23; project description 151; Record of Decision (ROD) for 12; setting portion of 151; as step in preparing or as final document of EIS 205; steps of 10–11, 52; Strategic Environmental Assessment (SEA) 13–14, 78, 151
Environmental Audit: definition of 42, 233
environmental concerns (EC) 53
Environmental Conservation Act 87
Environmental Impact Assessment (or Analysis; EIA) 1–19; air quality component 116, 118, 120, 121, 123, 125; archeological sites 133; burden of production 219; brief history of 1–2; Callahan Mining example 87; concepts and terms 16–17, 42–44, 62–63, 232; current US issues with 14–16; data collection and presentation 56–58; as decision-making process 6–11, 216–232; decision tools in 222; definition of 17; ecosystem services addressed in 177; energy component checklist 142; environmental justice and 174; environmental setting 92, 93, 104, 127; EPA and 174, 175; erosion factor 89; five principles for managing process of 66; hierarchy **25**; history of NEPA and 21–25; impact and mitigation component of report 80–81, 84, 85, 86; implementing 18–19, 65–79; land development process as related to 49–51; making and implementing the decision 216–232; mitigation "banking" 110; national/ global affairs and 16; noise impacts analysis 152, 154, 156; odor impact assessments 166; one-year timeline for **41**; players in process, roles of 51–54; potential and predicted impacts 129; process 21–44; project scoping 54–56; public participation in 58–62, *60*; as related to environmental planning 7; report completion and checklist 205, **206**, 209; report timeline **78**; as research 5–7; risk factors addressed by 178; sample job plan outline **39**; stakeholders **61**; stakeholder matrix **63**; outline stages in process of *9*; public awareness and 2–5; strategic environmental assessment compared to 13–14; three energy aspects for review of 138, *139*; traditional cultural practice (TCP) in, 128, 161; traffic impacts considered by 194; two-month timeline for project management of **68**; using EIA to improve the EIA process 232;

water resources impact checklist **99**; wetlands functions and values checklist **94**; wind energy projects 162; writing sample critique **210, 213**; writing up of 205–212

Environmental Impact Statement (EIS, DEIS, FEIS) 205; definition of 17; draft 19; first (Marsh) 2; program EIS 43 (definition); programmatic EIS 18 (definition; SEA); public participation in 60; review of draft of 19; strategic 232; *see also* DEIS, FEIS

environmental justice 33, 174–175; definition of 42; EIA's requirement to account for 219; environmental setting and 171–172; Green Book guidance on 206; judicial review and 26–27; mining operations and 86; National Environmental Justice Advisory Council 175; REA's implications for 228; White House Environmental Justice Advisory Council 175

environmental law: NEPA and 25–26

environmentally unsatisfactory (EU) 53–54

environmental management system: definition of 42

environmental media 68 (assignment), 70 (assignment)

environmental objection (EO) 53

environmental practitioners, preparation of students as 302–308 (Appendix)

environmental setting: definition of 17; environmental justice and 171–172

Environmental Protection Agency (EPA): Air Quality Index (AQI) 117; air quality regulated by 125; electricity generation and sulfur dioxide emissions statistics from 144; ENERGY STAR program 147; EPA Environmental Impact Statement Filing Guidance 47; former EIA/EIS results review system 52–54; National Environmental Justice Advisory Council 175; noise exposure levels set by 153, **154**; precautionary principle of 49; receptor models used by 121; Reference Guide to Odor Thresholds 118; Support Center for Regulatory Atmospheric Modeling (SCRAM) 120–121; Toxics Release Inventory (TRI) Program (EPA) 176 (assignment)

Environmental Situation Rapid Environmental Assessment Response Form **229–230**

EO *see* environmental objection

EOs *see* executive orders

EPA *see* Environmental Protection Agency

EPA/OSHA standards for permissible noise exposure limits on construction sites **154**

estoppel: definition of 43

ethnic and racial composition **172**

ethnicity: definition of 179

EU *see* environmentally unsatisfactory

executive directive 286

Executive Office of the President of the United States 285, 292

Executive Order 11991 281

Executive Order 12898 174, 175 (assignment)

Executive Order 13807 277, 281

Executive Order 14007 123

Executive Order 11514 xiv, 242, 244

Executive Order 12898, Federal Actions to Address Environmental Justice in Minority Populations and Low-Income Populations (1994) 174

executive orders (EOs) 28, 217, 282, 283; list of major EOs 140

executive summary 210; definition of 24

existing air quality conditions 115

existing conditions report 8, 80; definition of 43

ex parte: definition of 43

expert judgement 75

FedConnect 35

Federal Interagency Working Group on Environmental Justice & NEPA Committee (2016) 174

Federal Register 34

FEIS *see* Final Environmental Impact Statement

Final Environmental Impact Statement (FEIS): CEQ regulations regarding 205; commenting agencies' criteria applied to 106; definition of 17, 43, 63; Lake Champlain Lampricide example 211, 212; NEPA process and 12, 23, 216; North New Mining Project and Land Exchange example 87; PVT Land Company Integrated Solid Waste Management Facility Relocation Project example 218

Finding of No Significant Impact (FONSI): definition of 17, 43, 63, 78, 214; Environmental Assessment finding of 23, 213; level of detailed information required in 13; mitigated 214; NEPA analysis level two 22; public notice of 3, *4*, 10

Fish and Wildlife Coordination Act 100

Fish and Wildlife Service (FWS) (US) 27, 104, 106 (assignment), 107, 108, 211; biological opinion issued by 110

floodplain 96, 206; definition of 101
flow chart 74, 77 (assignment), *134*
flow rate 199
flow (traffic) 193, 194: definition of 202
flow (water) 100
form (aesthetic) **167**; definition of 168
fracking 3; **83**, **84**; definition of 89; hydrofracking 225, 226

genetic diversity 107, 177; definition of 112
GHGs *see* greenhouse gases
Glasson, John 30, 33
grass not always greener 308
Green Book (AASHTO; DOE) 203, 206
green building: definition of 147; *see also* LEED; USGBC
greenhouse effect 119
greenhouse gases (GHGs) 118, 123, 144; climate impact assessment and 125 (definition); social cost of carbon (SCC) and 191 (definition)
Green Mountains 201
green power **142**; definition of 147
green space 132, 187
Gulf Coast 174
Gulf Coast Extension 227
Gulf of Mexico 80, 220

Habitat Evaluation Procedure (HEP) 107, **107**; definition of 112
Habitat Evaluation System 107, **107**; definition of 112
Habitat Suitability Index **107**; definition of 112
habitats *see* species and habitats
handicap accessibility 195, 201; definition of 202
Hawken, Paul 15
hazard: definition of 179; environmental 86, 93; health 15, 115, 118; pollution 25
hazard evaluation 178
Hazardous Air Pollutants: definition of 125; odor impacts and 118, 166
hazardous materials 30, 36
hazardous waste 37, 38, 66, **84**
health hazard *see* hazard
health impact assessment (HIA): definition of 180
hearings: joint public 274; media-sharing of 226; NEPA related 276; preliminary 259; public 59, 212, 276, 277, 299; three types of 59
heating degree days (HDD): definition 147

Henning odor classification 167, **167**
HEP *see* Habitat Evaluation Procedure
historical records 93
historical site assessment **130**
historic preservation 127–135
historic property or resource 58; definition of 135
Historical Sites Act 128
hue **167**; definition of 168
HVAC system **145**; definition of 147
hydrocarbons 118, 119, 122
hydrology 92–102

illuminance: definition of 168–169
impact assessment: checklists and matrices for 76; Strategic Environmental Impact Assessment 8
impact mitigation 104
impact prediction 92, 104; definition of 101; models for 74–76
impact significance 81, 92, 104, 115; decision tree for use in Brazil 218, **218**; Strategic Environmental Impact Assessment 8
Impact Significance Assessment (ISI) 217; definition of 232
impacts, portable **40–41**, 47, 55, 66, 72, 80, 85–86, 92, 104; *see also* potential or predicted impacts
impacts, residual 12, 67, **78**, 80, 81, 88, 151, 168, 178; calculation 92, 104; EIA report **206**
impacts, types of xiii, 5, 22; air quality 118, 125; biological 101, 104, 107, **109**, **111**; chemical 101; cumulative 110, 115; development 3; EIA anticipation of 84; EIA review for 53–54; environmental 8, 32, **34**, 65–79; environmental/ energy 142–145, 146; fiscal and economic 184–189; harmful 3; indirect 180; minor 23, 48; mitigation of 9, 12, 13; principal 87; scoping of 56; social 171, 176–178; techniques of measuring 15
index and indices, definition of 78
indicator: definition of 78; environmental 69
indicator species **105**; definition of 112
indices, types of 69–71; EPA 52; noise 152; quality of life 33
indices, use of 69–71, 228, 300
indirect effects 285
indirect consequences 184

indirect emissions 122
indirect impacts: definition of 180
infrastructure 3, 128, 142, 182–191; building of 33; definition and discussion of 182, 191; EA planning exercise 67 (assignment); EIA planning exercise 44 (assignment); Executive Order 13807, Establishing Discipline and Accountability in the Environmental Review and Permitting Process for Infrastructure Projects 244, 277, 281; human settlement 150; master plan for 232; New Deal legislation and 25; public 86
infrastructure impact assessments 183
infrastructure relocations 172
infrastructure projects 15
in lieu fee mitigation 110
inorganic chemical parameters: definition of 101
inorganic chemicals **96**
integrated energy system: definition of 147
integrated pest management (IPM) 100
Integrated Noise Model (INM) 158
integrity (archeological): definition of 135
integrity (professional) 266
intensity: definition of 180
International Monetary Fund (IMF) 7, 16
International Organization for Standardization (ISO) 32; definition of 43
irrigation 25, 95
irrigation water: definition of 101
ISO *see* International Organization for Standardization
isochrone graphs 154

job plan 39
judicial hearing 59
judicial review: environmental justice and 26–27; NEPA compliance and 245
judicial system xiv

Kershner, Jim 21
Keystone Pipeline 226–227
keystone species **105**; definition of 112; *see also* umbrella species
Kunc, Hansjoerg P. 154

Lacey Act: definition 112
lack of objections 53

lamp efficiency: definition of 169
lamprey **97**
lampricide 211–212
land development: noise levels and 150
land development process 2–3; EIA and 22, 49–51; sustainable 7 (assignment)
land development process for a private sector project **50**
landfill 33, **34**, 83, 119, 162, 174 176, 231
landforms **83**, 93, 130, 162
Land Reclamation *see* United States Bureau of Land Reclamation
land use 138; mix 183; patterns **173**
Land Use Development Type **200**
lane (traffic): definition of 202
leaching 81; definition of 90
lead agency 7, 10–12; cooperating agency and 17, 213; cover sheet naming of 24; definition of 71, 214; regulatory agencies sought out by EIS preparers from 54, 105; ROD and 216; screening matrix including 49
lead agency as stipulated in Title 40 Code of Federal Regulations/NEPA § 1501.7 251
lead-based paint 36
Leadership in Energy and Environmental Design (LEED): checklist of energy-related emission reduction categories based on LEED green building certification **145**; definition of 147; EIA energy component checklist based on LEED green building certification **142**; green building certification program 141, **142**, 145, **145**
LEED *see* Leadership in Energy and Environmental Design
Leopold, Aldo 2
Leopold method 222
lethal concentration: definition of 112
level of service (LOS) 199–201
Life Cycle Assessment (LCA) 144, 146; definition of 43
life cycle cost methods 141
life cycle costing: definition of 147
life cycle of products 43
line (aesthetic) **167**; definition of 169
listening area 155; definition of 158
little NEPA 8, **29**; *see also* NEPA
loop of decomposition 67

LOS *see* level of service
lumens 168; definition of 169

Maine: Apple River 223 (assignment); Callahan Superfund Site 87; Department of Environmental Protection 94; EIA and public awareness: Portland Bayside 44 (assignment), 67 (assignment); Portland FONSI sample 4–5; Portland Jetport 190 (assignment); public record, definition of 59; traffic movement permit issued by 194; water flow rates 95; wetland checklist, sample **94**
Marcellus Shale 225
marginal abatement costs (MAC): definition of 191
marginal cost approach 189
marginal costing methods 188
marginal geographic locations 123
marginal topics 15
marsh 109; definition of 101
Marsh, George Perkins 2
master horizon *see* soil horizon
mastering Excel/Spreadsheets 308
mastering group interaction 305
master plan 62, 232
master plan stakeholder matrix **63**
matrices 56, 228; checklists and 76; usefulness of 56
matrix: EIA 175; impact assessment matrix 87 (assignment); impact matrix, categories of 72, **73**; impact trade-off analysis 217; Leopold matrix 22; MAP 178; master plan stakeholder matrix **63**; NEPA decision 217; Positive Matrix Factorization (PMF) 121; project and environment matrix **58**; Rapid Impact Assessment Matrix (RIAM) 233; sample matrix for offshore wind turbine project impact assessment **57**; sample risk matrix **220**; screening matrix for a project to construct an erosion control and stormwater barrier dike **48**
McHarg, Ian 2
methane 119
mitigated FONSI *see* FONSI
mitigation: amelioration and 78; definition of 78, 135, 214; documenting 72; impact mitigation, EIA report timeline **78**; infrastructure and community service 188, 189–190; lighting impact and **165**; management of 66; Modak and Biswas' model including 66; purchase of land 133; trade-off analysis and 146; Title 40 Code of Federal Regulations/NEPA § 1508.1 definition of 287; *see also* loop of decomposition

mitigation and monitoring 18, 24; aesthetics and visual impact analysis 168; air quality and climate change 121–125; archeology and historic preservation 132–133; biological species and habitats 109–111, **111**; energy management and 144–146; geology and topography 88–89; hydrology and water resources 100–101; noise impact analysis 155–158; social impacts and environmental justice 178–179; traffic and transport systems 201–202
mitigation bank 110, 112
mitigation measures: climate change and waterfront development **124**; Common types of biological impact mitigation measures **111**; definition of 17, 43; EAs and EIAS, concerns including 15; EIA process and 11; EIA review/concerns regarding 53; FONSI and 13; geology and topography in accounting for 80–90; monitoring plan for 9; remediation costs for 71; ROD for 10
Modak, Prasad, and Asit K. Biswas 66–67
models, types of 72; air quality *120*, 120–121; computer 118; dispersion 121; EIA 123; dose response 178; forecast 104; groundwater 93; habitat **73**; impact prediction 104, 106; mathematical 166; mechanistic 74; photochemical 121; physical 76, 162; receptor 121; simulation 75; statistical 193, 266; surface water 93
models for impact prediction 74–76
monitoring 104; definition of 17; *see also* mitigation and monitoring
Morris, Peter, and Riki Therivel 30
Muir, John 2

NAAQS *see* National Ambient Air Quality Standards
National Academies of Sciences, Engineering, and Medicine 144
National Ambient Air Quality Standards (NAAQS) 116; definition of 125
national certification 81
national conscience, awakening of 25
National Energy Conservation Policy Act (NECPA) 141
National Environmental Justice Advisory Council 175
National Environmental Policy Act (NEPA): air quality assessments and review by 115–116, 125; archeological assessments and 131 (assignment);

climate change and 123; concepts and terms 42–44; countries' environmental laws other than US compared to 30, 32; courts and xiv; cultural resources and 128, **129**; cumulative impacts and 232; current issues with EIA and 14–16; EPA and 22; definition of 17; EIA reports filed in response to 205; energy-producing projects and 138; environmental law and legal basis for 25–26; Executive Order 12898, Federal Actions to Address Environmental Justice in Minority Populations and Low-Income Populations (1994) and 174, 175 (assignment); fast decision-making and 227–232; Federal Highway Administration and 217; Federal Interagency Working Group on Environmental Justice & NEPA Committee 174; guidelines and screening 46–62; history of EIA and 21–25; little NEPAS 8, **29**; management of process 33–41; Nixon and xiii; passing and adoption of 2, 16; process 22; project planning and management 65; projects subject to 28; rational basis for 7; Section 106 and 105, *106*; state-equivalent ("Little NEPA") 2022 **29**; three intersecting federal laws involving cultural resources and **129**; trade-off analysis used by 223 (assignment); *see also* Council on Environmental Quality (CEQ); Finding of No Significant Importance (FONSI)

National Environmental Policy Act (NEPA), Title 40 Code of Federal Regulations Chapter V: agency planning and 247–256; implementing 237–301

national/ global affairs, EIA in 16

National Historic Preservation Act (NHPA) 131, 133, 206

national legislation 32

National Marine Fisheries Service (NMFS) 107

national park 184; potential noise 151, 157, **158**

National Pollutant Discharge Eliminating System (NPDES) 100

National Register of Historic Places 128, 135

National Water Use Information Program 100

Native Americans 128, 131, 161

natural capital: definition of 113, 232

NECPA *see* National Energy Conservation Policy Act (NECPA)

NEPA *see* National Environmental Policy Act

NEPA document 28, 220; definition of 214

NEPA pathway: definition of 214

NEPA process 7, 12, 16; definition of 214; managing 33–41

NEPA review 280; definition of 214

New York City Environmental Quality Review (CEQR) 47; state-equivalent "little NEPA" law **29**

NHPA *see* National Historic Preservation Act (NHPA)

nitrous oxide 119

NMFS *see* National Marine Fisheries Service (NMFS)

Noble, Bram 30

noise impact analysis 150–159

noise level isobar: definition of 158

noise pollution: receptors of 150–152, 155, 156, 158, **158**, 161; source of 150–152, 154, 156–159

NPDES *see* National Pollutant Discharge Eliminating System (NPDES)

objects: definition of 135

odor: categories of air quality impacts 115, 116, 118; Henning odor classification 167, **167**; odor impact assessments 166; Reference Guide to Odor Thresholds for Hazardous Air Pollutants (EPA) 118; *see also* physical impact variables

Omnibus Flood Control Act of 1939 25

Oso Landslide 89 (assignment)

overburden: acid rock drainage and 89 (definition); definition of 90

particulates (air) 119

party: definition of 43; lead agency 17 (definition); reviewing 12; third 12

payment in lieu of taxes (PLT) 190; definition of 191

peak flow 98

peak hour traffic 194, **195**, 197, **200**, 201; definition of 202

peak noise 152

pedestrian flow 201; definition of 202

Phase I Report 5; definition of 17; EA as equivalent to 23, 77

Phase I Site Assessment 129; definition of 113

Phase II Report 5; definition of 17

Phase II Site Assessment 129; definition of 113

Phase III Report 5; definition of 17

Phase III Site Assessment 129; definition of 113

Phases A–D *see* value tree analysis phases

physical impact variables: definition of 101
photochemical modeling 121
place (traffic): definition of 203
plume discharge 122
plumes 123; contaminant 101; well 97
PLT *see* payment in lieu of taxes
PMF *see* Positive Matrix Factorization (PMF) model
Positive Matrix Factorization (PMF) model 121
potential or predicted impacts; aesthetics and visual impact analysis 162–167; air quality and climate change 118–120; archeology and historic preservation 129–131; energy 142–144; geology and topography 85–88; hydrology and water supply 96–100; infrastructure and community service 184; noise impact analysis 152–155; social impact and environmental justice 176–178; species and habitat 106; traffic and transport 195
potential natural community: definition of 113
precautionary principle 49, 51 (assignment), 176; definition of 63
primary generation, cogeneration, and transmission 138, 140
primary succession: definition of 113
program EIS: definition of 43
programmatic EIS 14, 18, 110; *see also* Strategic Environmental Assessment (SEA)
project: erosion control 48; general transit questions to ask 199; modifying 217; offshore wind turbine 57; policy-focused xiii; private sector 50; sample threshold trip levels based on size of 200; significant earth resources, projects associated with 84; Smart Growth planning principles met by 198; variety of services engaged by 183; ways in which project can be modified 217
project alternatives 12, 13; definition of 18; NEPA decisions' considerations of 217
project and environment matrix 58
project decomposition 67
project description 11, 12, 13; aesthetics and visual impact analysis 162; air quality and climate change 118; archeology and historic preservation 128; definition of 18; energy 140; geology and topography 84–85; hydrology and water supply 95–96; infrastructure and community service 184; noise impact analysis 151; screening and 47, 49; social impact and environmental justice 172; species and habitat 106; traffic and transport 194

project management: EIA timeline 68; five principles for managing EIA process 66
project phase or life stage: EIA checklist of water categories 99; four phases of 33, 50–51, 58, 84, 92, 100, 106, 128, 162, 172
project review: fire and emergency department checklist 184; police department checklist 185
proposed action 3, 11, 13; agencies and 260–261, 265; alternatives on 262; CATEX list and 23; cover sheet and 24; cumulative effects of 42 (definition); definition of 43; EIA process and 26; EIA review and 53–54; EIS and 42 (definition); environmental assessment 250; environmental effect of, considered as variable 72; environmental impacts of 263; federal 260–271; FONSI 250–251; impact on climate change 123–124; impact topics and 213 (definition); matrices and checklists, selection 56; NEPA review and compliance for 29; NEPA review 249–250; NEPA thresholds and 248–249; scoping and 44 (definition); *see also* action; ROD
public archeology 127
public awareness: EIA and 2–5
public comments 5
public decision-making process 6; *see also* public meetings
public disclosure of information 58
public disbursement of documents 22
public domain 21, 219
public education 132
public environmental accountability 2
public funds 25
public health 26 (assignment), 85, 182
public hearings 59, 212, 226, 274, 276, 277, 299
public interest 124
public involvement, barriers 59
public involvement, reasons for 58–62; Title 40 Code of Federal Regulation/ NEPA § 1506.6 276–276
public lands 128, 182
public meetings 59, 183, 226; EIA decision-making and 226; EIA report timeline 78; *see also* public involvement, reasons for
public notice of Finding of No Significant Impact (FONSI) 4
public notices: EIA mandating of 3; EIA sample job plan and 52

public participation: accountability and 33, 65; EIA process and 10–11; global variation in 30, 32; long time frames demanded for 15; scoping process and 55
public policy decision-making 7
public review 205
public transportation 122
public trust 21
public water supply 95, 96
public works 86, 186; definition of 191
pure color 168
pure tone frequencies: definition of 158

Quechee analysis 165–166

race: definition of 180
radon 119
range of tolerance **97**; definition of 113
Rapid Environmental Assessment (REA) 176, 227; definition of 43, 233; Environmental Situation Rapid Environmental Assessment Response Form **229**, **230**
Rapid Impact Assessment Matrix (RIAM): definition of 233
REA *see* Rapid Environmental Assessment
real value 191; *see also* value
receptors (of noise pollution) 150–152, 155, 156, 158, **158**, 161
receptor models 121
reclamation 33, earth resources involved in 84; *Termination/closure/reclamation/long-term monitoring* 51; *see also* project phase or life stage
Reclamation Act of 1902 25
reclaimed water: definition of 101
Record of Decision (ROD) 10; definition of 18, 43, 215; EA and 12; EIA and 216, 226; EIS process and 24; *see also* Environmentally Preferable Alternative; Notice of Availability
redoximorphic features 81; definition of. 90
Reference Guide to Odor Thresholds for Hazardous Air Pollutants (EPA) 118
Request for Proposals (RFP) 34–38; EIA and 209–210; job plan and 39; definition of 43–44; respond to **41**, **68**; *see also* scope of work
Request for Qualifications (RFQ) 34–38; definition of 44; respond to **68**
residence time 93; definition of 125

resilience: definition of 113; forest 123; frameworks of 183; traffic impact analysis and 193
Refuse Act 25
reuse **142**, 263
revenues 184, 188, 189; definition of 191
risk 178, 219–221; definition of 180
risk profile **185**
road geometry: definition of 203
road profile: definition of 203
R-value: definition of 147; *see also* U-value

SA *see* Sustainability Assessment
scale (aesthetic) 164; definition of 169
SCC *see* social cost of carbon
Schmidt, Michael 74, 75
Schmidt, Rouven 154
Scope of Work (SOW) 36, 37; definition of 44
scoping 10, 11, 15; climate change considerations integrated into 124; definition and discussion of 8, 18, 44, 54, 215; EA as preliminary version of 23; EIA decision-making and 216; EIA process and 66; EIA requirements expanded for 32; NEPA and 54–56; RFPs and 35, 37; sample EIA job plan and timeline including **39**, **41**, **68**, **78**; screening and 35, 47; Title 40 Code of Federal Regulation/ NEPA § 1501.9 253–255
screening: definition and discussion of 8, 11, 18, 44; NEPA agency guidelines and 46–49; RFPs and 35, 37; sample EIA job plan and timeline including **39**, **41**, **68**, **78**; scoping and 35; vegetation 158–159, 168
screening matrix for project to construct an erosion control and stormwater barrier dike **48**
sea level rise 15, 38, **124**, 131
SEAs *see* Strategic Environmental Assessment
secondary succession: definition of 113
serial status or stage: definition of 113
shadow price: definition of 191
shorelines **94**, 162, 166
shores: near-shore aquaculture stations **177**; offshore drilling 49; offshore wind turbine project impact assessment **57**
sight distance **186**, **195**; definition of 203
signal warrant 197; definition of 203
significance 8, 15, 76; definition of 44, 135; impact significance assessment 217; statewide 165
significance of impacts 80, 85, 104, 115, 151

simulation models 75; computer 152, 162; hydrologic 93; *see also* impact prediction
simulation programs 74, 201
site: definition of 135
smokestack 121, 159
social cost of carbon (SCC): definition of 191
social impact assessment: definition of 180; UN good practices in impact mitigation **179**
social impacts: environmental justice and 171–180; mining operations 85
social justice 66; archeology and 131
social science 131
soil 81–82; basic properties of *82*; topsoil 84, **86**, **87, 89**
soil engineering report **85**
soil horizon 81; definition of 90
Soil Quality Index (SQI) 70
solar farm 138, 140, 143 (assignment); project 38 (list)
solar gain: definition of 148
solar panel **145**
solar panel fields 140
solar power *see* green power
solar radiation 120
Solicitation, Offer, and Award Form 38
space (aesthetic) 164; definition of 169
species and habitats 103–113
species diversity: definition of 113
"spheres" of US global influence/involvement 30–32
sprawl (urban) 189, 197, **199**; *see also* handicap accessibility
stability as project variable 71, **73**, 76, 222
stability, economic 10
stability, environmental 10; diversity and 112
stability, land 80, 89; seismic **85**, **86**; site **85**, 88; slope **83**, **84**
stability, project 30, 146
stability, residential **173**
stack emissions 122–123
State Historical Preservation Office (SHPO) 55; definition of 135
stopping distance: definition of 203
stopping sight distance (SSD): definition of 203
stormwater barrier dike **48**
stormwater discharge 128; definition of 101
stormwater management 95
stormwater runoff 75, **89**, **195**

Strategic Environmental Assessment (SEA) 32, 232; definition and discussion of 13–14, 18, 78–79, 233
Strategic Impact Statement: definition of 44
strip development 197
structure (soil): definition of 90; *see also* soil
structures 81, 119; buildings and 133 (definition); built 140; construction and 128; definition of 135; district and 134 (definition); fire and emergency services questions related to **185**; government 182; institutional 172, **173**; landscape checklist for **144**; LEED checklist for 141, **142**; market 32; physical 184; site and 135 (definition); *see also* public works
subcollector: definition of 203
sudden or startling noise: definition of 158
sulfur oxides and dioxide 117, 121, 122, 125, 144
sustainability 14, 183
Sustainability Assessment (SA): definition of 44

tailings 87; definition of 90
TCP *see* Traditional Cultural Practice
TCP *see* Traditional Cultural Property
texture (environmental or aesthetic) 163, 164, **167**; definition of 169
texture (soil) 81; definition of 90
Therivel, Riki 14, 30, 33
thermal value 148
THPO *see* Tribal Historic Preservation Office
threatened and endangered species 3, 27, 99, 103–109, 152; definition of 113
three horsemen of the treadmill 30, 33
tiering: definition and discussion of 13, 14, 18, 44, 215, 288; Title 40 Code of Federal Regulation/NEPA § 1501.11 256
timeline: EIA report **39**, **41**, **68**, **78**; internal 303; process 23
timeliness, issue of 66
toxic fumes 174
toxicity *see* lethal concentration
toxic levels 178
Toxics Release Inventory (TRI) Program (EPA) 176 (assignment)
toxic substances: Agency for Toxic Substances and Disease Registry (ATSDR) 225; *see also* air toxics
Traditional Cultural Practice (TCP) 128, 151, 161

Traditional Cultural Property (TCP): definition of 135
traffic and transport systems 193–203
traffic impact analysis: definition of 203
Transportation Control Measures (TCMs) 122
Transportation System Management (TSM) 122
treadmill of production 30, 33
Tribal Historic Preservation Office (THPO) 55, 136; definition of 135

umbrella process 217, 228
umbrella species: definition of 113
unavoidable adverse effects: definition of 44
undertaking agency 49
undertakings 11, 47, 188; categorical exclusion 62 (definition); definition of 135; federal agency 22, 37; federal or state 63 (EIS definition), 129; physical 81; project description (archology/land) 128; proposed action 43 (definition), 214 (definition); SEA done in advance of 79
United Nations: EIA use and 32; "good practices" 178 (assignment), **179**; *see also* Rapid Environmental Assessment (REA)
United Nations social impact assessment good practices in impact mitigation **179**
United States: "spheres" of global influence/involvement 30–32
United States Bureau of Land Reclamation 174
United States Constitution (US Constitution) 25, 138; Preamble 7
United States Department of Energy (DOE) 11, 27, 46
United States Green Building Council (USGBC) 141
United States Fish and Wildlife Service (FWS) 27, 104, 106 (assignment), 107, 108, 211; biological opinion issued by 110
UNMIX model 121
USGBC *see* United States Green Building Council (USGBC)
U-Value 148

value (color) 168; definition of 169
value: amenity 187; definition of 135; habitat unit values (HuV) **107**; loss of 219; magnitude 222; monetary 191; monitory 103; ranking value of importance 223; recreational **129**; scaled determinants of 217; socioeconomic 176, 177, 187, 188; real 191 (definition); R-value: 147 (definition); symbolic 111; U-Value 148 (definition)
values: core social 176; cultural 128, **129**, 168
values checklist for EIA on wetlands site **94**
value tree analysis for decisions *224*, 225, 226
value tree analysis phases *224*, 224–226; Phases A–D 116
vehicle traffic 69, 115, 118, 122
vehicle trip: definition of 203
Vermont: Act 250 8; public education component of archeological assessment 132; *see also* Quechee Analysis
Vermont Department of Fish and Wildlife 211
viewsheds 161, 162, *163*; definition of 169
visibility of noise source: definition of 158
visual impact analysis 161–169

water quality and water supply 92–101; definition of 101
Water Quality Index (WQI) 70
white collar jobs 188, 190 (assignment), 198 (assignment)
White House Environmental Justice Advisory Council 175
white list 47
White Mountains 201
Wild and Scenic Rivers Act 99
wildfire 30, **105**, 115, 123
wildlife: lighting impacts on **165**; noise disruptions to 150, 151; subsistence consumption of 174, 175
wildlife areas: fracking in 3
wildlife biology 5; Certified Wildlife Biologists 69
wildlife poaching 31, 103
wildlife resources 103
wildlife restoration 38
Williams Transco Pipeline 175
World Bank 7, 16, 33, 69, 227
WQI *see* Water Quality Index

xeriscape: definition of 148

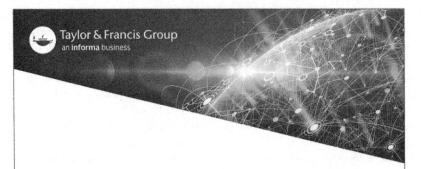

Made in the USA
Coppell, TX
22 January 2025

44782252R00188